W9-DEU-613

Risk, Resource, and Regulatory Issues

Remediation of Chlorinated and Recalcitrant Compounds

Editors

Godage B. Wickramanayake
Battelle

Robert E. Hinchee
Parsons Engineering Science, Inc.

The First International Conference on Remediation of Chlorinated and Recalcitrant Compounds

Monterey, California, May 18–21, 1998

Library of Congress Cataloging-in-Publication Data

International Conference on Remediation of Chlorinated and Recalcitrant Compounds
(1st : 1998 : Monterey, Calif.)
Risk, resource, and regulatory issues : remediation of chlorinated and recalcitrant compounds / editors, Godage B. Wickramanayake, Robert E. Hinchee : First International Conference on Remediation of Chlorinated and Recalcitrant Compounds, Monterey, California, May 18–21, 1998.
p. cm.
Includes bibliographical references and index.
ISBN 1-57477-056-X (alk. paper)
1. Organochlorine compounds--Biodegradation--Congresses. 2. Organic solvents--Biodegradation--Congresses. 3. Hazardous waste site remediation --Congresses. 4. Hazardous substances--Risk assessment--Congresses. I. Wickramanayake, Godage B., 1953– . II. Hinchee, Robert E. III. Title.

TD1066.073I58 1998
628.5′2--dc21

98-24757
CIP

Printed in the United States of America

Battelle Press
505 King Avenue
Columbus, Ohio 43201, USA
614-424-6393 or 1-800-451-3543
Fax: 1-614-424-3819
Internet: press@battelle.org
Website: www.battelle.org/bookstore

For information on future symposia and conference programs, write to:
Bioremediation Symposium
Battelle
505 King Avenue
Columbus, Ohio 43201-2693
Fax: 614-424-3667

CONTENTS

Innovative Site Characterization and Monitoring Approaches

Geostatistics/Data Quality Objectives

FOREWORD

Sites contaminated with chlorinated solvents and other recalcitrant compounds present formidable challenges in terms of risk assessment, resource allocation, and regulatory compliance. *Risk, Resource, and Regulatory Issues: Remediation of Chlorinated and Recalcitrant Compounds* provides a broad overview of these issues and presents case studies documenting successful risk-based site remediation strategies. Included in this volume are chapters on pathways analysis and modeling for risk assessment, human health and ecological risk assessment approaches, risk assessment in a Risk-Based Corrective Action (RBCA) context, resource allocation and cost considerations, regulatory issues, stakeholder involvement and technology acceptance, status and trends in technical impracticability, geostatistics and data quality objectives, and innovative site-monitoring approaches.

This is one of six volumes published in connection with the First International Conference on Remediation of Chlorinated and Recalcitrant Compounds, held in May 1998 in Monterey, California. The 1998 Conference was the first in a series of biennial conferences focusing on the more problematic substances—chlorinated solvents, pesticides/herbicides, PCBs/dioxins, MTBE, DNAPLs, and explosives residues—in all environmental media. Physical, chemical, biological, thermal, and combined technologies for dealing with these compounds were discussed. Several sessions dealt with natural attenuation, site characterization, and monitoring technologies. Pilot- and field-scale studies were presented, plus the latest research data from the laboratory. Other sessions focused on human health and ecological risk assessment, regulatory issues, technology acceptance, and resource allocation and cost issues. The conference was attended by scientists, engineers, managers, consultants, and other environmental professionals representing universities, government, site management and regulatory agencies, remediation companies, and research and development firms from around the world.

The inspiration for this Conference first came to Karl Nehring of Battelle, who recognized the opportunity to organize an international meeting that would focus on chlorinated and recalcitrant compounds and cover the range of remediation technologies to encompass physical, chemical, thermal, and biological approaches. The Conference would complement Battelle's other biennial remediation meeting, the In Situ and On-Site Bioremediation Symposium. Jeff Means of Battelle championed the idea of the conference and made available the resources to help turn the idea into reality. As plans progressed, a Conference Steering Committee was formed at Battelle to help plan the technical program. Committee members Abe Chen, Tad Fox, Arun Gavaskar, Neeraj Gupta, Phil Jagucki, Dan Janke, Mark Kelley, Victor Magar, Bob Olfenbuttel, and Bruce Sass communicated with potential session chairs to begin the process of soliciting papers and organizing the technical sessions that eventually were presented in Monterey. Throughout the process of organizing the Conference, Carol Young of

Battelle worked tirelessly to keep track of the stream of details, documents, and deadlines involved in an undertaking of this magnitude.

Each section in this and the other five volumes corresponds to a technical session at the Conference. The author of each presentation accepted for the Conference was invited to prepare a short paper formatted according to the specifications provided. Papers were submitted for approximately 60% of the presentations accepted for the conference program. To complete publication shortly after the Conference, no peer review, copy-editing, or typesetting was performed. Thus, the papers within these volumes are printed as submitted by the authors. Because the papers were published as received, differences in national convention and personal style led to variations in such matters as word usage, spelling, abbreviation, the manner in which numbers and measurements are presented, and type style and size.

We would like to thank the Battelle staff who assembled this book and its companion volumes and prepared them for printing. Carol Young, Christina Peterson, Janetta Place, Loretta Bahn, Lynn Copley-Graves, Timothy Lundgren, and Gina Melaragno spent many hours on production tasks. They developed the detailed format specifications sent to each author, tracked papers as received, and examined each to ensure that it met basic page layout requirements, making adjustments when necessary. Then they assembled the volumes, applied headers and page numbers, compiled tables of contents and author and keyword indices, and performed a final page check before submitting the volumes to the publisher. Joseph Sheldrick, manager of Battelle Press, provided valuable production-planning advice and coordinated with the printer; he and Gar Dingess designed the volume covers.

Neither Battelle nor the Conference co-sponsors or supporting organizations reviewed the materials published in these volumes, and their support for the Conference should not be construed as an endorsement of the content.

Godage B. Wickramanayake and Robert E. Hinchee
Conference Chairman and Co-Chairman

COST-EFFECTIVE SITE CHARACTERIZATION USING P450 REPORTER GENE SYSTEM (RGS)

Jennifer M. Jones (Columbia Analytical Services, Carlsbad, CA)
Jack W. Anderson (Columbia Analytical Services, Carlsbad, CA)

ABSTRACT: P450 Reporter Gene System (RGS) is a unique method to detect the transcriptional activation of the human CYP1A1 gene by PAHs and chlorinated organic compounds, including dioxins, furans, and coplanar PCBs. RGS utilizes a human cell line (101L) stably transfected with a plasmid containing firefly luciferase linked to the human CYP1A1 promoter. Solvent extracts (using EPA Method 3550) of sediment, soil, and tissue are applied directly to the cells. RGS responses, measured as relative light units in a luminometer, are expressed as RGS Toxic Equivalents (RGS TEQ), using Toxic Equivalency Factors (TEFs) already derived for a range of compounds. These equivalency values may then be used in assessing the hazard or risk of a sample, or the decrease in concentrations of contaminants resulting from remediation procedures. Because PAHs are readily metabolized, the RGS response to these compounds is greater at 6 than at 16 hours, whereas dioxins, furans, and coplanar PCBs produce a greater response at the 16 hour exposure period, enabling differentiation between these classes. This study demonstrates the cost-effectiveness of RGS in screening environmental samples to detect and characterize CYP1A1-inducing compounds, so that more expensive chemical characterization will be conducted on only a small subset of samples.

INTRODUCTION

Chlorinated organic compounds and PAHs are known to induce the CYP1A gene subfamily via the Ah-receptor (Nebert and Jones, 1989; Whitlock, 1990). This ligand-receptor complex is translocated to the nucleus of the cell, where it interacts with xenobiotic response elements in the promoter of the CYP1A1 gene and causes transcription of the P450 enzyme system. The resulting induction of the CYP1A1 gene and transcription of the P450 enzyme system is increasingly used as a measure of the toxic potencies of these compounds in *in vitro* systems (Anderson et al, 1995; Garrison et al, 1996; Murk et al, 1996; Richter et al, 1997).

The P450 Reporter Gene System (RGS) utilizes a cell line known as 101L, originally a human hepatoma cell line (HepG2) into which human CYP1A sequences are fused to the firefly luciferase gene and stably integrated (Quattrochi and Tukey, 1989; Postlind et al, 1993). RGS is used to assess human CYP1A induction by various environmental pollutants, especially the high molecular weight PAHs, coplanar polychlorinated biphenyls (PCBs) and 2,3,7,8-tetrachlorodibenzo-*p*-dioxin (TCDD) (Jones and Anderson, submitted). In the presence of CYP1A1-inducing compounds, the enzyme luciferase is produced, and can be detected by a simple assay that measures relative light units with a luminometer. RGS detection limits of CYP1A1-inducing compounds in a typical 40 g sample, are 0.02 ng/g for dioxins (TCDD), 0.12 ng/g for furans (TCDF), 10 ng/g for coplanar PCBs (PCB #126), and 100 ng/g for PAHs (Benzo[*a*]pyrene). Studies using a standard dioxin/furan mixture, a PAH mixture, and combinations of both mixtures indicate that the RGS response is additive, responding to the total amount of CYP1A1-inducing compounds present (Jones and Anderson, unpublished data). RGS has gained acceptance as a rapid and inexpensive *in vitro* approach to screening solvent extracts of environmental samples of soil, sediment, tissue, and water (APHA, 1996; ASTM, 1997).

Because PAHs and chlorinated compounds often co-occur in environmental mixtures, it is advantageous to be able to differentiate between contaminant types when screening environmental samples. Differences in RGS CYP1A induction by standard solutions of PAHs and chlorinated compounds at two exposure times (6 and 16 h) are used to develop criteria for determining the primary contaminant class in extracts of unknown composition.

This study presents the results of RGS testing of soil extracts from a contaminated site. Correlations between the chemical analyses (using GC/MS) and the calculated RGS TEQs for selected samples from this site will be shown to demonstrate the usefulness of RGS as an alternative to chemical analysis of all samples in an environmental assessment. Because it is inexpensive in comparison to chemical analyses, RGS is a cost-effective technique to measure the potential for the sample to cause harmful health effects.

MATERIALS AND METHODS

Test Solutions. The reference inducer, 2,3,7,8-tetrachlorodibenzo-*p*-dioxin (TCDD) was obtained from Ultra Scientific (North Kingstown, RI). All standard solutions were prepared in HPLC grade dichloromethane (DCM), except for TCDD, which was prepared in molecular biology grade Dimethylsulfoxide (DMSO) (Fisher, Pittsburgh, PA).

Cell Culture and Application of Test Solutions. The 101L cells were grown as monolayers in an atmosphere of 5% CO_2 and 100% humidity at 37° C. In preparation for testing, cells were subcultured into 6-well plates at a density of 2.5 x 10^5 cells/well and allowed to grow for 36 h in the environment described above to approximately 1 x 10^6 cells/well. Test solutions were applied at volumes of 2 µL to replicate wells each containing 2 mL of culture media so that the concentrations of solutions never exceeded 0.1% (v/v). Cell viability was assured using tryphan blue-dye exclusion.

Environmental Samples. Fourteen 40-g soil samples were collected from a site in December, 1997. Samples were extracted using EPA Method 3540, and the final volume was reduced to 1 mL DCM. Extracts were diluted and applied at a volume of 2 µL to replicate wells of the tissue culture plates as described above.

Four of the samples were selected for chemical characterization of dioxins and furans using EPA Method 8290 and PAHs using GC/MS SIM.

Luciferase Assay. The detailed methodology used in this study has been described elsewhere (APHA 1996, ASTM 1997). Briefly, after incubation with the test solutions, the cells were washed with salt solution and lysed with Cell Lysis Buffer. Cell lysates were centrifuged, and the supernatant was applied to a 96-well plate, followed by addition of a buffer containing ATP and $MgCl_2$. Reactions were initiated by injection of potassium phosphate buffer containing luciferin. Luminescence in relative light units (RLUs) was measured using a ML2250 Luminometer (Dynatech Laboratories, Chantilly, VA). Luciferase assay buffers were purchased from Analytical Luminescence Laboratory (Cockeysville, MD).

With each test run, a solvent blank (using a volume of DCM equal to the sample volume being tested) and 1 ng/mL TCDD were also applied to replicate wells. Mean fold induction of the solvent blank was set equal to 1, and the fold induction of each standard solution and TCDD was determined by dividing the mean RLUs produced by that solution by the mean RLUs produced by the solvent blank. The standard deviation and coefficient of variation were recorded for each test solution.

Equivalency Calculations. RGS Toxic Equivalents (RGS TEQ) are calculated using the equation below. According to previous concentration-response studies using a standard mixture of dioxins and furans, the RGS fold induction response is equivalent to the mixture TEQ in pg/mL (calculated using Toxic Equivalency Factors established by Safe, 1990). Dividing by 1000 yields TEQ in ng/g.

$$TEQ = (\text{fold induction}/1000) * ((V_e/V_a)/W_d)$$

Where V_e = total extract volume

V_a = volume of extract applied to cells

W_d = dry weight of sample

Using the concentration of each dioxin/furan analyte found using EPA Method 8290 and their TEFs (Safe, 1990), a Chemical Toxic Equivalency value (Chem TEQ) was calculated for the 4 soil extracts. In the same manner, a Chemical Benzo[*a*]pyrene (Chem B[*a*]PEq) was calculated for the same extracts using the concentrations of PAHs found by GC/MS and their TEFs derived from previous RGS studies (Jones and Anderson, unpublished data).

RESULTS AND DISCUSSION

The P450 RGS fold induction response represents a measure of the total CYP1A1-inducing compounds present in an environmental extract. This is quantified as equivalents based on RGS studies using standard mixtures of dioxins and furans as well as PAHs. Table 1 shows the range of RGS TEQs produced by the 14 samples. Of these, 4 extracts were also characterized for dioxins and furans, and the Chem TEQ calculated from these chemical analyses are shown in Table 1. Figure 1 illustrates the correlation of RGS TEQ and Chem TEQ for these 4 soil extracts. The correlation of RGS TEQ and Chem TEQ shown in Figure 1 (R^2 = 0.88) demonstrates the validity of using RGS in characterizing unknown environmental samples.

TABLE 1. RGS TEQ of all soil extracts and Chem TEQ and Chem B[a]PEq calculated from chemical analysis of 4 selected extracts.

Sample #	RGS TEQ (ng/g)	Chem TEQ (pg/g)	Chem B[a]PEq (µg/g)
1	10.6	14.4	4.2
2	1.0		
3	0.9		
4	0.3		
5	1.1		
6	59.1	51.8	1.1
7	41.0	47.4	2.6
8	47.0		
9	28.1	21.6	3.7
10	0.8		
11	0.1		
12	0.0		
13	0.2		
14	0.4		

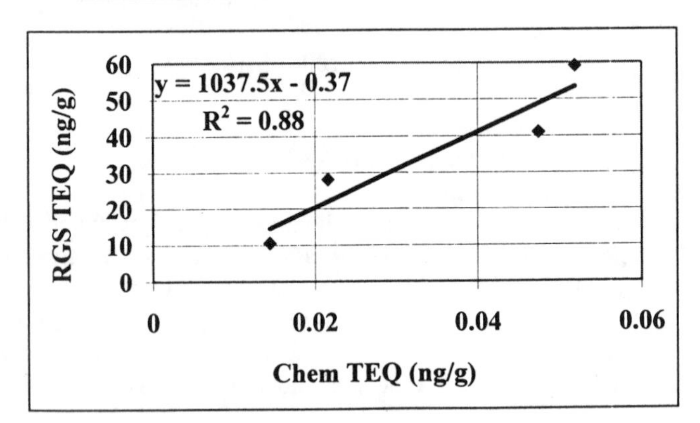

FIGURE 1. Regression analysis of RGS TEQ and Chem TEQ for 4 soil extracts chemically characterized using EPA Method 8290.

In addition, employing RGS at two time periods of exposure, 6h and 16h, enables the characterization of extracts as containing primarily PAHs or chlorinated compounds. From previous studies using standard solutions of individual PAHs, coplanar PCBs, dioxins and furans, as well as standard mixtures of these compounds, we have found that the RGS response to PAHs is earlier, producing a maximal response at 6h. In contrast, all chlorinated compounds tested produce maximal response at 16 h. These temporal differences in RGS response can be used to determine the class of contaminants (PAHs or chlorinated compounds) prevalent in environmental samples.

Figure 2 shows the results of testing to characterize the contaminants present in the 4 selected extracts (shown in Table 1). A greater response at 6h compared to 16h, as observed with Sample 1 and Sample 9, indicates the presence of significant levels of PAHs, while Sample 6 and Sample 7 produced a greater response at 16h, indicative of primarily chlorinated compounds.

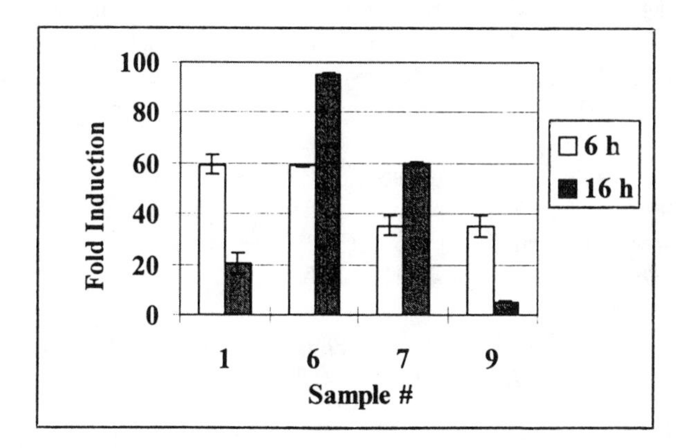

FIGURE 2. RGS fold induction responses to 4 soil extracts at 2 time periods.

The results of chemical characterization of these 4 extracts for PAHs, calculated as Chem B[*a*]PEq, are given in Table 1. These results confirm the presence of higher levels of CYP1A1-inducing PAHs in Sample 1 and Sample 9 as compared to Sample 6 and Sample 7.

The capability of not only detecting the total integrated response to CYP1A1-inducing chemicals, but also characterizing environmental samples using dual time period analysis, further demonstrates the usefulness of RGS as a tool for site assessment. In comparison to EPA Method 8290 for dioxins and furans, RGS is approximately one-tenth the cost. As an alternative to chemical characterization of all samples, RGS is a cost-effective screening method.

REFERENCES

A.P.H.A. 1996. "P450 Reporter Gene Response to Dioxin-like Organics. Method 8070." In *Standard Methods for the Exmination of Water and Wastewater*, 19th Ed. Supplement, pp. 24-25. American Public Health Association, Washington, DC.

Anderson, J. W., S. S. Rossi, R. H. Tukey, T. Vu, and L. C. Quattrochi. 1995. "A Biomarker, 450RGS, for Assessing the Potential Toxicity of Organic Compounds in Environmental Samples." *Environ. Toxicol. Chem.* 14:1159-1169.

A.S.T.M. 1997. "E 1853-96 Standard Guide for Measuring the Presence of Planar Organic Compounds which Induce CYP1A, Reporter Gene Test Systems." In *Biological Effects and Environmental Fate; Biotechnology; Pesticides, 1997 Annual Book of ASTM Standards,* Volume 11.05. Water and Environmental Technology, pp. 1392-1397. American Society for Testing and Materials, West Conshohocken, PA.

Garrison, P. M., K. Tullis, J. Aarts, A. Brouwer, J. P. Giesy, and M. S. Denison. 1996. "Species-specific Recombinant Cell Lines as Bioassay Systems for the Detection of 2,3,7,8-tetrachlorodibenzo-*p*-dioxin-like Chemicals." *Fund. Appl. Toxicol.* 30:194-203.

Murk, A. J., J. Legler, M. S. Denison, J. P. Giesy, C. Van de Guchte, and A. Brouwer. 1996. "Chemical-activated Luciferase Expression (CALUX): a Novel *In Vitro* Bioassay for Ah Receptor Active Compounds in Sediments and Pore Water." *Fundam. Appl. Toxicol.* 33:149-160.

Nebert, D. W., and J. E. Jones. 1989. "Regulation of the Mammalian Cytochrome P450 (CYP1A1) Gene." *Int. J. Biochem.* 21:243-252.

Postlind, H., T. Vu, R. H. Tukey, and L. C. Quattrochi. 1993. Response of Human CYP1-luciferase Plasmids to 2,3,7,8-tetrachlorodibenzo-*p*-dioxin and Polycyclic Aromatic Hydrocarbons." *Toxicol. Appl. Pharmacol.* 118:255-262.

Quattrochi, L. C., and R. H. Tukey. 1989. "The Human CYP1A2 Gene Contains Regulatory Elements Responsive to 3-Methylcholanthrene." *Mol. Pharmacol.* 36:66-71.

Richter, C. A., V. L. Tieber, M. S. Denison, and J. P. Giesy. 1997. "An *In Vitro* Rainbow Trout Bioassay for Aryl Hydrocarbon Receptor-mediated Toxins." *Environ. Toxicol. Chem.* 16:543-550.

Safe, S. H. 1990. "Polychlorinated Biphenyls (PCBs), Dibenzo-*p*-dioxins (PCDDs), Dibenzofurans (PCDFs) and Related Compounds: Environmental and Mechanistic Considerations which Support the Development of Toxic Equivalency Factors (TEFs)." *Crit. Rev. Toxicol.* 21:51-88.

Whitlock, J. P. 1990. "Genetic and Molecular Aspects of 2,3,7,8-tetrachlorodibenzo-*p*-dioxin Action." *Annu. Rev. Pharmacol. Toxicol.* 30:251-277.

PAHs AND RELATED COMPOUNDS AT A FORMER GASWORKS SITE

Staffan Lundstedt (Umeå University, Umeå, Sweden)
Bert van Bavel, Peter Haglund, Christoffer Rappe and Lars Öberg
(Umeå University, Umeå, Sweden)

ABSTRACT: Soil from a former gasworks site was characterized with respect to PAHs and related compounds. The soil was Soxhlet extracted and the extract cleaned up by passing it through a silica column. The concentrated samples were analyzed by GC/MS. More than 40 compounds and groups of compounds were identified, including PAHs, methylated PAHs, biphenyls, dibenzofuran and dibenzothiophene. The soil concentration of the 16 US-EPA priority pollutant PAHs was 159 mg/kg (dry weight). In addition, the PAH contamination was verified to be aged, according to PAH/alkyl-PAH ratios and high molecular weight PAHs-/low molecular weight PAHs-ratios.

INTRODUCTION

Soil from former gasworks sites is often highly polluted with coal tar and other by-products of coal pyrolysis. These pollutants consist mainly of polycyclic aromatic hydrocarbons (PAHs), of which several are known to be toxic and carcinogenic [Griciute *et al.* 1979]. To evaluate the soil toxicity it is important to estimate to what extent PAHs occur in the soil. Until recently, only 16 PAHs as listed by the US EPA have been studied at most contaminated sites, but reports show that many other toxic compounds can also be present [Dubourgier *et al.* 1997, Xiaobing *et al.* 1990], including other PAHs, methylated PAHs and heterocyclic PAHs containing oxygen, nitrogen or sulphur. To correctly estimate the environmental risk at former gasworks sites, it is desirable to measure the concentration of more polycyclic aromatic compounds than the 16 US-EPA priority-pollutant PAHs.

Within the Swedish national research program *Soil Remediation in a Cold Climate* (COLDREM), the research is focussed on a gasworks site in Stockholm, Sweden. The gasworks was relatively large and consumed about 400,000 ton coal annually, before it was shut down in 1972. The area is highly contaminated with PAHs and other hydrocarbons due to spill during production, clean up and transportation of by-products and gas, and due to leakage from underground tanks and piping. In order to find a suitable remediation technique for the area, soil from this site is treated with biological and chemical methods in laboratory and pilot scale studies. To better understand the results from these studies, the soil has been characterized with respect to PAHs and similar compounds.

MATERIALS AND METHODS

Sampling of soil. The area at the former coal gasification plant consists of excavated material including soil, blast stone and waste from the plant demolition. Therefore the ground is inhomogeneous and difficult to sample in a representative manner. However, at the beginning of December 1996 soil was excavated, mixed and put into barrels. Samples were taken and sent to Agro Lab (Kristianstad, Sweden) for analysis of dry substance, organic carbon content and pH. The soil contained 79 % dry substance of which 29 % was organic carbon. The soil pH was 6.2. Soil for PAH-analysis was sifted through a 4-mm sieve, homogenized and air-dried for three days. The dry soil was ground to a fine powder in a ball mill (Planetary Micro Mill "pulverisette 7" Fritsch, Germany).

Extraction and clean up. Three grams of the soil was put in a Soxhlet glass microfibre thimble and spiked with 630 µg perdeuterated internal standard (Cambridge Isotope Laboratory) containing naphthalene, acenaphthene, anthracene, fluorene, chrysene, pyrene and benzo(k)fluoranthene. The thimble was placed in a Soxhlet extractor and was extracted with 120 ml toluene for 18 h. The extract was evaporated to less than 1 ml and carefully applied on a silica column, consisting of 10 g activated silica gel and 1 g sodium sulphate at the top. The column was eluted with 25 ml hexane and 25 ml hexane/methylene chloride (3:2). The fractions were collected, pooled and then evaporated. The solvent was changed to toluene and the volume was reduced to approximately 1 ml by evaporation. Two individual soil samples were extracted and cleaned up by this procedure at different occasions.

GC/MS-analysis. The sample was analyzed by high-resolution gas chromatography / low-resolution mass spectrometry (HRGC/LRMS). The GC (Fisons GC 8000, 60 m DB-5 capillary column, 0.32 mm i.d., 0.25 µm film from J&W Scientific, CA) was operated in splitless mode. 1 µl aliquots of the extracts were injected using an autosampler. The MS (Fisons MD 800) was operated in the full-scan mode. PAHs and other compounds were identified according to retention time and mass spectra. The 16 US-EPA priority-pollutant PAHs were quantified by using a standard solution (QTM PAH Mix, Supelco, PA), containing 15 of the 16 PAHs (Table 2).

RESULTS AND DISCUSSION

Identification. A full-scan chromatogram of the soil extract is presented in Figure 1, and the identified compounds in Table 1. Most of the identified compounds are PAHs and methylated PAHs but also biphenyls, dibenzofuran and dibenzothiophene are detected. This is consistent with soil being contaminated with coal tar, which contains about 80 % creosote and anthracene oil. In 1989 Mueller *et al.* reported that creosote consists of 85 % PAHs, 10% phenolic compounds and 5 % other nitrogen-, sulfur- and oxygen-heterocyclics.

Quantification. The 16 EPA priority pollutant PAHs were quantified against a standard solution. The total concentration of the 16 PAHs was 159 mg/kg (d.w., Table 2). This PAH-concentration is comparable to that Xiaobing *et al.* [1990] found in soil contaminated with coal tar. On the other hand many other reports show much higher PAH-concentrations in soil from former gasworks sites [Dubourgier *et al.* 1997, Hollender *et al.* 1997]. Thomas *et al.* [1994] proposed that 10-20,000 mg/kg is the normal range for contamination of coal tar derivatives at former gasworks sites, and consequently the soil in this study is considered as relatively low contaminated.

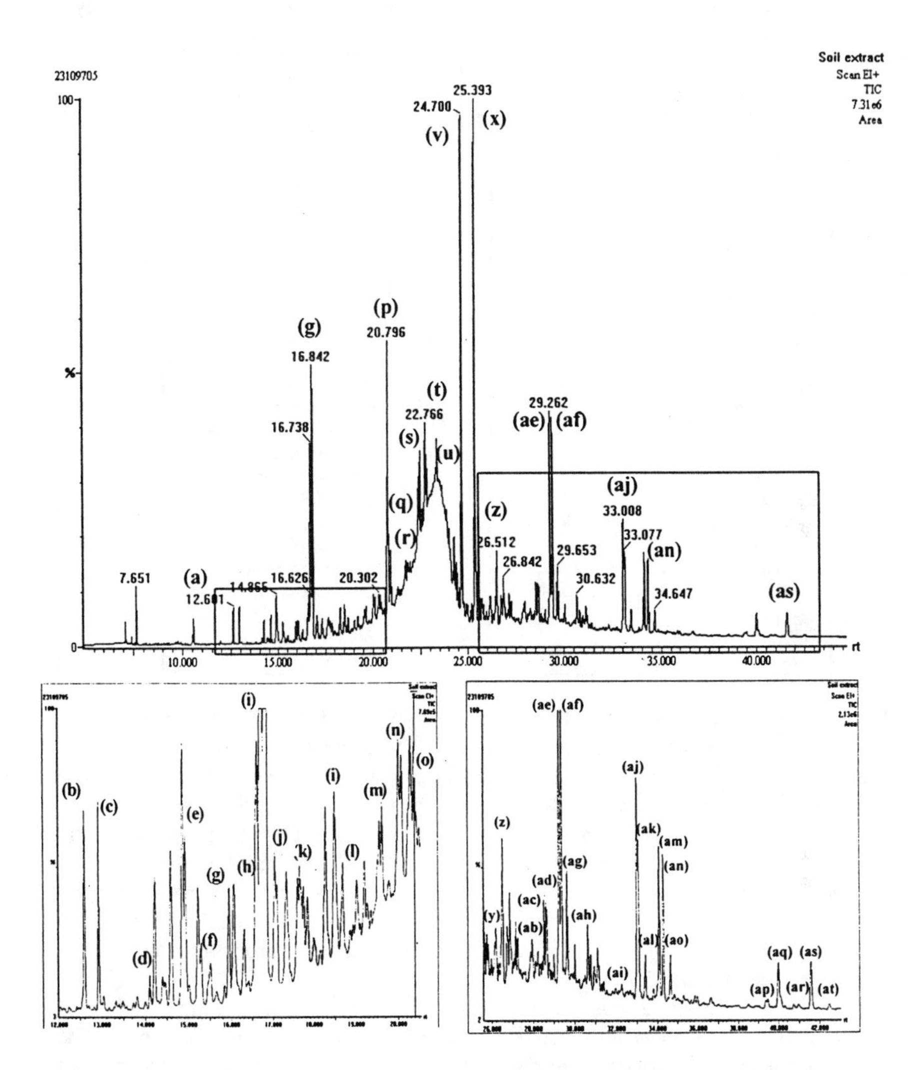

FIGURE 1. Full-scan chromatogram of soil extract from a former gasworks site. Amplified parts are marked in the total chromatogram.

However, a screening of the area indicated that a number of spots had soil concentrations of several thousands mg/kg of the 16 PAHs. This variation in the soil concentration emphasizes the importance of a careful site mapping and a representative sampling procedure.

The inhomogeneity of the soil is also indicated by the variation in PAH concentration between the two analyzed samples. Although, the soil was ground to a fine powder and carefully mixed, the resulting total concentration of the 16 PAHs varied between 196 and 122 mg/kg.

TABLE 1. Tentatively identified peaks in the full-scan chromatogram. The compounds were identified according to retention time and mass spectra.

Peak	Compound	MW
a	Naphthalene	128
b	2-Methylnaphthalene	142
c	1-Methyl naphthalene	142
d	Biphenyl	154
e	C_2-Naphthalene	156
f	Acenapthylene	152
g	Methylbiphenyl	168
h	Acenaphthene / Dibenzofuran	154 / 168
i	C_2-Biphenyl	182
j	C_3-Naphthalene	170
k	Fluorene	166
l	C_4-Naphthalene	184
m	Methylfluorene	180
n	C_3-Biphenyl	196
o	Dibenzothiophene	184
p	Phenanthrene	178
q	Anthracene	178
r	C_2-Fluorene	194
s	Methylphenanthrene / Methylanthracene	192
t	Phenyl naphthalene	204
u	C_2-Phenanthrene / C_2-Anthracene	206
v	Fluoranthene	202
x	Pyrene	202
y	C_3-Phenanthrene / C_3-Anthracene	220
z	Methylfluoranthene / Methylpyrene	216
ab	C_2-Fluoranthene / C_2-Pyrene	230
ac	C_4-Phenanthrene / C_4-Anthracene	234
ad	Benzo(c)phenanthrene	226
ae	Benzo(a)anthracene	228
af	Chrysene	228
ag	Triphenylene	228
ah	Methylchrysene	242
ai	C_2-Chrysene	256
aj	Benzo(b)fluoranthene	252
ak	Benzo(k)fluoranthene	252
al	Benzo(j)fluoranthene	252
am	Benzo(e)pyrene	252
an	Benzo(a)pyrene	252
ao	Perylene	252
ap	unknown PAHs	278 / 276
aq	Indeno(c,d)pyrene / Dibenz(a,c)anthracene	276
ar	Unknown PAHs	278
as	Benzo(g,h,i)perylene	276
at	Anthanthrene	276

Recovery estimation. To determine the recovery of the PAHs during the analytical procedure, PAH-quantification standard was added to an empty extraction thimble. The standard was extracted and cleaned up in the same way as the samples. Before analysis internal standard was added to the extract. The recovery was calculated by comparing the result with a standard solution containing both internal and quantification standard (Table 2). The recoveries of the PAHs varied between 10-100 %. The low molecular weight PAHs, *e.g.* naphthalene and acenaphthene, were recovered to 50 %, which is consistent with their relatively high volatility. During the clean up procedure there are several steps of evaporation that probably result in loss of the more volatile compounds.

More unexpected was the 60 %-recovery of some of the high molecular weight PAHs and the very low recovery of acenaphthylene and benzo(a)pyrene. The high molecular weight PAHs are not significantly volatilized during evaporation, but may be lost to some extent during the silica clean up, and also by adsorption onto glass surfaces. The much lower recovery of acenaphthylene and benzo(a)pyrene as well as the 100 % recovery of fluorene is more difficult to explain.

It should be noted that the internal standard (IS) only contained perdeuterated naphthalene, acenaphthene, anthracene, fluorene, chrysene, pyrene and benzo(k)fluoranthene, and that the other quantified PAHs were corrected for clean-up losses by the IS-compound eluting closest to it. However, the fluctuating recoveries of the PAHs are an indicayion that the concentration of some PAHs may be misestimated in this way.

TABLE 2. Concentrations (dry weight) of the 16 US-EPA priority pollutant PAHs in soil from a former gasworks site. Results from two individual soil samples, and the recovery for the analytical procedure.

PAH	Sample 1 Soil conc. (mg/kg)	Sample 2 Soil conc. (mg/kg)	Average Soil conc. (mg/kg)	Recovery (%)
Naphthalene	2.2	1.5	1.9	54
Acenaphthylene	0.2	0.2	0.2	53
Acenaphthene	0.7	0.5	0.6	11
Fluorene	0.9	0.7	0.8	100
Phenanthrene	16.8	10.9	13.9	63
Anthracene	3.1	1.9	2.5	43
Fluoranthene	23.5	14.5	19.0	68
Pyrene	20.4	13.0	16.7	70
Benzo(a)anthracene	16.0	9.7	12.9	65
Chrysene	17.0	10.3	13.7	77
Benzo(b)fluoranthene	20.6	15.0	17.8	79
Benzo(k)fluoranthene	13.9	7.1	10.5	79
Benzo(a)pyrene	18.2	11.1	14.7	18
Indeno(c,d)pyrene	15.7	10.2	13.0	65
Dibenz(a,c)anthracene	4.2	2.6	3.4	65
Benzo(g,h,i)perylene	22.3	13.1	17.7	59
Total	**196**	**122**	**159**	

Age of soil. The relative abundance of naphthalene and its methylated homologues (Figure 2), indicates that the contamination is originating from the past. In fresh coal tar the concentration of naphthalene is higher than the concentration of methylnaphthalene, which in turn is higher than the concentration of C_2-naphthalenes, which is higher than the concentration of C_3-naphthalenes. However, within a PAH homologous series both volatility and degradation rates generally are inversely proportional to the degree of alkylation [Douglas *et al.* 1992]. As a consequence highly alkylated PAHs are more persistent than the less alkylated. By comparing the peak areas in Figure 2 the concentration of the dimethylnaphthalenes seem to be highest in soil under study. Moreover, the concentration of the methylnaphthalenes seems to be at the same level as the concentration of naphthalene.

Furthermore the soil seems to contain relatively high amounts of the high molecular weight PAHs, *e.g.* pyrene, benzo(a)anthracene and chrysene. This also suggests that the contamination is old. Fresh coal tar contains relatively high amounts of the low molecular weight PAHs, such as naphthalene, methylnaphthalene, acenaphthylene, fluorene and phenanthrene [Hale *et al.* 1997]. These compounds are, however, more volatile and more easily degraded than the high molecular weight PAHs, which therefore will be more abundant in an aged soil.

ACKOWLEDGMENT

This study was carried out within the Swedish national research program COLDREM (Soil remediation in a cold climate). The authors wish to acknowledge MISTRA (Foundation for Strategic Environmental Research) for the financial support.

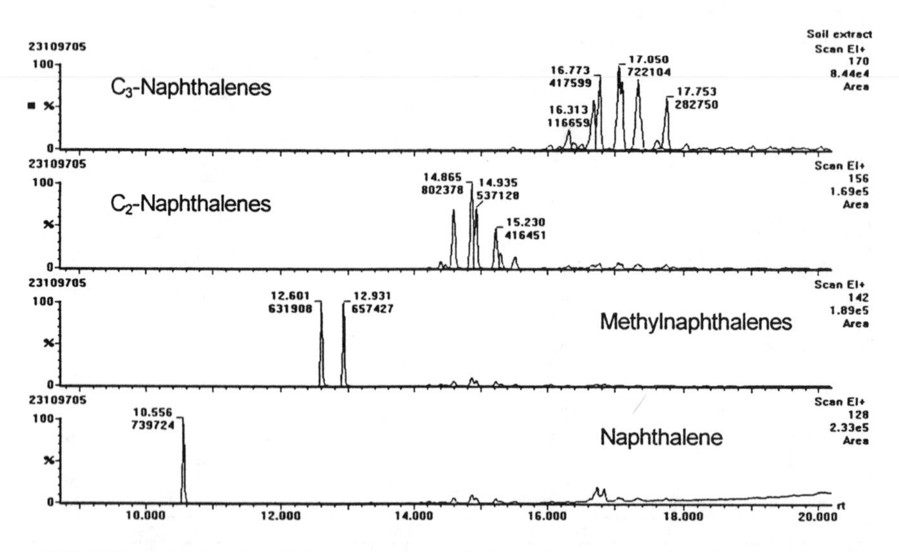

FIGURE 2. Selected ion chromatograms of the soil sample, showing the signals for naphthalene, methylnaphthalenes, C_2-naphthalenes and C_3-naphthalenes.

REFERENCES

Douglas, G. S., K. J. McCarthy., D. T. Dahlen., J. A. Seavey and W. G. Steinhauer 1992. "The use of Hydrocarbon Analyses for Environmental Assessment and Remediation." *Journal of Soil Contamination.* 1(3): 197-216.

Dubourgier, H. C., M. N. Duval., F. Cazier and P. Wijjfels 1997. "Bioremediation of Highly Polluted Soils from Former Coal Industries." In B. C. Alleman andA Leeson (eds.), *In situ and On.site Bioremediation,* pp.79-84. Battelle press, Columbus, Ohio.

Griciute, L. 1979. "Carcinogenicity of Polycyclic Aromatic Hydrocarbons". In M. Castegnaro, H. Kunte, P. Bogovski and E. A. Walker (eds.), *World Health Organization, IARC Publication No.29,* pp.3-11. Lyon, France.

Hale, R. C. and K. M. Aneiro 1997. "Determination of Coal Tar and Creosote Constituents in the Aquatic Environment." *Journal of Chromatography A.* 774: 79-95.

Hollender, J., J. Shneine, W. Dott, M. Heinzel, H. W. Hagemann and G. K. E. Götz 1997. "Extraction of Polycyclic Aromatic Hydrocarbons from Polluted Soils with Binary and Ternary Supercritical Phases." *Journal of Chromatography A.* 776: 233-243.

Mueller, J. G., P. J. Chapman and P. H. Pritchard 1989. "Creosote-contaminated Sites." *Environmental Science and Technology.* 23(10): 1197-1201.

Thomas, A. O. and J. N. Lester 1994. "The Reclamation of Disused Gasworks Sites: New Solutions to an Old Problem." *The Science of the Total Environment.* 152: 239-260.

Xiaobing, Y., W. Xiaobing, R. Bartha and J. D. Rosen 1990. "Supercritical Fluid Extraction of Coal Tar Contaminated Soil." *Environmental Science and Technology.* 24(11): 1732-1738.

EVALUATION OF ON-SITE ANALYTICAL METHODS FOR EXPLOSIVES IN COMPOST RESIDUES

Harry D. Craig (U.S. EPA Region X, Portland, Oregon, USA)
Ginger K. Ferguson and Andrew G. Markos (Black & Veatch Special Projects Corporation, Seattle, Washington, USA)
Phillip G. Thorne and Thomas F. Jenkins (U.S. Army Corps of Engineers Cold Regions Research and Engineering Laboratory, Hanover, New Hampshire, USA)
Carol A. Witt-Smith (U.S. EPA Region V, Chicago, Illinois, USA)

ABSTRACT: Composting is an emerging ex-situ solid phase biological treatment technology for degradation of nitroaromatic and nitramine explosives compounds in soil. A field demonstration was conducted to assess the performance of on-site analytical methods for explosives 2,4,6-trinitrotoluene (TNT) and hexahydro-1,3,5-trinitro-1,3,5-triazine (RDX) in compost residues during full scale remediation at the Umatilla Chemical Depot (Hermiston, OR) and U.S. Naval Submarine Base (Bangor, WA) Superfund sites and pilot scale composting at the Naval Surface Warfare Center Crane (Crane, IN) RCRA Corrective Action site. Compost samples were analyzed by each of the on-site methods and results compared to laboratory analysis using high performance liquid chromatography (EPA SW-846 Method 8330). The on-site methods evaluated include the EnSys TNT and RDX colorimetric methods (EPA SW-846 Methods 8515 and 8510) with and without an organic matrix cleanup step, and the DTECH TNT and RDX immunoassay methods (EPA SW-846 Methods 4050 and 4051). Accuracy of the on-site methods were evaluated using linear regression analysis, relative percent difference (RPD), and field analytical to laboratory analytical (FA/LA) ratio comparison criteria. Over the range of conditions evaluated, the colorimetric methods for TNT and RDX with organic matrix cleanup steps showed the highest accuracy. Significant differences were noted for samples run by the colorimetric methods with and without the organic matrix cleanup steps. The immunoassay TNT and RDX methods also showed reasonable accuracy for analysis of compost residues.

INTRODUCTION

Composting is an emerging ex-situ solid phase biological treatment technology for degradation of semi and non-volatile organic compounds in soil, particularly nitroaromatic and nitramine explosives compounds (EPA 1993, 1996a). Nine pilot scale treatability studies and three full scale remediation projects have been completed using composting to treat explosives contaminated soils in the U.S. and Germany. Due to the relatively quick reaction kinetics (i.e. > 99% degradation in 10 to 40 days), on-site analytical methods may be useful for full scale process monitoring and have the potential to reduce analytical turn around times. Inexpensive on-site methods for the most common explosives in contaminated soils have been developed and are now in common use (EPA 1996b). In contrast to soil, compost residues represent a complex organic matrix designed to enhance microbial degradation of explosives under thermophilic conditions. A field demonstration was conducted to assess the performance of soil on-site analytical methods for explosives 2,4,6-trinitrotoluene (TNT) and hexahydro-1,3,5-trinitro-1,3,5-triazine (RDX) in compost residues during full scale composting at the Umatilla Chemical Depot (Hermiston, OR) and U.S. Naval Submarine Base (Bangor, WA) Superfund sites and pilot scale composting at the Naval Surface Warfare Center (NSWC) Crane (Crane, IN) RCRA Corrective Action site undergoing EPA cleanup actions (EPA 1998).

METHODS

The on-site analytical methods evaluated include EnSys TNT and RDX colorimetric methods (EPA SW-846 Methods 8515 and 8510) and DTECH TNT and RDX immunoassay methods (EPA SW-846 Methods 4050 and 4051). Due to the quick reaction kinetics of compost degradation, initial (Day O) compost residues were analyzed for Phase I and II. The Phase I Umatilla trial utilized compost samples split by the TNT and RDX colorimetric methods without an organic matrix cleanup (OMC) step (EPA 1996b), and laboratory analysis by EPA SW-846 Method 8330 (EPA 1994). The Phase II compost mix designs are shown in Table 1.

TABLE 1. Phase II Soil/Compost Amendment Mix Design

Umatilla	SUBASE Bangor	Crane Mix #4	Crane Mix #5	Crane Mix #8
30% soil	25% soil	25% soil	25% soil	25% soil
21% cow, 1% chicken manure	24% cow manure	23% cow, 3% chicken manure	26% turkey manure	10% cow, 7% chicken manure
18% alfalfa	17% alfalfa	19% alfalfa	19% alfalfa	48% straw
18% sawdust	17% wood chips	19% sawdust	19% sawdust	
11% potato waste	17% potato waste/apple pomace	11% potato waste	11% potato waste	10% potato waste

The Phase II evaluation was conducted using a single acetone extract for all analyses to isolate analytical error from sample matrix heterogeneity. The Phase II colorimetric methods were evaluated with OMC steps, utilizing a styrene divinyl benzene (SDVB) column (Porapak) for the TNT method and a granular activated carbon (GAC) column for the RDX method (EPA 1998). The immunoassay methods for TNT and RDX were run in accordance with the manufacturer's recommendations. Several modifications to EPA Method 8330 for laboratory analysis of acetone compost extracts were conducted. A dilution (1:25) in water to achieve the correct solvent strength was required. The determative analysis was performed using the cyano phase (CN) Confirmation conditions described in Method

8330 (EPA 1994). The CN column used in the Confirmation analysis has the first peak come out three minutes later than the acetone peak and has no affect on the chromatograph. The samples required a dilution for RDX, and were made using HPLC grade water. Standard QC spikes and one set of sample matrix spikes were generated and a surrogate, 3,4-DNT was added to some samples (EPA 1998).

RESULTS

Bias, precision, and accuracy of the on-site methods were evaluated using 1) linear regression analysis, 2) relative percent difference (RPD) criteria, and 3) field analytical to laboratory analytical (FA/LA) ratio comparison criteria. Linear regression was evaluated with the on-site method as the y-axis and the laboratory method as the x-axis; slope and correlation coefficient (r) were calculated (Gilbert 1987). RPD and FA/LA ratio criteria were calculated as shown in equations 1 and 2 (Billets et al. 1996, Grant et al. 1996).

$$RPD = (FA - LA) \times 100 / [\,(FA - LA) / 2] \qquad (1)$$

FA = Field Analytical Method Conc. (mg/kg)
LA = Laboratory Analytical Method (HPLC) Conc. (mg/kg)

$$FA/LA\ Ratio = FA / LA \qquad (2)$$

FA = Field Analytical Method Conc. (mg/kg)
LA = Laboratory Analytical Method (HPLC) Conc. (mg/kg)

The results of Phase I and Phase II linear regression parameters, RPD, and FA/LA ratios are shown in Tables 2 through 4.

DISCUSSION/CONCLUSIONS

Accuracy is a combination of bias and precision in environmental measurement. Method bias is the tendency of the onsite analytical method to consistently overestimate or underestimate the true value. High bias (overestimating) is indicted by linear regression slopes greater than 1.0 and mean FA/LA ratios greater than 1.0. Low bias (underestimating) is indicated by linear regression slopes less than 1.0 and mean FA/LA ratios of less than 1.0. During the Phase I evaluation, both the colorimetric TNT and RDX methods exhibited low recovery when run without an OMC step (BSI 1995). During the Phase II evaluation the use of the OMC step on compost residues for the colorimetric methods significantly increased recoveries and is highly recommended for compost analysis.

Precision is a measurement of the size of closeness of agreement among individual measurements. Precision was assessed by the linear regression correlation coefficient (r) and FA/LA ratio standard deviation. The closer the (r) value is to 1.0, and the smaller the FA/LA standard deviation, the more precise the method. Based on (r) values and FA/LA standard deviation, colorimetric TNT results were consistently more precise than the immunoassay TNT results. In general, RDX

TABLE 2. Linear Regression Parameters

Compost Mix	TNT Colorimetric Slope	(r)	TNT Immunoassay Slope	(r)	RDX Colorimetric Slope	(r)	RDX Immunoassay Slope	(r)
Umatilla Phase I	0.32	0.81	---	---	0.24	0.75	---	---
Umatilla Phase II	0.69	0.94	0.82	0.78	0.90	0.86	0.82	0.52
SUBASE Bangor	1.03	0.96	0.74	0.56	---	---	---	---
Crane Mix #4	1.12	0.91	1.42	0.90	1.08	0.84	1.12	0.40
Crane Mix #5	1.05	0.96	1.64	0.36	0.70	0.39	0.90	0.42
Crane Mix #8	1.01	0.86	1.30	0.88	1.11	0.77	1.34	0.81
Phase II Mean	0.98	0.93	1.18	0.70	0.95	0.72	1.05	0.54

TABLE 3. Relative Percent Difference (RPD) Parameters

Compost Mix	TNT Colorimetric Mean	Med.	TNT Immunoassay Mean	Med.	RDX Colorimetric Mean	Med.	RDX Immunoassay Mean	Med.
Umatilla Phase I	126	112	---	---	132	131	---	---
Umatilla Phase II	34	33	46	48	20	14	33	36
SUBASE Bangor	53	51	56	46	---	---	---	---
Crane Mix #4	23	18	25	19	15	13	28	22
Crane Mix #5	6	4	50	44	35	28	24	23
Crane Mix #8	17	10	32	32	20	18	28	31
Phase II Mean	27	23	41	38	23	18	28	28

TABLE 4. FA/LA Ratio Parameters

Compost Mix	TNT Colorimetric Mean	S.D.	TNT Immunoassay Mean	S.D.	RDX Colorimetric Mean	S.D.	RDX Immunoassay Mean	S.D.
Umatilla Phase I	0.25	0.17	---	---	0.21	0.09	---	---
Umatilla Phase II	0.74	0.18	0.78	0.42	0.95	0.27	0.93	0.36
SUBASE Bangor	1.87	0.81	2.35	2.91	---	---	---	---
Crane Mix #4	1.17	0.23	1.25	0.35	1.07	0.17	1.15	0.34
Crane Mix #5	1.03	0.08	1.73	0.83	0.75	0.23	0.94	0.24
Crane Mix #8	1.16	0.33	1.28	0.35	1.12	0.24	1.34	0.20
Phase II Mean	1.19	0.33	1.48	0.97	0.97	0.23	1.09	0.29

precision based on (r) values and FA/LA standard deviation was similar between colorimetric and immunoassay methods, with the colorimetric method tending to be slightly better in some mix designs. Precision was further assessed by replicate and duplicate analyses. Quality assurance (QA) replicate and duplicate analyses were performed on all mix designs (data not shown). Based on replicate and duplicate analyses, colorimetric TNT results were significantly more precise than immunoassay TNT results. All of the colorimetric TNT replicate and duplicate results (100%) met QA/QC criteria. Only 38% of the immunoassay TNT replicate analyses and 80% of the duplicate analyses met QA/QC precision comparison criteria. All of the colorimetric RDX and immunoassay RDX replicate and duplicate analyses (100%) met QA/QC precision comparison criteria.

This study indicates that both colorimetric and immunoassay methods may be utilized in determination of explosives concentrations in compost residues during bioremediation treatment. There was reasonably consistent performance between the five compost mix designs tested in the Phase II evaluation. However, due to the large variation in nutrient amendments that could be utilized for composting, it is highly recommended that 20 to 30 samples be tested during pilot scale treatability studies to evaluate site specific correlation between the selected on-site methods and the laboratory method. It should be noted that sample heterogeneity for explosives in solid matrices (soils and compost) contributes significantly greater error than analytical differences between on-site and laboratory methods (EPA 1996b, 1998; Jenkins et al. 1996).

REFERENCES

Billets, S., G. Robertson, and E. Koglin. 1996. *A Guidance Manual for the Preparation of Characterization and Monitoring Technology Demonstration Plans.* U.S. Environmental Protection Agency, Characterization Research Division, Las Vegas, NV.

Bioremediation Service Inc. (BSI). 1995. *Data Analysis Report for the Composting Trial Test, Contaminated Soil Remediation, Explosives Washout Lagoons, Umatilla Depot Activity, Hermiston, Oregon.* Prepared for U.S. Army Corps of Engineers, Seattle District, Contract No. DACA67-C-0031, Seattle, WA.

Gilbert, R.O. 1987. *Statistical Methods for Environmental Pollution Monitoring.* Van Nostrand Reinhold Company, NY.

Grant, C. L., T.F. Jenkins, and A.R. Mudambi. 1996. *Comparison Criteria for Environmental Chemical Analyses of Split Samples Sent to Different Laboratories*, U.S. Army Corps of Engineers, Cold Regions Research and Engineering Laboratory, CRREL Special Report 96-9, Hanover, NH.

Jenkins, T.F., C.L. Grant, G.S. Brar, P.G. Thorne, T.A. Ranney, and P.W. Schumacher. 1996. *Assessment of Sampling Error Associated with the Collection and Analysis of Soil Samples at Explosives Contaminated Sites.* U.S. Army Corps of Engineers, Cold Regions Research and Engineering Laboratory, CRREL Special Report 96-15, Hanover, NH.

U.S. EPA. 1993. *Handbook: Approaches for the Remediation of Federal Facility Sites Contaminated with Explosive or Radioactive Wastes.* Office of Research and Development, EPA/625/R-93/013, Washington, D.C.

U.S. EPA. 1994. *SW-846 Method 8330, Nitroaromatics and Nitramines by High Performance Liquid Chromatography (HPLC), Revision 0.* Office of Solid Waste, Washington, D.C.

U.S. EPA. 1996a. *Engineering Bulletin: Composting.* Office of Research and Development, EPA/540/S-96/502, Cincinnati, OH.

U.S. EPA. 1996b. *Field Sampling and Selecting On-Site Analytical Methods for Explosives in Soil.* Federal Facilities Issue Paper, Characterization Research Division, EPA 540/R-97/501, Las Vegas, NV.

U.S. EPA. 1998. *Compost Field Screening Technologies, UMDA, SUBASE Bangor, and Crane NSWC, Final Report.* Prepared for U.S. EPA Region X, Project No. 71370, Seattle, WA.

FIELD MONITORING OF AEROBIC BIODEGRADATION OF 1,2-DICHLOROBENZENE

Rick Gillespie, Bruce Alleman, and Albert J. Pollack (Battelle, Columbus, OH)
Catherine M. Vogel (USAF AFRL/MLQ, Tyndall AFB, FL)
Jon Ginn (USAF OO/ALC, Hill AFB, UT)
R. Ken Crowe (BDM Management Services, Tyndall AFB, FL).

ABSTRACT: This study was performed in conjunction with a pilot-scale bioventing system located at a former chemical disposal pit at Hill AFB, Utah. The purpose of this study was to compare the efficacy of ex situ and in situ monitoring methods for determining O_2 utilization rates in the vadose zone. The field monitoring techniques implemented at the site included in situ O_2 sensors, an ex situ On-Line Environmental Monitoring System (OEMS), and manual sampling with gas analysis by hand-held instruments. Quarterly respiration tests were used to compare the precision and accuracy of the different methods. The results from the three methods of monitoring O_2 utilization rates were relatively comparable. Overall, the in situ O_2 sensors were the most effective method for recording reliable data during the respiration tests performed in this study.

INTRODUCTION

Bioventing is a conventional, cost-effective technology for the treatment of vadose-zone contaminated soils, through aeration of the soils to stimulate in situ biological activity and promote bioremediation. Bioventing typically is applied in situ to the vadose zone and is applicable to any chemical that can be aerobically biodegraded. Because the technology application for fuel hydrocarbons is well understood, the research emphasis has changed to focus on broadening the list of contaminants that can be remediated by bioventing.

A pilot-scale study is being conducted to examine the effectiveness of bioventing for treating non-fuel hydrocarbons that can be directly metabolized, such as chlorinated aromatic hydrocarbons (CAH) and polycyclic aromatic hydrocarbons (PAH), as well as compounds that are degraded co-metabolically such as trichloroethylene (TCE) and other chlorinated solvents (Battelle, 1997).

The pilot-scale bioventing system is located at a former chemical disposal pit at Hill AFB, Utah. The primary contaminant present at the site was 1,2-dichlorbenzene (1,2-DCB). Secondary contaminants present at the site included perchloroethylene (PCE), *cis*-1,2-dichloroethylene (1,2-DCE), chlorobenzene, 1,1,1-trichloroethane (1,1,1-TCA), and toluene. Data from a successful demonstration will be combined with other studies being conducted by the Air Force to expand the candidate contaminant list, and to produce an addendum to the Air Force's Principles and Practices of Bioventing Manual.

O_2 utilization rates are defined as the rate that O_2 is depleted in the soil gas over time following the shutdown of the O_2 source (i.e., system blower). O_2 utilization rates are used to calculate microbial degradation of contaminants at bioventing sites. A standard method for determining O_2 utilization requires manual extraction of gas from the vadose zone, followed by gas analysis with hand-held instruments. The purpose of this study was to compare the efficacy of ex situ and in situ monitoring methods for determining O_2 utilization rates in the vadose zone at the Hill AFB site.

Six in situ O_2 sensors were installed at multiple depths within the radius of influence of the bioventing system. A total of 24 soil-gas monitoring probes were installed to allow removal of gas from the vadose zone by both manual sampling and analysis, and automated sampling and analysis by the OEMS. The in situ O_2 sensors were placed adjacent to a soil-gas monitoring probe to allow direct comparisons. Quarterly respiration tests were performed with all three technologies to compare the precision and accuracy of the different methods.

Objective. The objective of this study was to compare the effectiveness of three different field methods and to determine the most effective method for monitoring aerobic biodegradation of 1,2-DCB. The three field methods used to monitor aerobic degradation of 1,2-DCB included: 1) in situ O_2 sensors, 2) automated sampling and analysis with an OEMS, and 3) manual sampling with analysis by hand-held meters. Several factors were used to evaluate each method, including precision and accuracy of measurements, system reliability, and relative cost.

MATERIALS AND METHODS

The bioventing study included monitoring respiration and removal of contaminants in an actively vented plot and a non-vented control plot. The results from the non-vented control plot will be used to assess the background reduction of contaminants without venting. The bioventing system was a single vent system designed for air injection. Eight tri-level soil-gas monitoring points were installed at discrete distances from the vent well. The system was powered by a 0.5-horsepower (hp) regenerative air blower that required 115-volt, electrical service.

Once the monitoring points were constructed, a tubing and sensor wire conduit was installed from each monitoring point to a field trailer near the center of the active treatment plot. The tubing and sensor wire conduit was plumbed to the interior of the field trailer. A panel was constructed to connect tubing from each individual monitoring point to quick-connect fittings. A total of 24 monitoring points were connected to the sampling panel. This design allowed for simple, quick sampling in a controlled environment. In addition, the 6 in situ O_2 sensors and the 24 type K thermocouples were connected to a 30-channel datalogger inside the field trailer.

In Situ O_2 Sensors and Data Acquisition System. In situ O_2 sensors produced by DataWrite™ were used to monitor O_2 concentrations in the vadose zone. These commercially-available sensors used a galvanic cell that produced a millivolt signal proportional to the O_2 concentration in the soil gas. An optional sideport was installed on the sensors to allow in situ calibration during the project. The sensors were calibrated during each shutdown test by injecting high-purity air (21% O_2) and nitrogen (N_2) gas (0% O_2) directly into the sensor housing.

The in situ O_2 sensors were connected to a Datataker 505™ 30-channel datalogger (DT 505). The DT 505 is a microprocessor-based datalogger capable of measuring a multitude of sensor types. The DT 505 allowed the manipulation and calibration of the six in situ O_2 sensors. Type K thermocouples were installed on the remaining 24 single-ended inputs on the DT 505. The thermocouple data were used to evaluate biodegradation rates versus seasonal soil temperatures; however, the data from the thermocouples are not presented in this paper.

OEMS System: The OEMS system, developed by Battelle, permitted automated soil-gas sampling and analysis. In addition, the microprocessor-based OEMS system allowed remote operation by modem. The OEMS employed a vacuum-assisted sampling train that pulled soil gas from monitoring points and delivered it to a common cell, containing a suite of sensors. The sensors responded to concentrations of O_2, CO_2, and TPH in the soil gas. The O_2 sensor was a galvanic cell that produced a millivolt signal proportional to the percent O_2 in the sample. The CO_2 and TPH sensors were infrared units that relied on the principle that most gases absorb energy at a particular wavelength.

During remote operation, the OEMS was automatically calibrated with a prepared, high-pressure cylinder gas standard at the beginning and end of each sampling event. The cylinder contained 5.09% O_2, 9.98% CO_2, and 506 ppmv hexane with N_2 balance gas. The OEMS was programmed to sample ambient air that contained 20.9% O_2, < 0.1% CO_2, and low TPH. This calibration step permitted ongoing validation of each sensor's response during soil-gas sampling.

Manual Soil-Gas Sampling for Field Analysis. Soil-gas samples were collected from each monitoring point directly into a Tedlar™ gas sampling bag using an evacuated chamber device. After an initial purge, the sampling process was repeated and a second sample bag was collected for field analysis. Manual samples were analyzed for O_2 and CO_2 concentrations with a GasTechtor™ 32520X. A Trace-Techtor™ was used to quantify TPH concentrations.

Meters were calibrated prior to each sampling event. A three-point method was used to calibrate the O_2 sensor on the GasTechtor™. A 20.9%, 10%, and a 0% O_2 standard were used to ensure accuracy of the measurements recorded during manual sampling. A 0% (ambient air) and 10% CO_2 standard (balance N_2) were used to calibrate the CO_2 meter. A hexane 510 ppmv standard was used to calibrate the TraceTechtor™ TPH sensor. It should be noted that CO_2 and TPH data are not presented in this paper.

In Situ Respiration Tests. Three quarterly in situ respiration tests were conducted between July 1997 and January 1998. The in situ respiration tests were performed to measure O_2 utilization rates to calculate biodegradation rates (Leeson and Hinchee, 1997). The O_2 utilization rates were calculated for each monitoring probe location by performing a regression analysis on the data that fell within the zero-order part of the curve. Contaminant biodegradation rates were calculated from the O_2 utilization rates using the stoichiometric relationship for the oxidation of a theoretical molecule $C_{4.2}H_{7.5}Cl_{.59}$. This molecule was determined using the mass-weighted average composition of the primary contaminants at the site.

Based on the O_2 utilization rates (% per day), the biodegradation rate (mg of contaminant per kg of soil per day) was estimated using the following equation.

$$K_{\beta} = \frac{-K_o A D_o C}{100} \quad (1)$$

where:

K_{β}	=	biodegradation rate (mg/kg/day)
K_o	=	O_2 utilization rate (percent per day)
A	=	volume of air/kg of soil (L/kg)
D_o	=	density of O_2 gas (mg/L)
C	=	mass ratio of contaminant to O_2 required for mineralization (mg contaminant/mg O_2).

Substituting a porosity value of 0.3, a soil bulk density of 1.4 g/cm^3 , an O_2 density of 1,330 mg/L, and a contaminant-to-O_2 mass ratio of 1:2.943 yields equation 2 (Leeson and Hinchee, 1997).

$$K_{\beta} = (-.97) K_o \quad (2)$$

RESULTS AND DISCUSSION

Respiration Tests. The three in situ respiration tests were conducted in July 1997, October 1997, and January 1998. Figure 1 presents results from the three methods at a common monitoring point during a single respiration test.

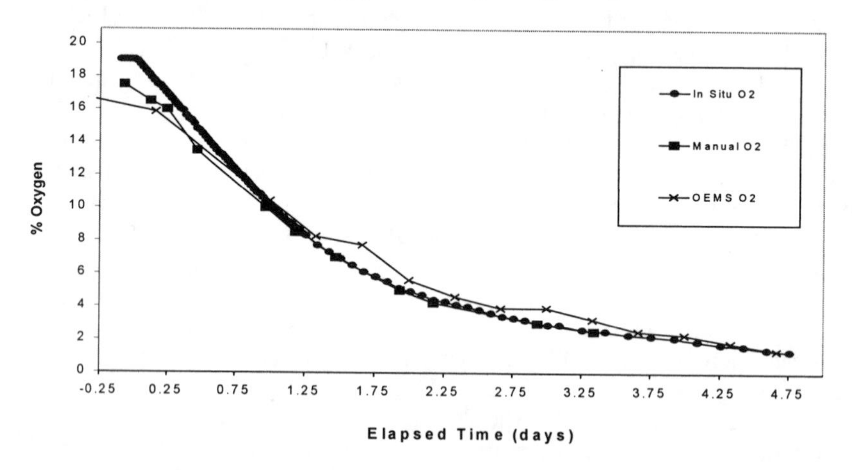

Figure 1. Selected O_2 Utilization Rates Measured with the DT 505, Manual Sampling, and the OEMS During a Quarterly Respiration Test at Hill AFB

The O_2 utilization rates were calculated for each method by performing a regression analysis of the zero-order part of the O_2 utilization curve. In addition, the coefficient of determination (r^2) was determined for each method at each monitoring point to interpret the proportion of observed O_2 variation versus time (Devore, 1991). The r^2 value was used to estimate the precision of each method. Table 1 presents the results of the three different methods for each respiration test.

Table 1. Quarterly Respiration Test Results.

SAMPLING POINT ID	IN SITU O_2 SENSORS			OEMS			MANUAL SAMPLING		
July 1997	K_o	K_b	r^2	K_o	K_b	r^2	K_o	K_b	r^2
MPA-12'	-5.20	5.04	.991	(b)	(b)	(b)	-5.54	5.37	.992
MPB-17'	(a)	(a)	(a)	(a) (b)	(a) (b)	(a) (b)	(a)	(a)	(a)
MPC-12'	-12.99	12.60	.965	(b)	(b)	(b)	-12.92	12.53	.998
MPD-17'	-12.80	12.42	.995	(b)	(b)	(b)	-11.70	11.35	.999
MPE-7'	-3.71	3.60	.987	(b)	(b)	(b)	-2.83	2.75	.978
MPG-12'	-1.03	1.0	.994	(b)	(b)	(b)	-.984	0.95	.954
Oct 1997									
MPA-12'	-5.47	5.31	.999	-2.98	2.89	.816	-6.93	6.72	.928
MPB-17'	-23.94	23.22	.981	-5.32	5.16	.238	-22.20	21.53	.926
MPC-12'	-5.06	4.91	.979	-3.79	3.68	.904	-4.78	4.64	.969
MPD-17'	-16.09	15.61	.999	-6.67	6.47	.631	-14.41	13.98	.980
MPE-7'	-2.94	2.85	.979	-2.38	2.31	.966	-2.32	2.25	.977
MPG-12'	-2.21	2.14	.992	-1.92	1.86	.933	-2.10	2.04	.897
Jan 1998									
MPA-12'	-2.37	2.30	.987	-1.91	1.85	.921	-2.02	1.96	.987
MPB-17'	-8.47	8.22	.994	-4.34	4.21	.920	-7.48	7.26	.992
MPC-12'	-13.46	13.06	.961	-2.94	2.85	.920	-12.75	12.37	.989
MPD-17'	-9.60	9.31	.995	-6.08	5.90	.962	-8.22	7.97	.994
MPE-7'	-2.53	2.45	.984	-2.14	2.08	.976	-2.20	2.13	.988
MPG-12'	-1.93	1.87	.997	-1.41	1.37	.953	-1.38	1.34	.996

(a) – insufficient O_2 concentrations to perform respiration test
(b) – OEMS not installed prior to start of respiration test

During the initial respiration test performed in July 1997, O_2 utilization rates were measured with manual sampling and in situ O_2 sensors. The OEMS began measurements approximately 1.25 days after the beginning of the respiration test. The second and third respiration tests conducted in October 1997 and January 1998 allowed comparison of all three methods. Overall, the three methods resulted in O_2 utilization rates that were relatively comparable. Manual sampling and the in situ O_2 sensors provided results that were similar at every point. The OEMS measurements were similar to the other methods at most locations, with the exception of those points with higher O_2 utilization rates.

The in situ O_2 sensors with the DT 505 performed well during the three respiration tests. This method presented several advantages over the other two methods, including near limitless data collection and precise measurements. The

512 kilobyte (Kb), non-volatile memory card used with the datalogger provided safe storage for recorded data. One of the disadvantages associated with using the in situ O_2 sensors with DT 505 was the inability to perform remote calibration and download of recorded data. However, the DT 505 can be designed to be operated remotely with a modem. Another disadvantage of the in situ O_2 sensors is that retrieval of the sensors is impracticable at project completion. This results in a significant capital cost for sensors that have to be replaced at each new site.

The advantages of the OEMS included automated validation of the gas-constituent sensors by its on-line calibration capability; collection of CO_2, TPH, and soil-gas pressure data in addition to the O_2 readings; remote control and data transfer via modem; and the ability to relocate the entire monitoring system to another remediation site after completion of the project. The major disadvantage of using the OEMS at Hill AFB was the 6.5-hr time period needed to complete a sampling event for all 24 soil-gas monitoring points, which resulted in less O_2 readings, 3 per day, than either the manual or in situ O_2 sensor methods.

The advantages of manual sampling included real-time data collection, a well-established method, and frequent calibration. The use of hand-held meters allowed personnel to observe O_2 utilization rates in the field and to focus sampling on monitoring points with higher rates of aerobic microbial activity. The disadvantages of manual sampling and analysis with the GasTechtor™ and TraceTechtor™ at Hill AFB included labor required to monitor O_2 utilization rates, difficulty with calibration, and frequent operation and maintenance problems. Manual sampling required staff to be present at the site for extended periods of time during the respiration test. This labor requirement can result in higher costs compared to the two automated methods of sampling and analysis.

Overall, the in situ O_2 sensors with the DT 505 was the most effective method for measuring O_2 utilization rates at Hill AFB. These automated sensors allowed collection of reliable data at frequent intervals. Figure 1 shows readings every 30 minutes for the first 24 hours of the respiration test. This frequency facilitated the determination of the zero-order part of the O_2 utilization curve, and would have been unattainable using the other two methods. In addition, the in situ sensors consistently had the highest r^2 values for the recorded data.

REFERENCES

Battelle. 1997. *Final Work Plan for Demonstration of Bioventing of Non-Petroleum Hydrocarbon Contamination at Hill Air Force Base, Utah.* Report Prepared for the Environics Directorate of the Armstrong Laboratory, Tyndall AFB, FL, June.

Devore, J. 1991. *Probability and Statistics for Engineering and the Sciences.* 3rd edition. Brooks/Cole Publishing Co., Pacific Grove, CA.

Leeson, A. and R. Hinchee. 1997. *Soil Venting.* CRC Press, Inc. Boca Raton, FL.

AN INTEGRATED LANDFILL MODELING SYSTEM FOR EVALUATING REMEDIATION ALTERNATIVES

Paul J. Van Geel (Carleton University, Ottawa, Ontario, Canada)
Grant R. Carey (Environmental Software Solutions, Ottawa, Canada)
Edward A. McBean and Frank A. Rovers (Conestoga-Rovers & Associates, Waterloo, Canada)

ABSTRACT: An Integrated Landfill Modeling System (ILMS) has been developed to provide an important predictive tool for evaluating the effects of various design parameters on the performance of landfill remediation alternatives. The ILMS combines the HELP model for predicting the hydrologic performance of various landfill cap configurations, a leachate composition model, and a groundwater biodegradation-redox model (BIOREDOX) to predict the redox-dependent biodegradation of chlorinated aliphatic hydrocarbons (CAHs) in landfill-impacted aquifers. This study presents an application of the ILMS for a hypothetical landfill site, to determine the effects of landfill cap permeability on the extent of PCE, TCE, DCE, and vinyl chloride plumes below the landfill. Model input parameters are based on field conditions reported in the literature. Simulation results indicate that an engineered permeable landfill cap may be more effective than a low permeability cap at some sites because it maintains favourable conditions for the bioattenuation of CAHs.

INTRODUCTION

As late as 1990, approximately 25% of the sites on the Superfund National Priority List were municipal landfills that accepted hazardous waste, including CAHs. A principal concern associated with these landfills is the impact to underlying groundwater resources. An innovative remediation strategy under consideration for unlined municipal landfills, where natural attenuation is shown to be effective in neutralizing risks to downgradient receptors, involves the placement of an aerobic, permeable landfill cap. The permeable cap must be carefully engineered to control the moisture percolation rate into the refuse and the resulting leachate flux through the base of the landfill, so that landfill-derived pollutants such as CAHs will continue to be effectively attenuated within and below the landfill.

A common remediation strategy for unlined landfills is to place a low permeability cap in order to reduce the total flux of leachate to the underlying aquifer. A net reduction in the mass flux to the underlying aquifer will result in lower concentrations immediately below the landfill due to mixing or dilution. A low permeability cap will also extend the contaminating life span of the landfill. The net reduction in mass flux to the underlying aquifer will alter the distribution of redox zones in the underlying aquifer. The magnitude of the anaerobic zone is likely to decrease as a result. However, the change in redox conditions below the landfill may adversely affect the degradation of landfill-derived contaminants such as CAHs.

REDOX-DEPENDENT BIODEGRADATION

The oxidation of landfill leachate and the corresponding reduction of various electron acceptors results in the formation of various redox zones in the underlying aquifer below landfills. The sequence of redox zones of increasing redox potential that may be generated downgradient from a landfill are as follows: methanogenic zone, sulfate-reducing zone, iron-reducing zone, nitrate-reducing zone, and aerobic zone.

Reductive dechlorination is one of the principal mechanisms involved in the degradation of CAHs in the anaerobic zone. The anaerobic reduction of PCE, TCE, DCE and vinyl chloride to ethene has been observed to varying degrees at many sites. McCarty (1996) provides a general review of the redox-dependent biodegradability of chlorinated ethenes. In general, the complete transformation of PCE and TCE to ethene will only occur in the methanogenic zone. PCE and TCE will be reduced to DCE in the sulfate-reducing and iron-reducing zones; however, DCE will not be further transformed to vinyl chloride or ethene. Vinyl chloride may be oxidized under aerobic conditions or under anaerobic conditions where iron is the electron acceptor (Bradley and Chapelle, 1996).

MODEL SCENARIO

A hypothetical landfill scenario was developed to illustrate the application of the ILMS to evaluate various landfill remediation alternatives. The landfill scenario is based on an old, unlined landfill overlying a shallow, unconfined aquifer. Contaminants were assumed to have leached from the landfill over a period of 15 years at which time two remedial alternatives were proposed: an engineered permeable cap that maintains a controlled rate of infiltration into the landfill, and a low permeability cap which significantly reduces the infiltration rate. Both scenarios were modeled for an additional 5 years after placement of the respective caps.

The hydrogeochemical setting for this hypothetical scenario is similar to conditions reported for the Vejen Landfill Site in Vejen, Denmark (Lyngkilde and Christensen, 1992; Heron et al., 1995).

INFILTRATION MODEL

The Hydrologic Evaluation of Landfill Performance (HELP) model was used to estimate the landfill infiltration rates for the different landfill cap configurations. The soil types and design configurations for the HELP model simulations are provided in Figure 1. Based on the HELP model predictions, 30 cm and 2.5 cm of percolation enter the landfill each year for the permeable and low permeability caps, respectively.

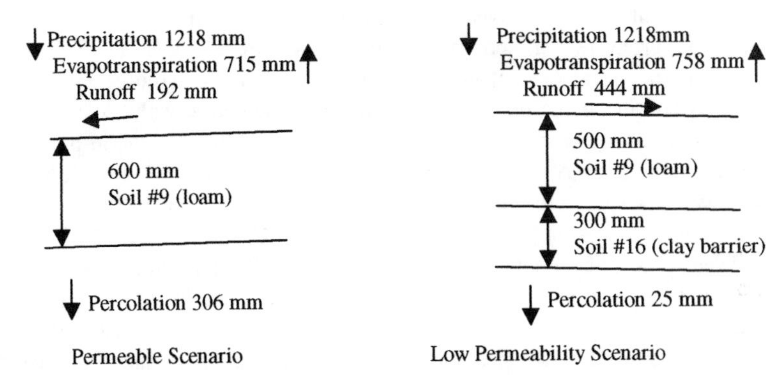

FIGURE 1. Permeable and Low Permeability Landfill Cap Scenarios

LEACHATE MODEL

The landfill total organic carbon (TOC) concentration was represented in the leachate composition model because TOC serves as the primary substrate for microbial reactions in the underlying aquifer. Based on the leachate model presented by Lu et al. (1981), the initial TOC concentration was assumed to be 14,000 mg/L, and the TOC concentrations were assumed to decline exponentially with a first order rate of 0.26 yr^{-1}. The decline in concentrations was assumed to occur over a ten-year period at which time the TOC concentration was assumed to remain constant at 1,000 mg/L. The long term TOC concentrations reflect conditions seen at the Vejen Landfill (Heron et al., 1995). The leachate concentrations for PCE, TCE, DCE, and vinyl chloride were assumed to remain constant at 200 µg/L for the entire simulation. This simplified landfill leachate model was assumed for this study, however, additional research is required to develop a leachate composition model that more accurately reflects the processes occurring within the landfill. Specifically, the change in leachate concentrations after the placement of a low permeability cap requires further investigation.

GROUNDWATER FLOW MODEL

A two-dimensional cross-section of the unconfined aquifer was modeled using MODFLOW. The groundwater flow simulation was based on an average aquifer thickness of 10 metres, an average linear groundwater velocity of approximately 200m/year, and a uniform regional recharge rate of 30 cm/year. The recharge rate below the landfill was specified to be consistent with the infiltration rates determined for each landfill cap configuration. The landfill was assumed to be 100m in length.

BIOREDOX MODEL

BIOREDOX was used to simulate the fate and transport of the leachate plume in the shallow aquifer, including a representation of the coupling between the oxidation of landfill-derived TOC and the reduction of available electron acceptors in groundwater and in the soil. TOC, PCE, TCE, DCE, and vinyl chloride are the organic leachate constituents included as solutes in the model simulation. The inorganic electron acceptors represented as solutes in the model include dissolved oxygen, nitrate, and sulfate, as well as mineral-phase ferric hydroxide coatings on soil particles. It was assumed that carbon dioxide was non-limiting with respect to methanogenic activity. The BIOREDOX governing equations and verification program are described by Carey et al. (1998a,b)

The redox-dependent biodegradability of the chlorinated ethenes was specified in the model using biodegradation rates presented in Table 1. It was also assumed that the reductive dechlorination rates for the chlorinated ethenes decreased when the low-permeability cap was placed over the landfill, because of the corresponding decrease in TOC concentrations below the landfill. Further study is warranted to define a general relationship between reductive dechlorination rates and TOC concentrations in leachate plumes.

TABLE 1. Biodegradation Rates

	Redox Zone				
Organic	CH_4	SO_4	*Fe(III)*	NO_3	O_2
TOC	k_{TOC}	k_{TOC}	k_{TOC}	k_{TOC}	k_{TOC}
PCE	k_{RD}	0.75	0.5 k_{RD}	0	0
TCE	k_{RD}	0.75	0.5 k_{RD}	0	0
DCE	k_{RD}	0	0	0	0
Vinyl	0.75	0	k_{OXD}	0	k_{OXD}

Scenar	K_{TOC} (d^{-1})	k_{RD} (d^{-1})	k_{OXD} (d^{-1})
no cap	7.6E-	7.6E	7.6E
high-K	7.6E-	7.6E	7.6E
low-K	7.6E-	1.9E	7.6E

Figure 2a presents the redox zone distribution that was simulated using BIOREDOX, based on the initial 15 years of TOC migration from the landfill. The general sequence of redox zones simulated using BIOREDOX is consistent with what was expected, and compare favorably with the redox zones observed in the shallow aquifer underlying the Vejen landfill site (Lyngkilde and Christensen, 1992a). Figures 2b and 2c illustrates the simulated redox zones for the permeable cap and low-permeability cap scenarios corresponding to a time of 5 years after placement of the respective caps. Figure 2c illustrates how the placement of the low-permeability cap, and the corresponding decline in TOC concentrations in the aquifer, resulted in a significant shift in the redox zone distribution in the aquifer.

To illustrate the potential effects of the shift in the redox zones on the degradation of CAHs in groundwater, TCE concentration contour corresponding to 5 ug/L is also plotted on Figure 2. For the purpose of this discussion, the extent of the TCE plume will be defined by the 5 ug/L concentration contour. For the permeable cap scenario, the TCE plume remains within the iron-reducing zone where TCE will still degrade based on the conceptual model defined earlier. After the placement of the low permeability cap and the shift in the redox zones, a portion of the TCE plume downgradient of the landfill is no longer contained within the strongly reducing zones of the anaerobic zone where conditions are favorable for TCE degradation. Hence, this portion of the plume is allowed to continue to migrate downgradient of the site. Attenuation of the TCE will still occur as a result of dilution via dispersion, however, no further degradation will occur in the aerobic zone.

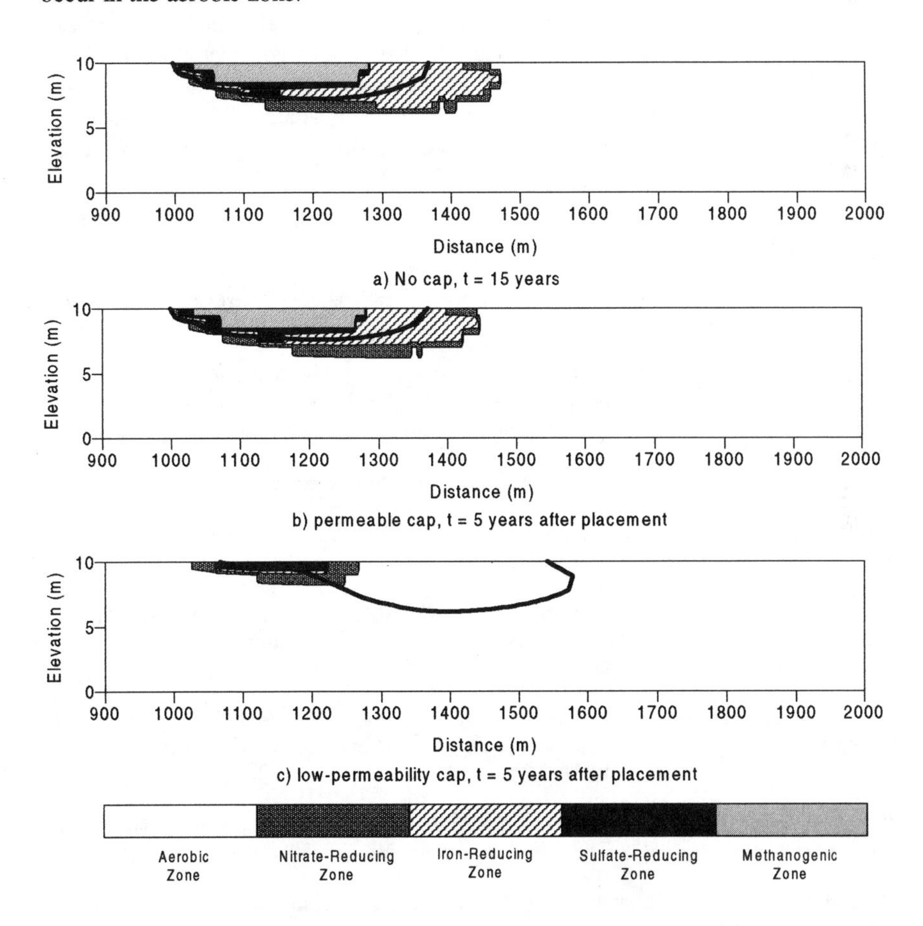

FIGURE 2. Simulated Redox Zones and 5 ug/L TCE Concentration Contour

SUMMARY

This study demonstrates the application of an Integrated Landfill Modeling System (IMLS). Linking the cover design (HELP model) and leachate model with groundwater flow and transport codes (MODFLOW and BIOREDOX), provided an effective tool for predicting the effects of various design parameters on the performance of different landfill remediation strategies. The hypothetical landfill scenario presented is an example of how the ILMS might be applied at an actual site.

The ILMS was used to simulate two landfill remediation alternatives. This example illustrates conditions for which the placement of a low-permeability cap may cause a detrimental impact to groundwater quality. Although this initial assessment is based on a specific setting and several simplifications and assumptions, it provides important evidence that serious consideration should be focused on the risk and cost implications of low-permeability caps. Particular attention should be paid to alternative remedies (such as aerobic, permeable caps) that can be engineered to suit site-specific conditions and to provide more cost-effective remedial solutions.

REFERENCES

Bradley P.M. and F.H. Chapelle. 1996. "Anaerobic Mineralization of Vinyl Chloride in Fe(III)-Reducing Aquifer Sediments." *Environmental Science & Technology*, 30, 2084-2086.

Garey, G.R., P.J. Van Geel, and J.R. Murphy, 1998a, *BioRedox: A coupled Biodegradation-Redox Model for Simulating Natural and Enhanced Bioremediation of Organic Pollutants – V1.0 User's Guide,* Conestoga-Rovers & Associates, Waterloo, Ontario, Canada.

Garey, G.R., P.J. Van Geel, and J.R. Murphy, 1998b, *BioRedox: A coupled Biodegradation-Redox Model for Simulating Natural and Enhanced Bioremediation of Organic Pollutants – V1.0 Verification Manual,* Conestoga-Rovers & Associates, Waterloo, Ontario, Canada. In preparation.

Heron G., P.L. Bjerg, and T.H. Christensen. 1995. "Redox Buffering in Shallow Aquifers Contaminated by Leachate." *Intrinsic Bioremediation: Third International In Situ and On-Site Bioreclamation Symposium,* Batelle Press, Columbus, OH, 143-151.

Lu J.C.S., R.D. Morrison, and R.J. Stearns. 1981. "Leachate Production and Management From Municipal Landfills: Summary and Assessment." In: *In-Land Disposal Municipal Solid Waste Proc. Seventh Ann. Res. Symp.*, EPA/600/9-81-002a.

Lyngkilde J., and T.H. Christensen. 1992. "Redox Zones of a Landfill Leachate Pollution Plume (Vejen, Denmark)." *Journal of Contaminant Hydrology*, 10, 273-289.

McCarty P.L. 1996. "Biotic and Abiotic Transformations of Chlorinated Solvents in Ground Water." In: *Symposium on Natural Attenuation of Chlorinated Organics in Ground Water,* EPA/540/R-96/509, 5-9.

LATERAL AND VERTICAL CHARACTERIZATION OF VOCs IN ALLUVIAL AQUIFERS

Mark A. Trewartha (SECOR International Inc., Tualatin, Oregon)

ABSTRACT: By utilizing depth-specific discrete groundwater sampling techniques and multilevel well equipment, multiple water-bearing zones within alluvial aquifers can be laterally and vertically characterized and monitored with significant reductions in time and cost. This approach was used to characterize the distribution of dissolved phase volatile organic compounds (VOCs) in thick sequences of unconsolidated sediments at two sites in Oregon. During site investigation activities, depth-specific groundwater samples were collected during drilling operations utilizing a discrete sampler. By reviewing analytical results of depth-specific samples prior to well installation, optimum placement of groundwater monitoring well screen interval or MGMS well sampling ports was achieved. In addition, by installing MGMS wells with multiple sampling ports, groundwater samples and hydraulic head measurements can be collected from multiple aquifer zones within one borehole location. By utilizing these techniques and equipment, the magnitude and lateral and vertical extent of VOCs was successfully delineated. The MGMS wells also allow continued monitoring of VOC concentrations in groundwater and groundwater gradient and flow direction in multiple aquifer zones. In addition, by monitoring aquifer conditions from multiple zones in one borehole location, the number of required borehole locations was reduced and subsequently the costs for site investigation activities were reduced.

INTRODUCTION

Two sites located in Oregon overlie thick unconsolidated sequences of alluvial deposits forming the Linn Gravels and Troutdale Formation. These formations are the result of thick alluvial sequences of sand, gravel, silt, and clay deposited from early Pliocene time to the present. During Pleistocene glacial epochs, alluvium was continually deposited in fans at the mouths of mountain canyons and as gravel trains in the southern part of the area, and sediment-laden glacial meltwater from the Columbia River system inundated the Willamette Valley. Within the areas of the two sites, these formations range in thickness from 120 to 260 feet.

The two sites were suspected of containing dissolved phase VOCs in the groundwater due to historical site activities. Based on analytical results of groundwater samples collected from onsite or nearby drinking water supply wells, the vertical extent of VOC contamination was suspected to be approximately 100 feet below the water table, approximately 110 feet below ground surface (bgs).

Objective. The primary objectives were to further assess the magnitude and lateral and vertical extent of VOC contamination in the unconsolidated alluvial deposits beneath each site. Depth-specific discrete groundwater sampling techniques and multilevel well equipment were utilized to characterize the sites and provide a network of wells that could adequately monitor VOC concentrations in groundwater and groundwater gradient and flow direction.

EQUIPMENT AND METHODS

Boreholes were drilled using dual-wall, reverse air circulation, percussion hammer drill rig or an air-rotary drill rig with outer steel casing. These techniques utilize an outer steel casing which is driven downhole as the borehole is drilled. Based on the drilling techniques available, these types of drilling operations were chosen for the following reasons:

- representative formation samples could be obtained;
- drilling through cobble-laden formations was relatively quick;
- groundwater samples representative of depth-specific aquifer zones were easily obtained as single and dual-walled drill stems temporarily cased the boring during drilling operations.

Discrete groundwater samples were collected from selected depths as boreholes were drilled. The purpose of collecting these samples was to provide analytical results of multiple water-bearing zones within the alluvial aquifer. These results would provide a detailed vertical profile of contaminant distribution.

During borehole advancement, drilling was stopped at 10- or 20-foot intervals. As groundwater entered the well casing through the bottom of the borehole, an electronically controlled discrete sampler was lowered down the drill casing to near the bottom of the hole. The sampler remained closed as it was lowered to the desired sampling depth. After the sampler reached the designated sampling depth, a solenoid was activated from the surface which opened the sampler. The sampler remained open for approximately one minute and was then closed and brought to the surface. A quick disconnect metering valve was utilized to transfer the groundwater sample from the discrete sampler into 40 milliliter volatile organic aromatic (VOA) vials for analysis.

Well Construction. One cluster of three groundwater monitoring wells and subsequent multi-level groundwater monitoring system (MGMS) wells were installed to characterize and monitor the lateral and vertical extent of dissolved VOC compounds in the water-bearing zones of the unconsolidated alluvium. Because the MGMS wells allow monitoring and sampling of several aquifer zones from a single borehole, MGMS wells were selected instead of additional well cluster configurations.

Groundwater monitoring wells were constructed of 4-inch-diameter Schedule 80 PVC blank casing and 0.020-inch slotted PVC well screen in the bottom ten feet. MGMS wells consisted of 2-inch inside diameter (ID), Schedule 80 PVC outer casing, with sample ports set at specific intervals and depths which were selected based on analytical results of discrete groundwater samples and formation lithology.

The sample ports are 6-inch long enclosed sections of PVC with a single row of machine-cut holes. Due to the large diameter of holes, and the possibility of the annular sand filter pack clogging them, an inert geotechnical fabric was secured over each sample port using plastic fasteners. Subsequent to 1996, samples ports were encased in 2½-foot sections of 0.010-inch slotted PVC. A dedicated pump was installed at each sample port interval. Two small diameter hard poly tubes are attached to each pump. The tubes run to the surface enclosed within the outer casing. One tube serves as the drive line for the pump, while the other tube serves as the sample line. In addition to these two tubes, a larger "openhole" tube is attached at each sample port. This "openhole" tube, which also runs to the surface enclosed within the outer casing, serves as an access for groundwater level measurements and as a backup sample collection point should the dedicated pump fail.

Groundwater Sampling Methods. Prior to sample collection, each groundwater monitoring well was purged to evacuate oxidized water from the well column. As soon as the well recovered sufficiently, a sample was collected and tested in the field for groundwater parameters including pH, temperature, and specific conductivity. The well was then retested for groundwater parameters after sampling as a measure of purging efficiency and as a check on the stability of the water samples over time. Groundwater samples were collected from the well screen interval utilizing the discrete sampler or submersible pump at a low flow rate (0.5 to 1.0 gallon per minute).

Two procedures were utilized to purge and sample the MGMS ports: sampling by the in-line dual-valve pump or sampling by utilizing a single-valve sampler. The pumps are designed to be operated from the surface utilizing compressed nitrogen. The nitrogen is applied to drive water up the sample tube, and a vent cycle is used to allow the sample port to recharge. The drive and vent cycles are timed such that a continuous stream of water is discharged to the surface, without any air bubbles entrained in the sample tube.

When a dedicated pump malfunctioned, groundwater samples were collected from the larger open hole tubes with a single valve system utilizing the openhole hard poly tube. The single valve system consists of a dedicated floating Teflon ball and dedicated ¼-inch OD hard poly tubing. The Teflon ball was inserted into the open hole tubing and the ¼-inch OD hard poly tubing advanced to the level of the sample port. A manifold was connected to the open hole tubing, with the ¼-inch OD hard poly tubing passing through it. The open hole tubing served as the drive line, and the ¼-inch OD tubing served as the sampling line. Purging procedures were conducted as with the in-line dual valve pumps.

Analytical program. Groundwater samples collected from the groundwater monitoring wells and MGMS well sampling ports were placed on ice in an insulated cooler and submitted to an onsite mobile laboratory or an offsite laboratory. Groundwater samples were analyzed for halogenated volatile organic compounds (HVOCs) by Environmental Protection Agency (EPA) Method 8010.

RESULTS AND DISCUSSION

Hydrostratigraphic Units. The main aquifer zones in the study area are the sediments of the Linn Gravels and Troutdale Formation which consist of interfingered and layered unconsolidated to semi-consolidated sediments deposited by fluvial and lacustrine processes. Thin clay and silt horizons were observed in several of the deep soil borings. These zones were not encountered in all deep soil borings at each site and are considered discontinuous.

Vertical Distribution of Contaminants. VOC concentrations from depth-specific discrete groundwater samples collected during the drilling of boreholes and analytical results of groundwater samples collected from subsequently installed groundwater monitoring wells and MGMS well sampling ports are presented graphically on Figures 1 through 3.

Initially at the first site, a cluster of three groundwater monitoring wells with 10-foot slotted screen sections were installed based on field observations (i.e., lithology, sample screening techniques). The groundwater monitoring wells were installed in the shallow, intermediate, and deep zones of the underlying alluvial aquifer. Depth-specific discrete groundwater samples collected during drilling were submitted to an offsite laboratory and analyzed under standard turn-around time. Review of analytical results of depth-specific discrete groundwater samples collected during drilling indicated that the highest concentrations of VOCs existed between the screen intervals of the shallow and intermediate groundwater monitoring wells. Subsequently, depth-specific discrete groundwater samples were submitted to an onsite mobile laboratory or an offsite laboratory and analyzed under an expedited turn-around time.

Once boreholes had reached total depth and analytical results of depth-specific discrete groundwater samples were reviewed, a groundwater monitoring well or MGMS well was installed. The location of the well screen interval and locations of the MGMS sampling ports were based analytical results of the depth-specific discrete groundwater samples. Although a few anomalies were noted, comparison of analytical results of groundwater samples collected from groundwater monitoring wells and MGMS well sampling ports and analytical results of depth-specific discrete groundwater samples collected from similar depths during drilling indicated similar VOC concentrations and trends in VOC concentration distribution.

By reviewing analytical results of depth-specific samples prior to well installation, optimum placement of groundwater monitoring well screen interval or MGMS well sampling ports was achieved. In addition, by installing MGMS wells with multiple sampling ports, continued monitoring of VOC concentrations in groundwater and measuring of hydraulic head can be conducted from multiple aquifer zones within one borehole location.

CONCLUSIONS

By utilizing these techniques, the magnitude and lateral and vertical extent of VOCs can be successfully delineated and optimum monitoring of aquifer conditions in various discrete zones is possible. In addition, by using the MGMS well equipment, monitoring aquifer conditions from multiple zones in one borehole location is possible. An overall reduction in the number of required boreholes and cost for site investigation can thus be achieved.

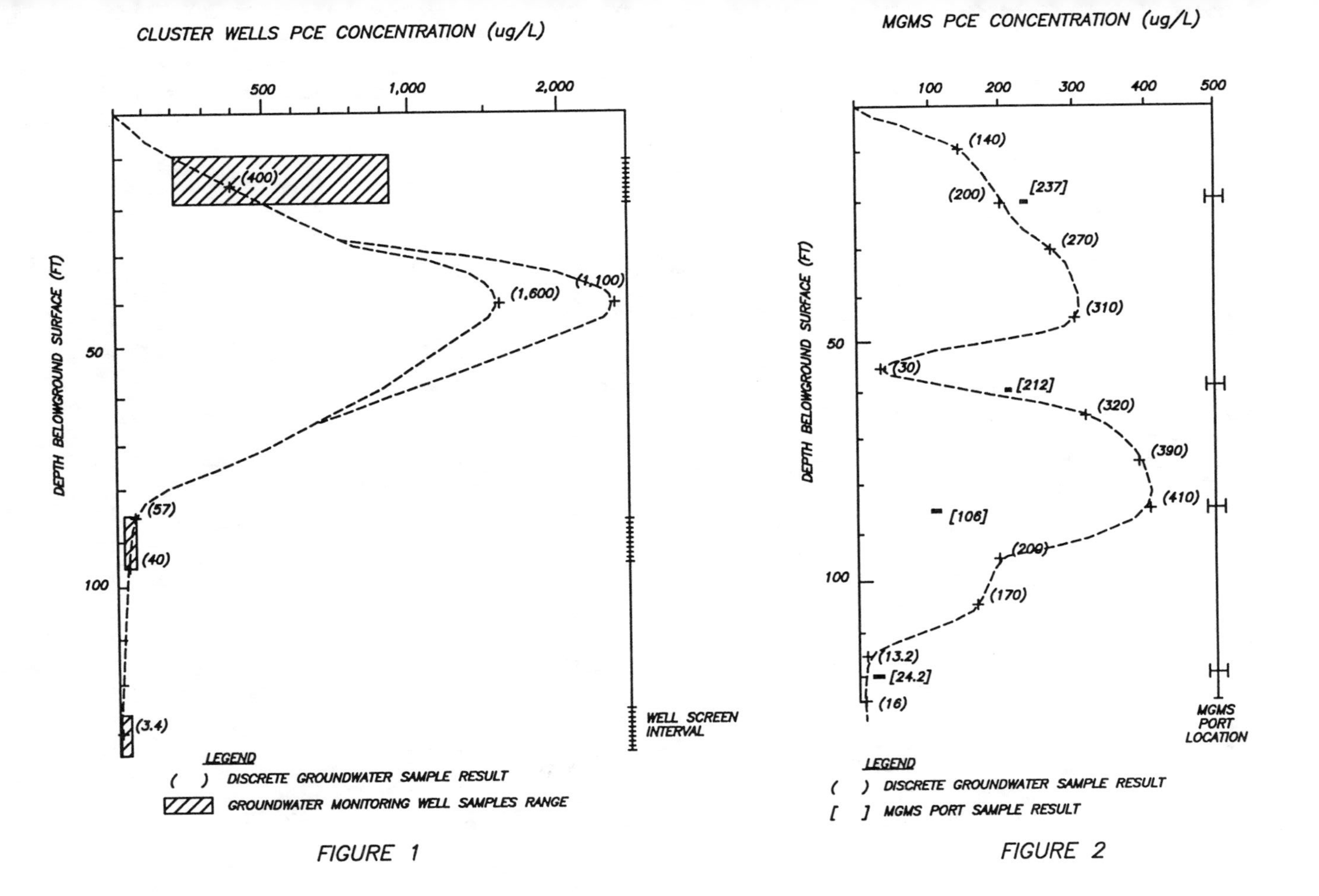

FIGURE 1

FIGURE 2

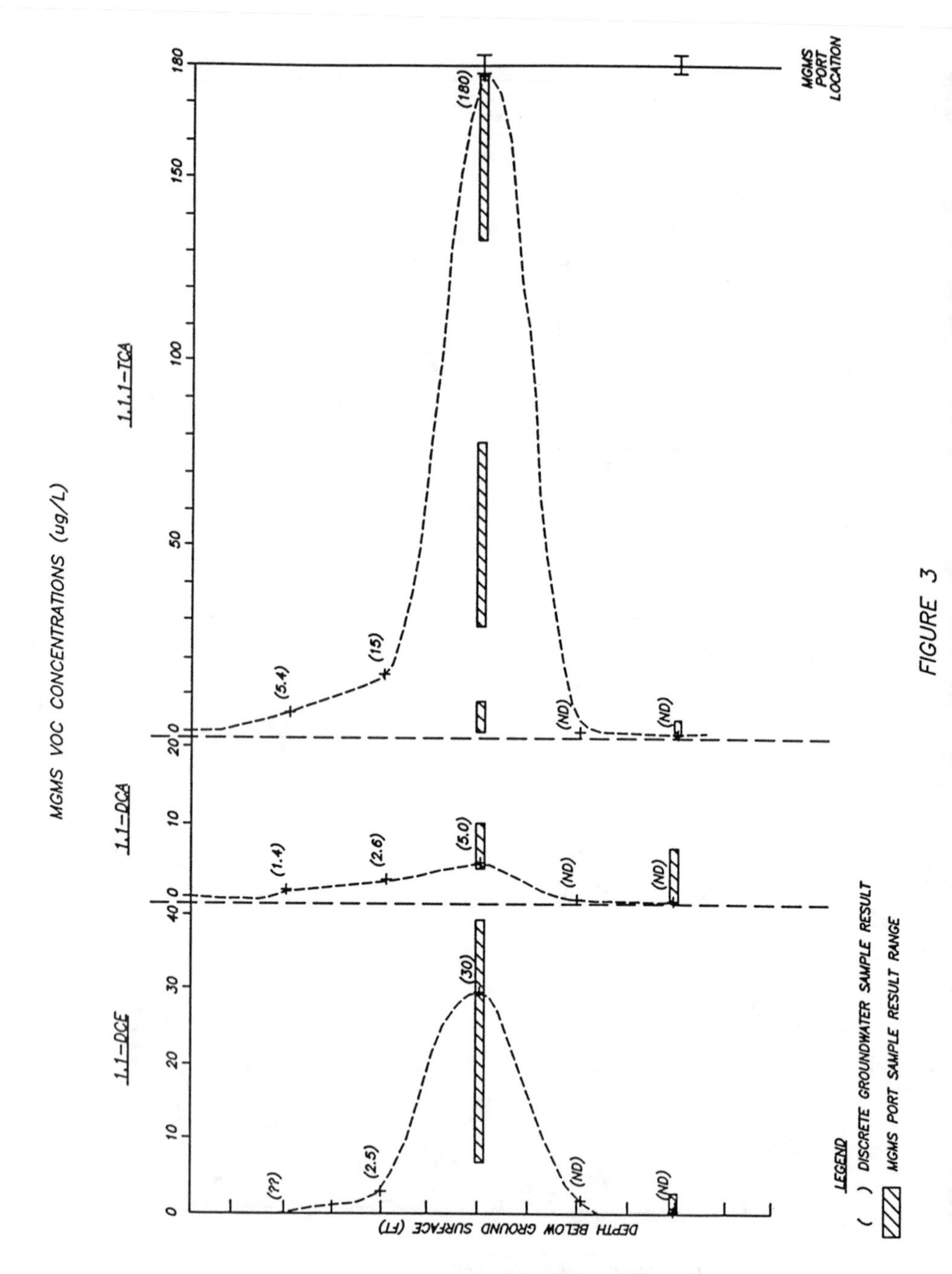
MGMS VOC CONCENTRATIONS (ug/L)
1.1.1-TCA
1.1-DCA
1.1-DCE
DEPTH BELOW GROUND SURFACE (FT)
MGMS PORT LOCATION
(180)
(15)
(5.4)
(ND)
(5.0)
(2.6)
(1.4)
(30)
(2.5)
(??)
0
10
20
30
40
50
100
150
180
LEGEND
( ) DISCRETE GROUNDWATER SAMPLE RESULT
MGMS PORT SAMPLE RESULT RANGE

FIGURE 3

REFERENCES

Burns, S. 1998. "Environmental, Groundwater, and Engineering Geology: Applications in Oregon." In R.E. Trewartha and Hunt, Characterization of Dissolved Organic Compounds in the Linn Gravels Aquifer, Salem, Oregon, pp. 673-689. Star Publishing Company, Belmont, California.

Piper, Arthur M., 1942, Ground-Water Resources of the Willamette Valley, Oregon, U.S. Geological Survey Water-Supply Paper 890.

Price, Don, 1967, Ground Water in the Eola-Amity Hills Area, Northern Willamette Valley, Oregon, U.S. Geological Survey Water Supply Paper 1847.

COMBINING CHEMOMETRIC AND GRAPHICAL METHODS AS TOOLS IN HYDROGEOLOGIC CHARACTERIZATION

Thomas A. Delfino (Geomatrix Consultants, Inc., San Francisco, California)
Scott D. Warner (Geomatrix Consultants, Inc., San Francisco, California)
Scott L. Neville (Aerojet General Corporation, Sacramento, California)
Craig A. Stewart (Geomatrix Consultants, Inc., Newport Beach, California)

ABSTRACT: A chemometric method (principal component analysis) was used in combination with two graphical methods (mapping a ratio of major cation concentrations [as percent milliequivalents per liter (meq/L)] and plotting Piper trilinear diagrams [Piper, 1944]) to assess groundwater movement in a key region of a site affected by volatile organic compounds (VOC) and perchlorate. All three methods of interpretation showed a likely relationship between groundwater discharging from a spring and from two other sampling points, and suggested that mixing of water of at least two hydrochemical types was occurring in the area. These results were used to develop the conceptual model for groundwater flow at the site and to assist planning for subsequent additional characterization and groundwater management activities.

INTRODUCTION

Decisions regarding remediation of volatile organic compounds (VOCs) or other chemicals in groundwater require an understanding of groundwater movement. For sites with complex lithologic and hydrogeologic conditions, chemometric and geochemical methods can be used to improve the interpretation of groundwater movement.

We combined geochemical and chemometric methods to interpret groundwater movement at a site with complex lithology and hydrogeology situated within the western Sierra Nevada metamorphic belt. Two intermittent creeks drain the site. A veneer of soil less than 0.5 meters has developed over much of the site. The highly fractured bedrock underlying the site consists of a Jurassic-age slate, a Jurassic-age volcanic unit, and a Mesozoic-age intrusive termed the "feldspar porphyry."

The locations of the monitoring wells and surface water sampling stations are shown in Figure 1. Trichloroethylene (TCE) and perchlorate have been detected in groundwater samples from monitoring wells near the center of the site. TCE and perchlorate have not been found in groundwater collected from a manmade spring structure located within the drainage of the creek that drains most of the central portion of the site. Understanding the source of groundwater to the manmade spring structure was a primary goal of the work described here.

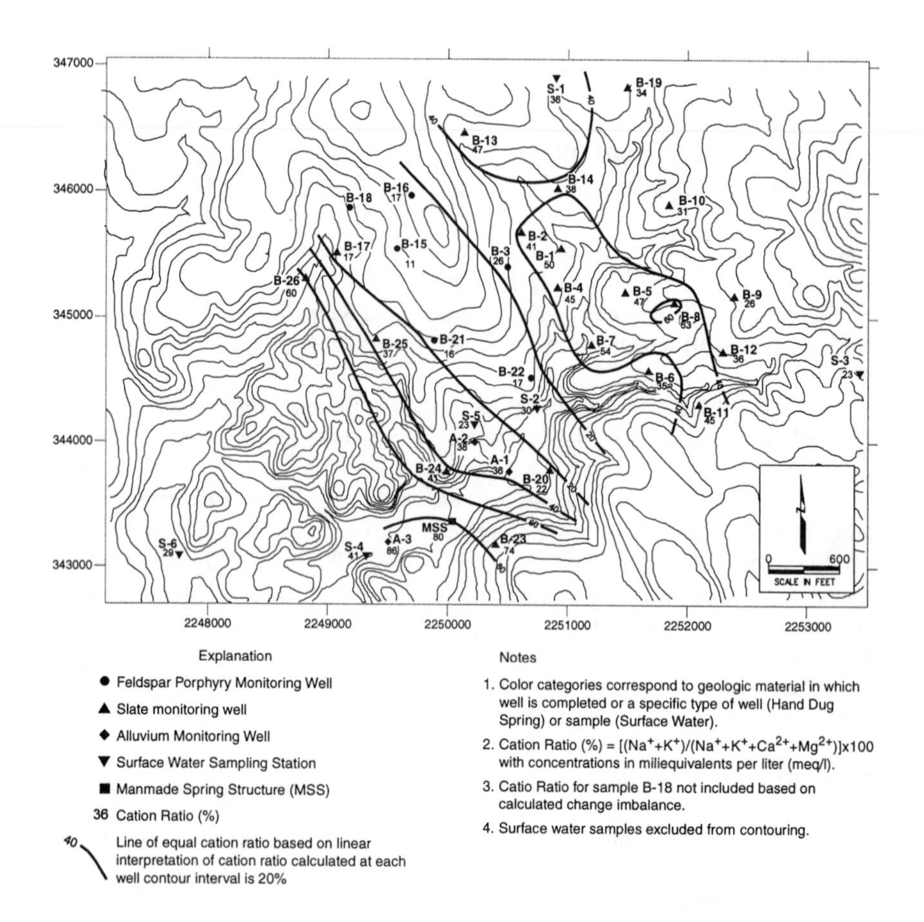

Figure 1. Site map and cation ratio distribution

METHODS

We used chemometric and graphical geochemical methods to assess groundwater and surface water data from a January 1997 monitoring event. For the assessment, we used a Piper trilinear diagram, mapping of major cation ratios, and principal component analysis.

Piper Trilinear Diagram. The Piper trilinear diagram is a useful tool for visually representing the major-ion chemistry of multiple water samples. This type of representation is convenient for showing the effect of mixing of groundwaters from two different sources. The concentrations of major cations (calcium, manganese, sodium, and potassium) and anions (chloride, sulfate, carbonate, and bicarbonate) in units of meq/L are plotted on two triangles as percentages. The intersection of lines extended from the two sample points on the triangles to a central "diamond" results in a single point that represents the major-ion chemistry of the sample.

Mapping Major Cation Ratios. Assessing trends in the ratio of major cationic constituents $[(Na^{+}+K^{+})/(Na^{+}+K^{+}+Ca^{2+}+Mg^{2+})]$ is a useful method for evaluating hydrochemical evolution and mixing of groundwater in crystalline rock aquifers. Similar to the construction of the Piper trilinear diagram, the cation ratio is calculated using concentrations of the major cations in units of meq/L and expressed as a percentage.

Principal Component Analysis. Principal component analysis is a means of segregating and classifying samples on the basis of similarity or differences within variables. Principal component analysis treats each variable as a dimension in a multidimensional space. A transformation of axes attempts to capture as much information in as few dimensions as possible. The first transformed axis (principal component 1 or PC1) is aligned to capture the greatest possible variability in the samples. The second transformed axis (PC2) is aligned so that it is perpendicular to PC1 and captures the greatest possible remaining variability. Subsequent transformed axes (PC3, PC4, et seq.) are aligned so that they are perpendicular to all prior axes and capture the greatest possible remaining variability. Prior to transforming the axes, the variables are scaled so that all variables have equal weight and the origin is located at the center of gravity of the data. When plotted, samples having similar component measurements cluster together. Dissimilar samples either form separate clusters or are outliers. Results are evaluated by visual interpretation of clustering and/or outliers.

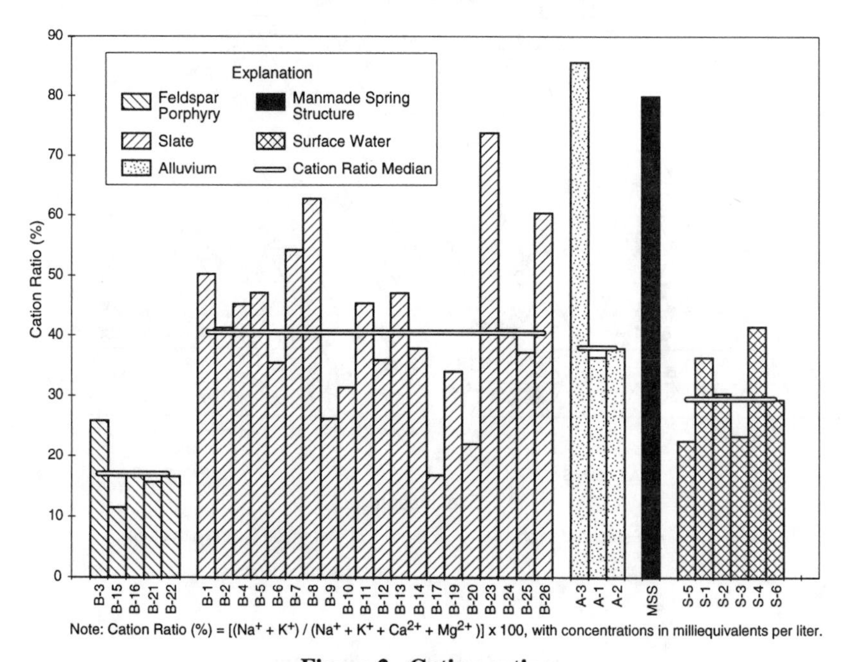

Note: Cation Ratio (%) = $[(Na^{+} + K^{+}) / (Na^{+} + K^{+} + Ca^{2+} + Mg^{2+})] \times 100$, with concentrations in milliequivalents per liter.

Figure 2. Cation ratios

RESULTS AND DISCUSSION

Data from the January 1997 monitoring event consisted of analytical results for water samples from 29 monitoring wells, 6 surface water stations, and the manmade spring structure. Before performing the assessment, we checked the anion-cation balances for each sample. We excluded monitoring well B-18 from the assessment because of a significant imbalance.

Major cation ratios revealed differences between the groundwaters in monitoring wells A-3, B-23, and the manmade spring structure and groundwaters and surface waters from the rest of the site, see Figures 1 and 2. The differences were more distinct between A-3, B-23, and the manmade spring structure and near by monitoring wells A-1, A-2, B-20, and B-24. Similar distinctions are revealed in the Piper trilinear diagram, Figure 3.

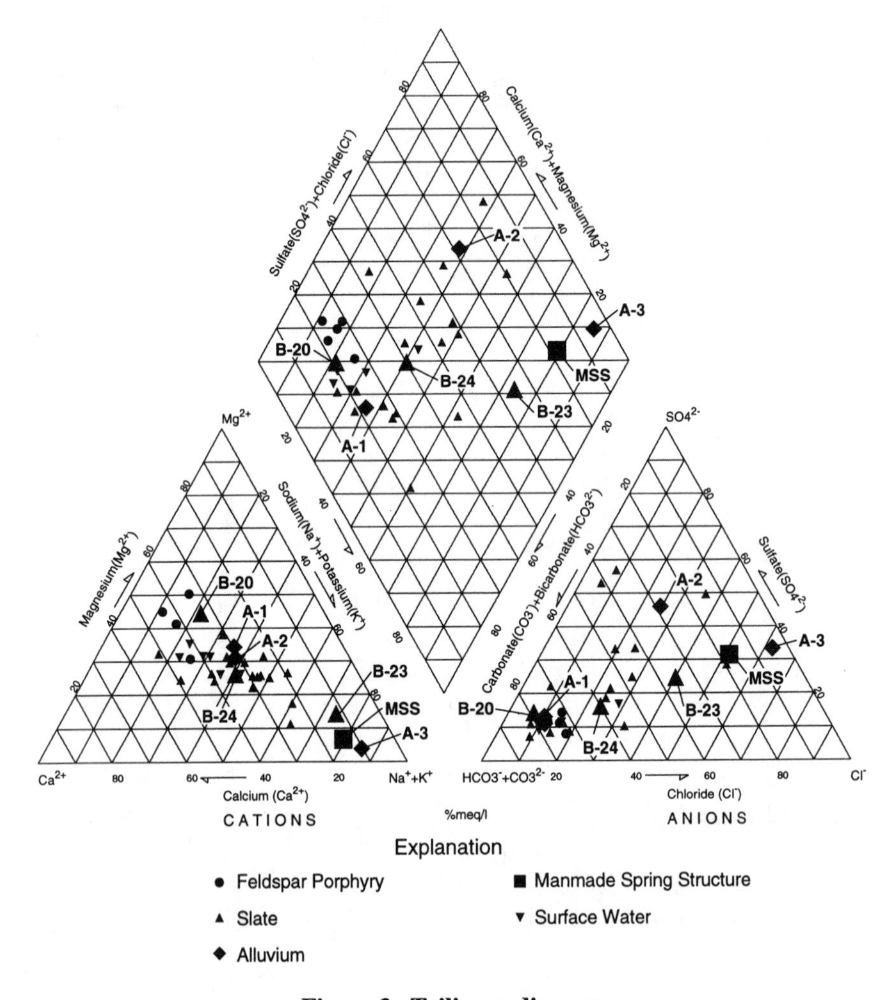

Figure 3. Trilinear diagram

Principal component analysis revealed that A-3 was a significant outlier from all other wells and surface waters, see Figure 4. The manmade spring structure also was an outlier and was located along a line drawn from the origin to A-3. B-23 is separated from the three main clusters representing the majority of the data and also along a line drawn from the origin to A-3.

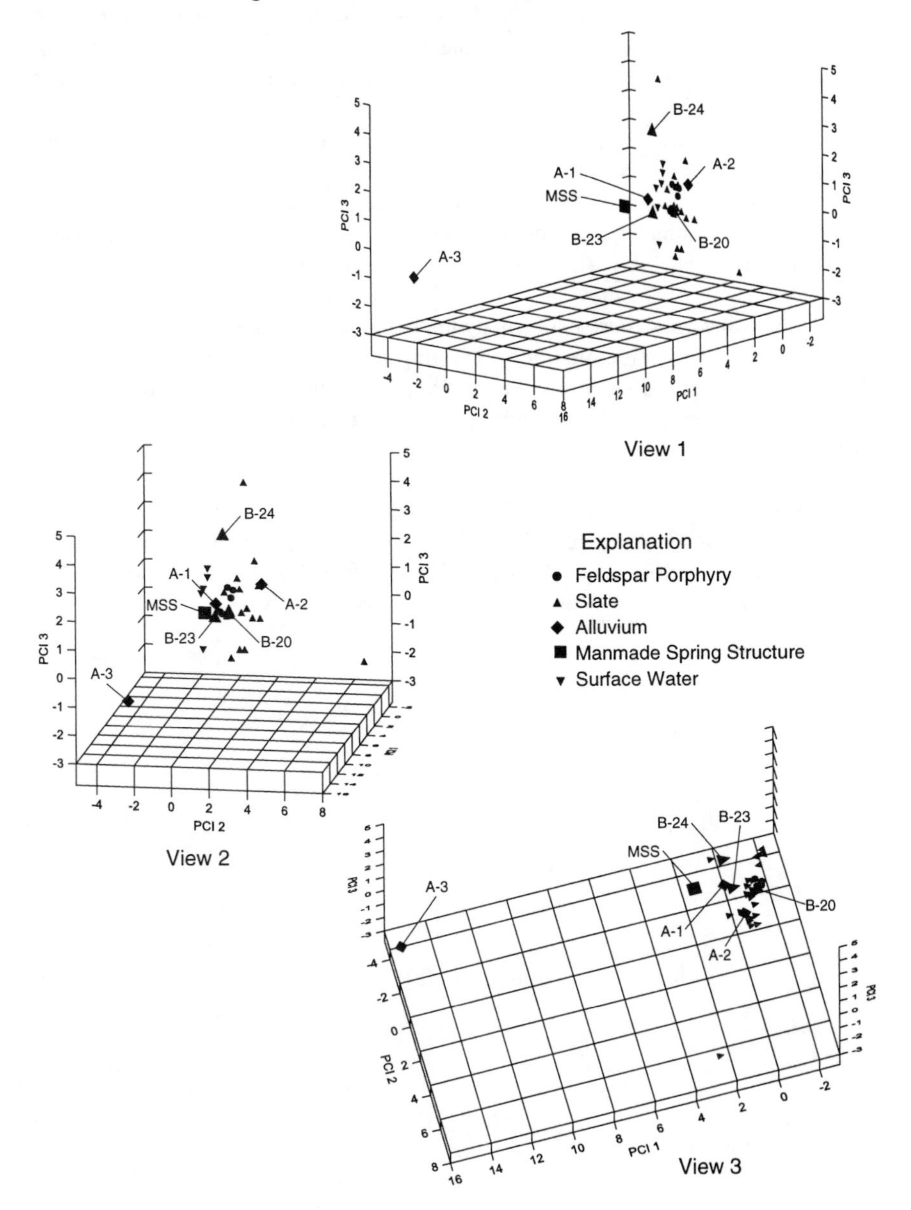

Figure 4. Plots of principal components 1, 2, and 3

INTERPRETATIONS

The cation ratios of groundwater samples collected from the manmade spring structure, alluvial monitoring well A-3, and bedrock monitoring well B-23 are similar to one another, and not typical of less evolved groundwater from a crystalline rock aquifer source. The locations at which the samples from these wells plot on a trilinear diagram indicate that the water at the manmade spring structure is intermediate in composition between the water at well B-23 and that at well A-3. The water may represent a mixture of more evolved water (possibly from bedrock at greater depth or with different mineralogic composition) and very shallow, more brackish alluvial water.

Principal component analysis shows groundwater samples from alluvial well 850 as a pronounced outlier from the cluster defined by most groundwater samples. This analysis also suggests a compositional relationship between groundwater samples from wells A-3 and B-23 and the manmade spring structure.

The relatively more-evolved hydrochemical character of groundwater at the manmade spring structure and monitoring wells B-23 and A-3, coupled with the water-level conditions at the manmade spring structure, suggest the potential presence of upward vertical hydraulic gradients and upward groundwater flow in this area.

REFERENCES

Piper, A. M. 1944. "A Graphic Procedure in the Geochemical Interpretation of Water Analysis." *Trans. Amer. Geophys. Union. 25*: 914-923.

SIDEWALL SENSORS FOR MEASURING IN SITU PROPERTIES

Larry Murdoch, Clemson University, South Carolina; and FRx Inc., Cincinnati, OH
William Slack, FRx Inc., Cincinnati, Ohio, USA
Bill Harrar, Geological Survey of Denmark and Greenland, Copenhagen, DK
Robert Siegrist, Colorado School of Mines, Golden, Colorado, USA
Timothy Whiting, Fuss & O'Neill, Longmeadow, Massachusetts, USA

ABSTRACT

Important subsurface parameters often vary markedly over short vertical distances, but sensors placed in vertical boreholes may be unable to resolve these variations. One problem is that conventional methods only allow one, or perhaps a few, sensors to be placed in each borehole. To address this limitation we have developed a method for accessing the *sidewall* of a borehole (Fig. 1). The method uses a device that pushes sensors or sediment samplers laterally into the sidewall to distances slightly less than the diameter of the borehole. The device can obtain a core sample 15 cm long and 4 cm in diameter, and then insert a permeable sleeve for extracting water samples. The same device has been used to insert several types of electrodes capable of measuring water content (using TDR waveguides), Eh (using platinum electrodes), or electrical resistivity (using a miniature Wenner-type array). As many as 22 water samplers and 19 resistivity electrodes have been installed in a single borehole at vertical spacings as close as 10 cm. This approach can be used to install horizontally oriented TDR waveguides at virtually any depth, thereby extending the TDR technique to the study of deep vadose zones. At a Region I Superfund site, TDR wave guides were installed to a depth of 12 m in glacial till. Other applications include measurement of Eh at a site where in situ chemical oxidization was used, and the in situ sensors provided results that are similar to data obtained from soil cores.

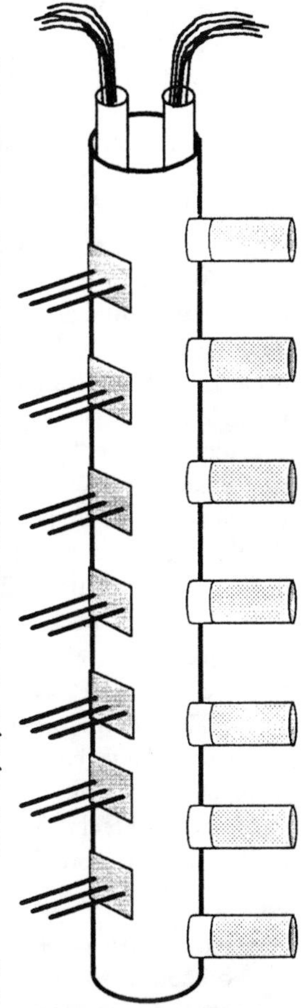

Figure 1. Piezometers and electrode sensors in the sidewall of a borehole.

INTRODUCTION

The need for a more detailed understanding of the subsurface has led to the development of a remarkable array of new methods for sampling and measuring properties at different depths in a borehole (e.g. special issue of the Journal of Hydrology, October, 1995). Some of the methods are designed to be placed in an existing well, and they typically obtain water samples from selected points along the screen. Other methods make use of special well completions with individual tubes that extend to different depths.

In some cases, a cluster of individual boreholes is used with each borehole containing a single access tube terminated in a short screen (Lapham and others, 1997). This approach is particularly attractive in cohesive formations or where vertical hydraulic gradients are a concern. An alternative is to bundle the access tubes together and place them in a single borehole. This approach is attractive in clean sands that will collapse around the bundle of tubes.

Those methods have dramatically improved our understanding of hydrology and environmental processes, but they have shortcomings for some applications. Methods that use a well with a long screen and gravel pack are limited to areas with small vertical gradients; steep vertical gradients would cause fluid to flow along the gravel pack and distort the results. Moreover, methods that use a bundle of tubes are best suited to formations that collapse to seal the annulus between the bundle and the wall of the bore. This excludes many fine-grained formations and may cause some concern when applied to areas where vertical hydraulic gradients are important and where uncertainty exists regarding the borehole seal. Methods that use multiple wells with short screens avoid some of the problems outlined above, but they can be costly to implement where detailed vertical resolution is desired.

We have developed a method for accessing the sidewall of a borehole to both increase resolution and provide new types of data for assessing and monitoring subsurface conditions. The method uses a tool that pushes sediment core samplers, or sensors such as piezometers, laterally into the sidewall to distances slightly less than the diameter of the borehole. This allows as many as several dozen piezometers to be embedded in the formation adjacent to a borehole (Fig. 1), but it is by no means limited simply to increasing the resolution of head measurements. Horizontally oriented core samples can be obtained to facilitate measurements of anisotropic aquifer properties. Moreover, horizontal core samples provide a new perspective on vertical fractures or macropores. This approach also provides a platform for placing arrays of many other types of sensors, including TDR waveguides, Eh electrodes, or electrical resistivity probes. Sensors and water samplers embedded in a host formation may provide more representative data than similar devices contained in a well bore.

The sidewall method has been used by us at four sites in the U.S. and one site in Denmark. The technique for placing sensors in the subsurface has been developed in some detail, but particular applications involving specific sensor packages are still evolving. The equipment and techniques for creating sidewall monitoring wells, and several applications are described in more detail below.

METHOD

The method that we have developed is based on a tool (Fig. 2) for accessing the sidewall of a borehole in sediment or soft rock. To obtain a core sample or place a sensor requires lowering the access tool to a desired depth, activating a pushing mechanism, retracting the mechanism, and recovering the access tool. To create a piezometer in a cohesive formation, however, currently requires two trips into the borehole. One trip is used to obtain a sediment sample and create a hole in the sidewall, and a second trip registers the access tool and advances a porous sleeve that holds the hole open. Sensors are emplaced from the bottom to the top of the borehole and several types of sensors can be placed at nearly the same depth by varying the angular position. The boreholes are sealed after the sensors have been emplaced. We have used portland cement with an additive to prevent shrinkage to seal boreholes in the vadose zone, or bentonite to seal boreholes in the saturated zone. The current process has been conducted using either an open borehole, or a borehole held open by steel casing. In the latter application, the steel casing was jacked up and the borehole sealed incrementally as the samplers were installed.

Sidewall Access Tool. The sidewall access tool is a device capable of pushing samplers or sensors laterally out of a borehole. We have constructed a variety of such devices, each with a slightly different application in mind. A design theme that is central to all of them is to develop a stroke that is as long as possible. This is done in order to place the sensors as far from borehole as possible. The most reliable devices push sensors normal to the axis of the borehole and are capable of penetrating distances approximately 4 cm (1.5 in) less than the diameter of the borehole. Most of the recent devices are designed to push a sensor 16 cm into the wall of a borehole 20 cm (8 in) in diameter.

One design that is particularly versatile makes use of a fluid-powered cylinder and chain to push the sampler (Fig. 2). The cylinder is mounted parallel to the axis of the borehole and the chain is fastened to either end of the rods from the cylinder and passes over small-diameter sprockets at the bottom of the tool. A sliding block fastened to the chain at the bottom of the tool is used to hold either sample tubes or sensors. The block slides forward to push a device into the sidewall by pressurizing one side of the cylinder and it is retracted by pressurizing the other side of the cylinder (Fig. 2).

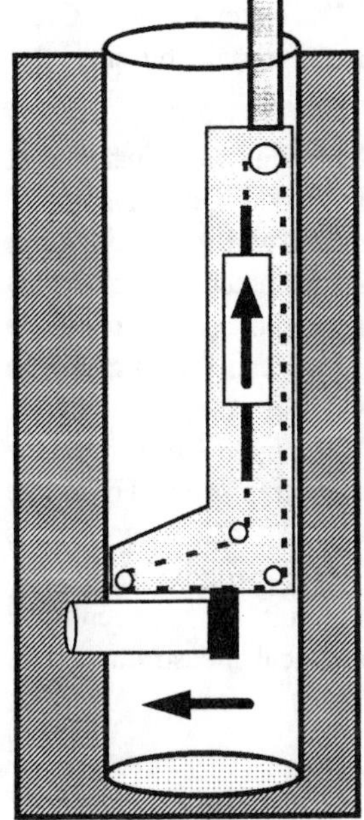

Figure 2. Sidewall access tool.

Sidewall access tools based on the design outlined above have been constructed using both air and hydraulic cylinders. The air-powered device can be driven with a small compressor to produce about 350 kg force (750 lbs). The hydraulically-powered device requires a hydraulic power pack above ground, and it can develop approximately 1500 kg (3300 lbs) force. The pneumatic device is light weight and uses a convenient power source, but the power it produces is limited and it is best suited to pushing relatively thin sensors into soft formations. The hydraulic device is slightly more cumbersome, but it develops enough force to drive core samplers through stiff clay till or weathered rock.

Sensors. Piezometers are the most widely used sensors in the sidewall system. The current piezometer design makes use of a porous polyethylene filter, 4 cm in diameter x 7 cm long. The piezometers are accessed by two tubes, one that reaches to the upper side of the distal end of the filter, and another one that goes to the lower side of the proximal end (relative to the borehole wall). This configuration allows air to be purged from a piezometer, and it allows water to be circulated through the device. Heads are measured either using a micro-bore electrical water level indicator that fits into a 6-mm-diameter access tubing, or using a small piezo-resistive transducer mounted permanently in the piezometer.

Water samples can be obtained from the piezometers. Current applications use suction to lift the water samples through 3-mm-diameter tubing. This results in relatively small stagnant volumes (20 to 40 ml) that should be purged prior to retaining a sample for analysis.

Several types of electrode sensors have been utilized. The electrodes consist of 15-cm-long, narrow rods, either mounted singly or in groups of three or four onto base plates. Electrodes for measuring Eh were created by placing a small Pt wire at the tip of a 3 mm-diameter stainless steel tube. A single Ag-AgCl electrode was installed near the ground surface as a reference for Pt electrodes in a nearby borehole. TDR waveguides were constructed using 3 parallel rods fixed to a solid base and wired to a coaxial cable (e.g. Fig. 1; Zegelin and White, 1989). The waveguides produced signals that could be readily analyzed using standard TDR methods to estimate electrical conductivity, dielectric constant, and moisture content. Wenner-type electrodes were constructed using 4 parallel rods fixed to a base and pushed into the sidewall. Wires were attached to each rod and a current was applied between the outer two rods. The voltage difference between the inner two rods was measured and the ratio of applied current:measured voltage was calibrated to electrical conductivity.

APPLICATIONS

We have used sidewall samplers for environmental applications at three locations in Ohio, one in Connecticut, and one in Denmark. All the locations are underlain by silty clay glacial deposits with downward hydraulic gradients.

Piezometers and Water Samplers. Piezometers and Wenner electrodes were installed at several locations to monitor the migration of an ionic tracer at Flakkebjerg, Denmark. The piezometers were used to measure hydraulic head during the test and to obtain water samples analyzed for the tracer. The electrodes were used to detect the electrical conductivity changes accompanying changes in tracer concentration. Each borehole contained approximately 20 piezometers and 20 electrodes between 2 and 9 m b.g.s. The completion of each borehole took between one and two days.

The site is underlain by silty clay till to a depth of 10 m. The upper 3 to 4 m of the till is highly fractured, whereas below 4.5 m, the till is only sparsely fractured. Hydraulic conductivity of the till between 2 and 4 m depth is approximately 10^{-7} m/s, whereas it is 10^{-8} or less below 5 m, according to data from slug tests. Interestingly, slug tests conducted using the small sidewall piezometers give similar results to conventional methods, suggesting that the technique may be suited to making local measurements of hydraulic conductivity in situ.

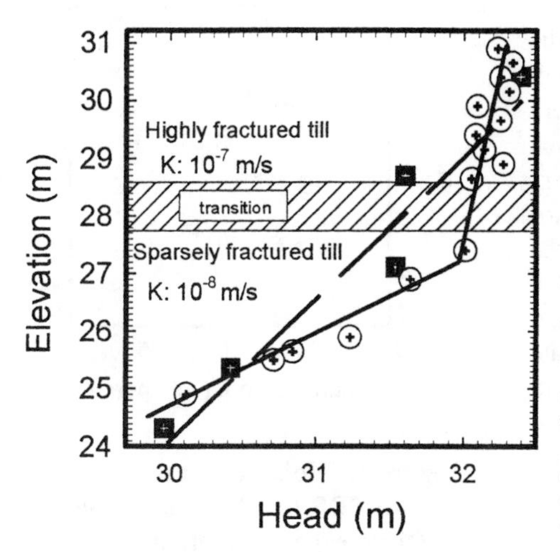

Figure 3. Head measurements at conventional cluster (squares) and sidewall piezometers (circle).

Several clusters of conventional wells with short (25 cm) screens completed at different depths indicated a downward vertical hydraulic gradient. Five of these wells were available in the depth range of 2 to 7 m near one of the sidewall wells, and head measurements were made to compare to the results from the sidewall piezometers. Head measurements from the cluster of wells are somewhat variable, but probably can best be described by a vertical head gradient that is uniform from 2 to 7 m and equal to 0.43 (dashed line in Figure 3). This suggests that the vertical flux decreases by an order of magnitude from the highly fractured to sparsely fractured zone (because the conductivity decreases). This conclusion is hydrologically untenable, however, indicating that this cluster of 5 wells provides insufficient resolution of the hydraulic gradient at this site.

In contrast, data from the sidewall samplers show that the head gradient varies with depth. The heavy solid lines shown in Figure 3 indicate that the gradient is from approximately 0.083 in the highly fractured till to 0.82 in the sparsely fractured material. This implies a vertical flux of 8.3×10^{-9} m/s in the upper zone and 8.2×10^{-9} m/s in the lower zone. In this case the data indicate that the vertical flux is uniform with depth, which is the expected result.

Deep Vadose Monitoring. A Superfund site in Connecticut is underlain by 12 m of cobble-rich silty clay glacial sediment overlying fractured bedrock. Chlorinated solvents in the silty clay create a contaminant source zone for a ground water plume in the fractured rock. It should be feasible to remove some of mass from the source zone using vapor extraction. However, the water content of the sediments must be reduced before vapor extraction is a viable process.

Several methods for measuring the moisture content at this site were evaluated, but most were eliminated because they were unable to provide data to 12 m in cobble-rich silty clay. Time domain reflectometry appeared to be a viable method for measuring changes in moisture content if the waveguides were installed without undue deformation at the desired depths. A total of 30 waveguides were installed in the sidewalls of 2 boreholes between 1 and 12 m bgs. Most of the waveguides produced waveforms that resemble our above-ground tests, but two gave erratic waveforms probably because the component rods bent upon encountering cobbles during insertion. A Trase Moisture Content Measurement System was used to automatically interpret the TDR waveform and estimate moisture content. The results indicate that volumetric moisture content varies between 25 and 37 percent (Fig. 4). The variation is erratic with depth and may be due to a variable volumetric percentage of rock as cobbles in the vicinity of the waveguides. Volumetric moisture contents determined in the lab from large intact samples ranged from 0.27 to 0.37, a similar span as observed in the TDR data.

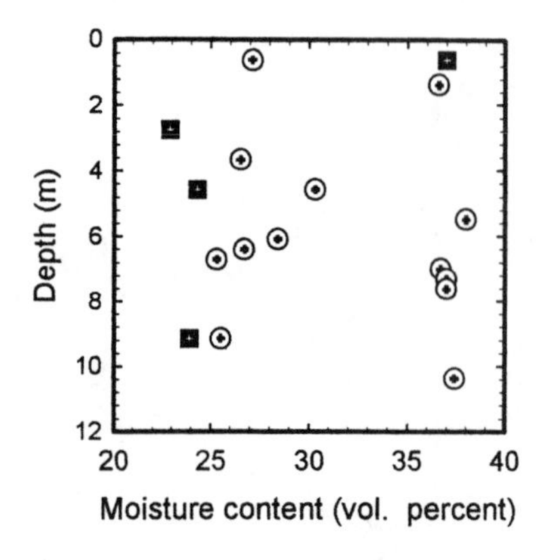

Figure 4. Moisture content measured with TDR (circles) and in the lab (squares) as a function of depth.

Conclusions

A method for obtaining horizontal core samples and installing sensors in the sidewall of boreholes in sediment has been developed to address some of the shortcomings of current sampling and monitoring technology. The sidewall method is capable of installing several dozen or more piezometers to markedly increase the resolution of vertical profiles of hydraulic head and concentration. In addition, electrodes or waveguides can be placed to provide real-time monitoring of a variety of parameters, including Eh, moisture content, and electrical resistivity. In certain hydrologic conditions, these capabilities should improve the resolution, quality, and type of data available for studying subsurface processes.

References

Lapham, W.W., F.D. Wilde, and M.T. Koterba. 1997. Guidelines and standard procedures for studies of ground water quality; selection and installation of wells and supporting documentation. *USGS Water Resources Investigations Report* 96-4233. 110 p.

Zegelin, S.J., I. White, and D.R. Jenkins. 1989. Improved field probes for soil water content and electrical conductivity measurement using time domain reflectometry. *Water Res. Research*, (25)11, 2367-2376.

INNOVATIVE TOOLS TO EXPEDITE SITE CHARACTERIZATION

Dev Murali, Raveendra Damera, Mike Hill, Tom Rose, and Mark Lewis (General Physics Corporation, Columbia, Maryland), Raymond McDermott and Naren Desai (U S Army Garrison, Aberdeen Proving Ground (APG)), Maryland

ABSTRACT: Historic degreasing operations at the Assembly Shop site located at APG have resulted in dissolved contamination of groundwater with chlorinated solvents. A suite of investigating tools was utilized to thoroughly characterize the plume and design a remedy to protect the ecologically sensitive wetlands. Tools include direct push sampling combined with field gas chromatography (GC) to screen and obtain real time data for geologic and contaminant characterization. The data resulted in optimum placement of soil borings and monitoring wells. Down-hole logging supplemented geologic details and aquifer testing determined the hydrogeologic characteristics. Pollen analysis and radiocarbon dating confirmed the continuity/homogeneity of the confining lithostratigraphic units. Seepage velocity measurements in the creek yielded groundwater discharge rates. The study concluded that the plume is confined to the shallow aquifer and solvents such as trichloroethylene (TCE); 1,1,1-trichloroethane (1,1,1-TCA); tetrachloroethene (PCE) and their degradation products exceeded federal maximum contaminant levels (MCLs). Groundwater monitoring for three events revealed decreasing trends for TCE, 1,1,1-TCA, and PCE while increasing trends for degradation products indicating natural/intrinsic biodegradation. Low levels of TCE and 1,1,1-TCA were observed in the creek indicating direct impacts from site contaminated groundwater discharges. Feasibility Studies are currently underway to select appropriate methods for remediation.

INTRODUCTION

The Assembly Shop occupied at Building 525 ("the site"), is located in the Aberdeen Area of APG. The 3-acre site as seen in Figure 1 is delineated by Mulberry Point Road on the north-northwest, and the Woodrest Creek and its associated wetlands on the southeast. The facility was used for assembly, processing, and painting of artillery components and weapon systems. Two vapor degreasers existed within the facility. The vapor degreasers used chlorinated solvents to degrease gun barrels. It is suspected that the generated waste solvents were disposed of in the boiler room. Chlorinated solvents are no longer used at the facility. In addition, vehicles and gun barrels were steam cleaned and applied with chemicals at a washrack. The wastewater was discharged into the sanitary sewer after passing through oil/water separators. The site contained two underground storage tanks (USTs): a 15,000-gallon (56,775 liter) for #2 fuel oil storage and a 1,000-gallon (3,785 liter) UST for solvent storage. The fuel oil UST was abandoned in place, and the solvent UST was removed in 1993.

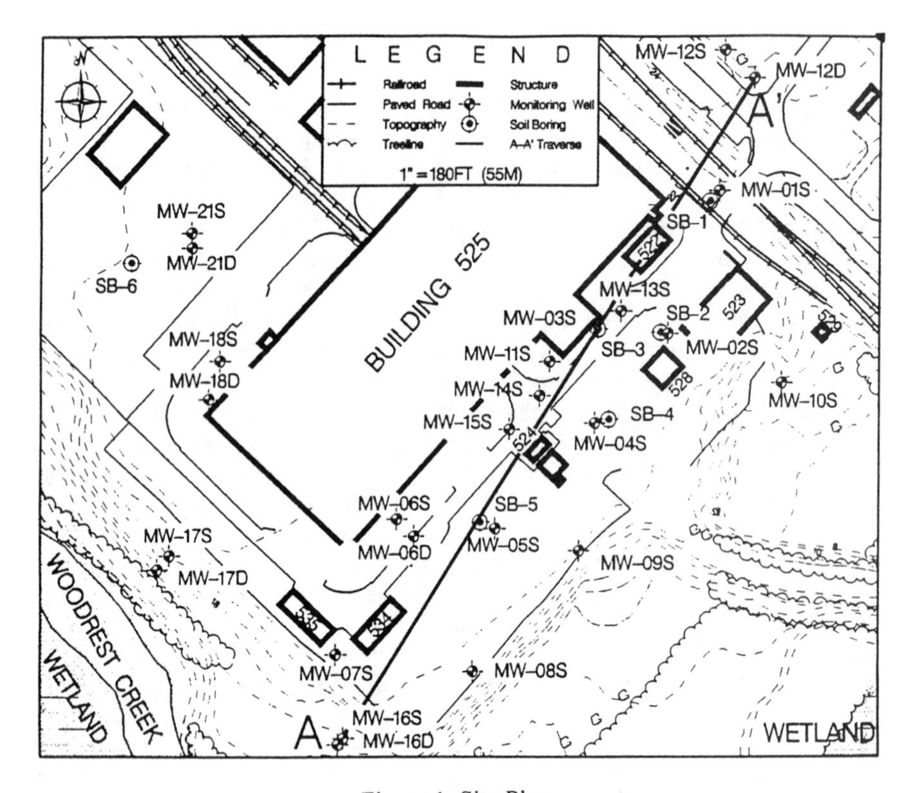

Figure 1. Site Plan

A site assessment conducted as part of the UST closures resulted in the discovery of a chlorinated solvent plume along with localized petroleum contamination in the UST vicinity. The discovery of chlorinated solvents triggered a detailed site characterization.

Objective. The objectives of the site characterization were to utilize innovative tools to expedite hydrogeologic characterization of two aquifers and determine the vertical and horizontal extent of dissolved chlorinated solvent contamination to support a risk assessment and feasibility study; and provide design parameters for an effective interim remedy.

FIELD METHODS

Several innovative USACE-approved field methods were implemented to achieve the project objectives. Initial investigations focused on the southern side of the building, followed by the northern side and included: direct push surveys to obtain soil and groundwater samples from 47 locations, and a field GC to analyze the samples on site to provide real-time data. Only 20 percent of the samples underwent laboratory analysis. This data were used to place six stratigraphic borings, and 29 monitoring wells at strategic locations varying in depths from 20 to 100 ft (6.1 to 30.48 m) to encounter three aquifers represented by various lithostratigraphic units. The geologic data was further confirmed by geophysical logging through use of Electro-Magnetic Induction (e-m), conductivity, resistivity, and natural gamma techniques. Representative soil samples were analyzed for pollen counts and radiocarbon dating to gather geochronometric data in an effort to correlate stratigraphic units for comparison to geologic samples. Select samples were also analyzed for geotechnical parameters to characterize hydraulic properties, and bacteriological parameters to evaluate bacterial populations present in subsurface soils and predict the effectiveness of *in-situ* biodegradation. Three rounds of groundwater sampling provided the concentration trends and slug testing provided the aquifer characteristics. Additional sampling included sanitary sewers, surface water and sediments to determine the influence of site contaminants in the Woodrest Creek. Seepage velocity measurements were taken in the creek to calculate groundwater discharge rates.

DISCUSSION OF RESULTS

Geology and Hydrogeology. The site is underlain by Talbot formation which is the most widespread coastal plain Pleistocene sedimentary unit present in Harford County, Maryland (Southwick et al., 1969). Lithologic, geotechnical, pollen, and radiometric data collected from the investigation provided fairly detailed geologic information up to 100 ft (30.48 m). In general, the site geology is characterized by gravels, silty sands, silts, silty clays, and clays of varying thickness and lateral extent. The lithologies have been grouped into six groups based on the field observations from the subsurface investigations. A representative cross section shown in Figure 2 illustrates the horizontal and vertical variation of the lithofacies across the site. Lithofacies one (LF1) consists of silts, silty clays, and clays extending up to 16 ft (4.88 m) underlain by LF2 represented by silty sand mixed with fine gravel. This facies extends from 9-25 ft (2.74-7.62 m). Facies LF1 and LF2 represent the surficial aquifer. LF3 consists of silty clay to clay extending from 21 to 26 ft (6.41-7.92 m) defined as the first confining unit. Underlying this unit is LF4 extending from 26-70 ft (7.92-21.34 m) comprised of silty, gravely sand representing the first semi-confined aquifer. LF5 consists of clayey silt, and clayey sand extending from 58-62 ft (17.68-18.90 m) defined as the second semi-confining unit. LF6 from 62 to 110 ft (18.90-33.53 m) consists of laminations of organic fragments in a silt/sand matrix and represents the second semi-confined aquifer. The pollen analysis and radiocarbon data confirm the lithologic and age relationship with respect to individual lithofacies. Geophysical logging confirmed the geology of the subsurface, especially the confining units.

LF 1 and LF2, representing the surficial aquifer, are present throughout the site. LF2 is continuous as evidenced by the facies thickness, grain size, and uniform gamma peaks. Groundwater flow is southwest towards the Woodrest Creek, and

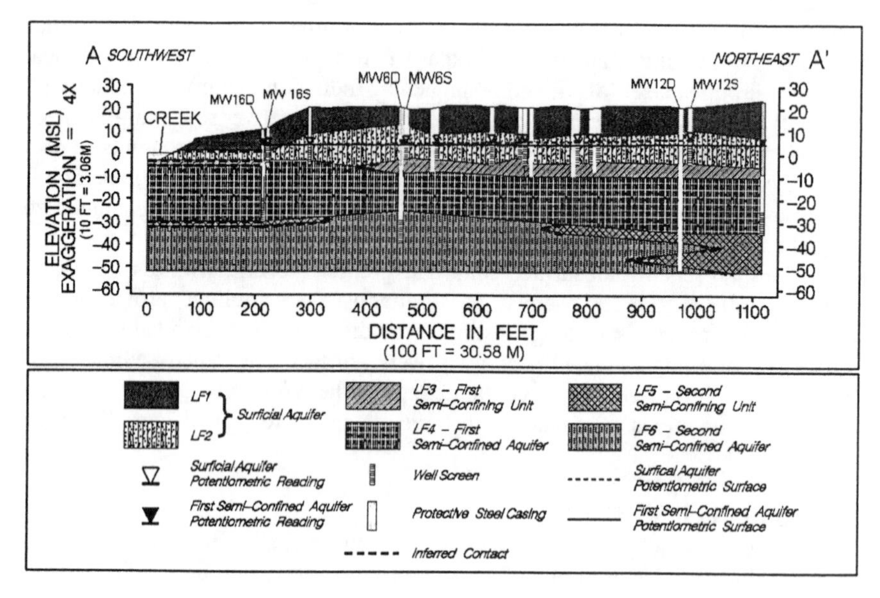

Figure 2. Cross Section A – A'

hydraulic conductivity is 8.11×10^{-4} cm/sec. LF3, the first confining unit, exists throughout the site except breaches around MW-07S. LF4, the first semi-confined aquifer, is present across the site and is underlain by discontinuous semi-confining unit (LF5) as evident in MW-06D. Groundwater flow follows the same direction as the surficial aquifer. There are currently no wells in LF6. The horizontal gradient in the surficial aquifer ranges from 0.0004 to 0.006 ft/ft.

Soil Results. Analysis of soil samples indicated localized presence of low levels of adsorbed contaminants with a maximum of 34 μg/kg of TCE, and 44 μg/kg of 1,1,1-TCA in shallow soil borings. It is an indication that a majority of the compounds are present in the dissolved form in the groundwater. Standard plate count results in representative soils indicated a healthy population of heterotrophic bacteria in the shallow boring at MW-03S (2.27×10^{-7} CFU/g) with no apparent toxicity problems. This boring is located in the suspected source area. In addition, low phosphorus concentrations (an order of magnitude less than the background concentration) and depleted nitrogen levels detected in the source area indicate a significant biological activity in the subsurface.

Groundwater Results. Multiple rounds of groundwater sampling data are available for the first 10 wells (MW-01S through MW-10S), with only one set for the remaining 19 wells. The primary dissolved contaminants detected in the surficial aquifer are the chlorinated solvents such as PCE, TCE, 1,1,1-TCA, and their degradation products 1,1-Dichloroethene (1,1-DCE), *cis*-DCE, and vinyl chloride. These compounds exceeded their respective MCLs. Low levels of PCE was detected in the only deep well screened in the first semi-confined aquifer. The discussion is restricted to the southern side of the site and shallow aquifer due to the amount of data available for presentation. Figures 3a and 3b depict the average concentration contours for TCE, 1,1,1-TCA, PCE, and 1,1-DCE for the baseline and the most recent sampling. The baseline sampling results in Figure 3a display an anomalous area around MW-3 with a maximum concentration of 2666 μg/l and most likely indicates the source area for the plume possibly originating from the vapor degreasers within the building. The plume tends to follow the hydraulic gradient towards the Woodrest Creek based on the

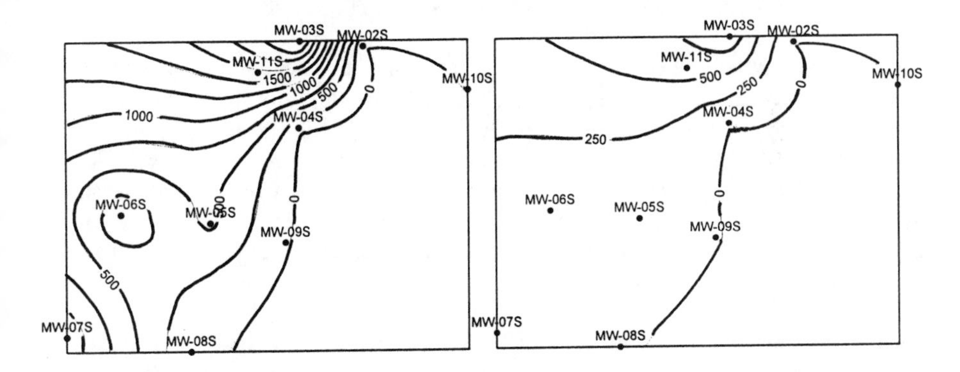

Figure 3a. March 1995
Isoconcentration Contours

Figure 3b. February 1997
Isoconcentration Contours

results from the down gradient wells (MW-04S, 05S, 06S, 07S, and 08S). The solvent concentrations across the site have distinctly decreased with time. In some cases (for example MW-07S), the concentrations have decreased by an order of magnitude. An increase in the concentration of degradation products such as 1,1-DCE, *cis*-DCE and vinyl chloride has been observed with time (Figure 4). A combination of robust bacterial populations, presence of less halogenated ethenes, and depleted nutrient levels indicates anaerobic reductive dehalogenation (ARD) of TCE and PCE has been occurring in the subsurface. Dissolved oxygen concentrations of 0 to 1.0 mg/L at several locations and reduced subsurface conditions with a redox potential of –30 to 250 mV further support this conclusion. Common electron acceptors in ARD process include nitrate, sulfate, and carbon dioxide, although the use of petroleum hydrocarbons such as toluene as a substrate are also reported (Cookson, 1995). Nitrate and sulfate concentrations in the source area dropped by an order of magnitude with a slight decrease in petroleum hydrocarbons.

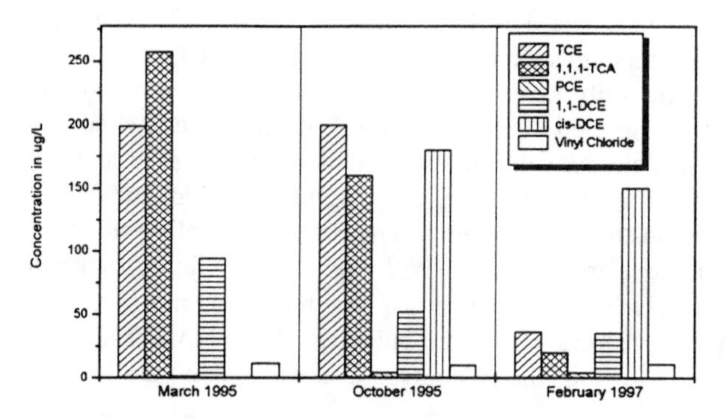

Figure 4. VOC Concentration vs. Time in MW-05S

Surface Water/Sediments. Comparison of the surface water samples from two rounds show a decrease in TCE, and 1,1,1-TCA from 50µg/l to 5µg/l. The latest round also showed the presence of *cis*-DCE, and vinyl chloride in the samples. This data is consistent with the groundwater data trends. A seepage velocity study revealed that groundwater discharges to the Woodrest Creek at an overall average velocity of 5.74 X 10^{-6} cm/sec. Sampling of sediments in the Woodrest Creek detected chlorinated compounds concentrations below the EPA Region III Risk Based Concentrations (RBCs).

Risk Assessment: The preliminary risk assessment indicated that the chemical concentrations in surface water are not likely to pose any risks to ecological receptors. However, toxicity tests of benthic organisms may be necessary to quantify risks to ecological receptors. Even though the groundwater concentrations are above the MCLs, there are no known current or future uses of groundwater in the vicinity of the site. A baseline risk assessment is underway to quantify risks to human health and the environment.

CONCLUSION

The systematic use of innovative and appropriate field methods has expedited the characterization of two aquifers thus completing entire field investigation within one year. The following conclusions are drawn from the studies:

1. In general, the site geology grouped into six facies is characterized by gravels, silty sands, sands, silts, and silty clays of varying thickness and lateral extent. The hydrogeology of the site characterized by three aquifers and two semi-confining units.
2. Water table is approximately 12.5-19.5 ft below ground surface (3.81-5.94 m) with groundwater flowing in a southwesterly direction towards the Woodrest Creek.

3. Groundwater sampling in the surficial aquifer has resulted in detection of chlorinated solvents and their degradation products. A zone of ARD was delineated based on measured reductions in concentrations of chlorinated solvents over time; slight increases of concentrations of degradation products over time; depletion of nutrient concentrations below background levels; and low dissolved oxygen and redox potential measurements.
4. A feasibility study is currently being performed to evaluate various options to include natural attenuation, in-well stripping, in-situ reactive walls, etc. Appropriate cost-effective technology will be selected to mitigate the site.

REFERENCES

Cookson, J.T., Jr. 1995. “Bioremediation Engineering, Design and Application”. McGraw-Hill, Inc., New York, NY., 525pp.

Southwick, D.L., Owens, J.A., and Edwards, J. 1969. “The Geology of Harford County, Maryland”. Maryland Geological Survey, Baltimore, Maryland.

A VAPOR EXTRACTION MONITORING SYSTEM FOR PERFORMANCE ASSESSMENT

Lee A. Brouillard (Duke Engineering & Services, Albuquerque, NM)
James E. Studer P.E., P.G., and Cythia P. Ardito (DE&S, Albuquerque, NM)
Richard E. Fate (Sandia National Laboratories, Albuquerque, NM)

ABSTRACT: During 1996, three intermediate depth (150 ft) and three deep (485 ft) vapor extraction (VE) wells were installed to complement pre-existing shallow VE wells. To reduce monitoring system installation costs, each of the new VE wells were constructed with three to five subsurface soil gas monitoring ports. These monitoring ports provide the mechanism with which to assess *in-situ* performance of the VE system. The VE monitoring system allows the collection of soil gas samples for use in documenting changes in *in-situ* soil gas composition over time and the measurement of extraction vacuum or air injection pressure at critical depths.

INTRODUCTION

The Chemical Waste Landfill (CWL) at Sandia National Laboratories, New Mexico (SNL/NM) occupies approximately 2 acres and is underlain by a 490-ft thick vadose zone composed of alluvial silt, clay, silty to clayey sand and gravel. Liquid waste disposal into unlined pits at the CWL occurred from approximately 1961 to 1982. Vapor-phase transport of volatile organic compounds (VOCs), such as trichloroethene, has resulted in the development of a soil gas plume and groundwater contamination.

Remediation of the VOC soil gas plume underlying the CWL was initiated in May of 1997. A vapor extraction (VE) Voluntary Corrective Measure (VCM) was implemented to reduce the mass and extent of the VOC plume and prevent further degradation of the groundwater. Remediation performance objectives for the reduction in VOC concentrations within specific depth intervals beneath the site were defined in 1996. Multi-use vadose zone wells were installed to meet the following project requirements: 1) create a subsurface air extraction and injection system to reduce the concentration of VOCs within the soil gas plume, 2) establish a fixed soil gas monitoring system with which to measure temporal changes in VOC concentrations at selected depths, and 3) have the capability to monitor subsurface vacuum/pressure values and determine zones of influence.

Included in this paper are the design specifications for a multiple-use, vapor extraction/soil gas monitoring well. General measurement and analysis procedures used to monitor and assess the VE system's operational performance, and the method used to document progress towards meeting remediation performance goals, are also presented.

DESIGN AND INSTALLATION OF A VAPOR EXTRACTION/ SOIL GAS MONITORING WELL

Air Rotary Casing Hammer (ARCH) drilling was used to create a thirteen inch diameter borehole in which each vapor extraction/soil gas monitoring well was constructed. Well completion depths ranged from 140 ft below ground surface (bgs) to 486 ft bgs. Soil gas monitoring port construction specifications were modified after design parameters for a combination groundwater and gas sampling well patented by Hubbell et al (in press). Well construction materials included standard items such as five-inch diameter schedule 80 PVC threaded flush-joint casing, 10/20 mesh sand for filter pack around the well screen and soil gas sampling ports, hydrated bentonite pellets for the filter pack seal, and bentonite grout to isolate the well screen and each sampling port interval (Figure 1). Five hundred foot coils of 0.25 in OD stainless steel tubing were cut to pre-determined lengths to obtain continuous sections of access tubing extending from the ground surface to each soil gas sampling port. A two-foot section of 0.375 in OD stainless steel tubing, which comprises the soil gas port, was attached to the access tubing using swaglock ™ fittings. Stainless steel hose clamps secured the sampling port tubing to the outside of the well casing. Duct tape secured the access tubing to the outside of the well casing between port intervals (Figure 1).

Three deep vadose zone wells with five soil gas monitoring ports, and three intermediate depth wells with three monitoring ports (a total of 24 soil gas monitoring ports), were installed to augment existing shallow VE and SEAMIST ™ soil gas monitoring wells at the site. A simplification of the vadose zone lithologic sequence, and the monitoring port completion depths for the wells located along a northwest-oriented transect of the site, is shown on cross-sectional view A-A' (Figure 2).

Compared to an alternate 24-port soil gas monitoring system requiring six separate boreholes, such as the multi-level monitoring port SEAMIST ™ membrane, installation of a multiple-use system resulted in a cost savings of approximately $60,000. The majority of the additional time required for construction of a multi-use well, when compared to a standard vapor extraction well, is associated with installing the bentonite seals and filter pack interval around each soil gas monitoring port, and the curing time needed for grouted sequences between soil gas monitoring port intervals.

DATA COLLECTION USING SOIL GAS MONITORING PORTS

The soil gas monitoring ports provide dedicated sampling locations from which *in-situ* measurements of the pressure/vacuum field and samples of soil gas can be taken under controlled conditions before, during, and after the period of system operation. Changes in soil gas concentration and the pressure/vacuum field can be monitored for effects caused by air extraction or injection, mass transport from VOC source areas, and changes in barometric pressure. These dedicated monitoring locations avoid many of the errors associated with the

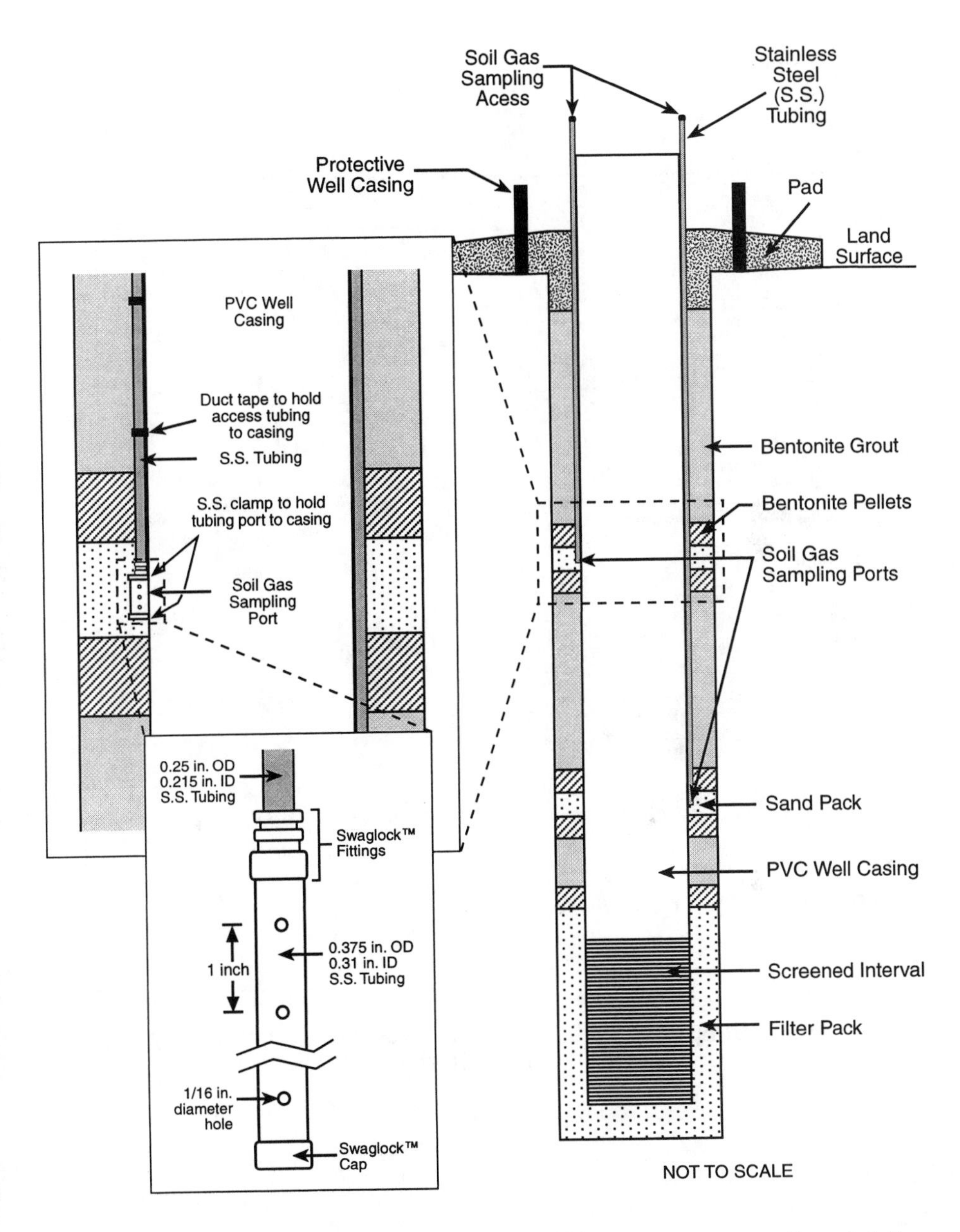

FIGURE 1. Schematic of soil gas monitoring port construction details.

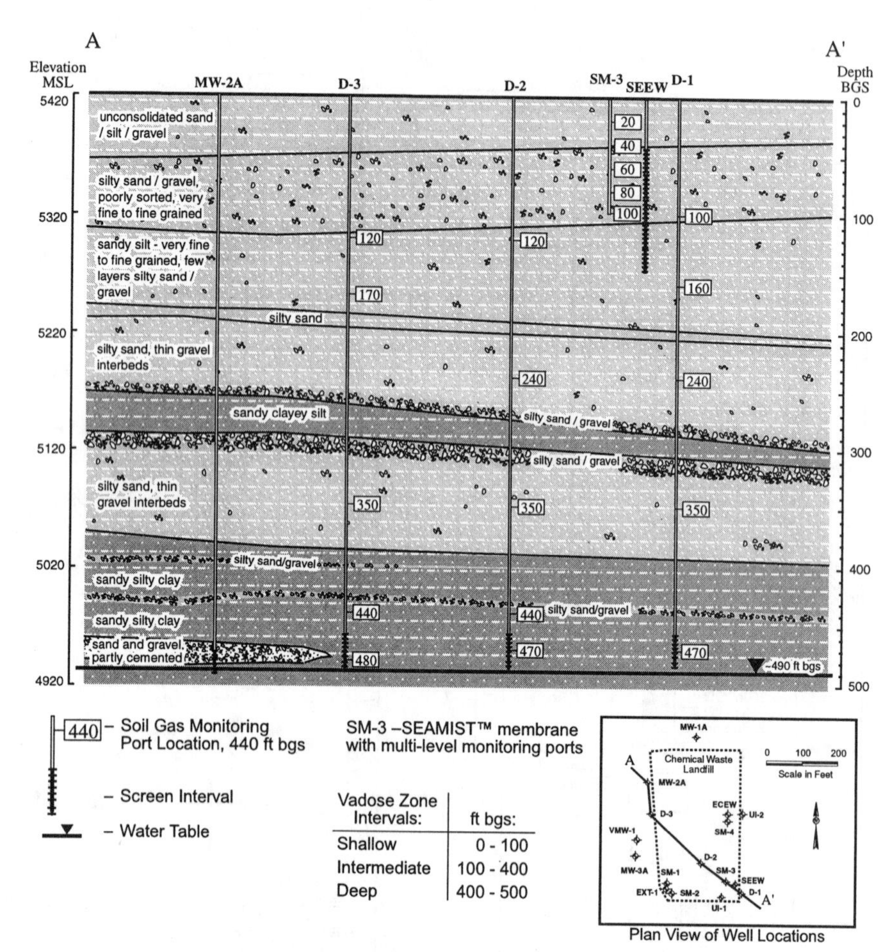

FIGURE 2. Simplified vertical profile of vadose zone lithology, well screen and soil gas monitoring port locations

analysis of *in-situ* soil gas concentrations that occur when samples are collected during drilling.

Soil gas monitoring ports placed within the shallow, intermediate, and deep vadose-zone intervals underlying the site (Figure 2), are being used to periodically measure the response of the soil gas plume as the VE system operates through time. Both air injection and extraction are currently being used to enhance removal of VOCs directly above the water table within the deep vadose zone. Monitoring ports attached to the well screen permit the measurement of the induced vacuum or pressure occurring in the filter pack and sediments adjacent to the screen, as well as VOC concentrations between periods of operation.

Vacuum/Pressure Monitoring Data. *In-situ* injection and/or extraction pressures are measured through the soil gas monitoring ports. Magnehelic gauges capable of measuring relatively high [up to a 100 in water column (w.c.)], moderate (1 - 10 in w.c.), and low (0.01 - 0.5 in w.c.) pressures/vacuums are used. Swaglock ™ fittings at the ends of the monitoring port access tubes allow for easy, fast connection to the gauges. Barometric pressure values obtained from a local weather monitoring station are used when calculating the absolute *in-situ* pressure/vacuum values and each well's area of influence.

Soil Gas Analytical Data. A 1/10 horse-power electric rotary vane compressor is used to establish a vacuum within the soil gas monitoring port access tubing. Stagnant air is first purged from the tubing. VOC levels are monitored continuously using an organic vapor analyzer (OVA) equipped with a photo ionization detector to determine when *in-situ* soil gas is being extracted. A valve located near the end of the tubing is used to switch flow to a SUMMA ™ canister and collect the soil gas sample for laboratory analysis. As part of three sampling events completed to date, EPA Method TO-14 laboratory analytical results have been compared to the OVA measurements taken in the field and a correlation equation has been developed. A linear equation regression coefficient of 0.92 has been established between total VOC concentrations indicated by laboratory analytical results and OVA field readings. Establishing this correlation has resulted in the capability to evaluate changes in the soil gas plume using OVA measurements between laboratory analysis sampling events. Laboratory analytical data, augmented with numerous inexpensive OVA measurements of soil gas concentrations, has permitted the compilation of a more comprehensive data set with which to document VE VCM performance.

EVALUATING VOC REMOVAL PERFORMANCE

Laboratory analytical results of soil gas samples collected from the soil gas monitoring ports are being provided to regulators to document changes in VOC concentrations and configuration of the soil gas plume underlying the site. These data provide the basis for reporting progress toward meeting cleanup performance goals. The spatial distribution of the soil gas sampling ports permits plotting of analytical results and creation of two-dimensional illustrations used for assessing changes in VOC concentrations and configuration of the soil gas plume. Figure 3 (cross-sectional view a) shows the concentration distribution of VOCs in the soil gas plume based on the limited sampling data available prior to installation of dedicated soil gas monitoring ports. The interpreted configuration of the soil gas plume, based on the samples collected from soil gas monitoring ports shortly after startup of the extraction/injection system, is shown on cross-section view b (Figure 3). The distribution of the soil gas plume is much better defined using the data set that was available after installation of the soil gas monitoring ports. Soil gas analytical data gathered using the monitoring ports after six months of operation, are shown on cross-section view c (Figure 3).

CONCLUSIONS

With the considerable time and money often invested in drilling projects, it is essential to optimize the uses that any remediation or monitoring system can provide. In this case, combining a vapor extracion/air injection well with an *in-situ* soil gas monitoring system resulted in project cost savings because of the reduction in well materials and drilling, and also created a monitoring system which meets all the performance reporting requirements for the project.

REFERENCES

Hubbell, J.M., T.R.Wood, B. Higgs, and A.L. Wylie, (in press). "Design, Installation, and Uses of Combination Ground Water and Gas Sampling Wells." *Ground Water Monitoring & Remediation, Spring 1998.*

(Sandia National Laboratories is a multiprogram laboratory operated by Sandia Corporation, a Lockheed Martin Company, for the United States Department of Energy under Contract DE-ACO4-94AL85000)

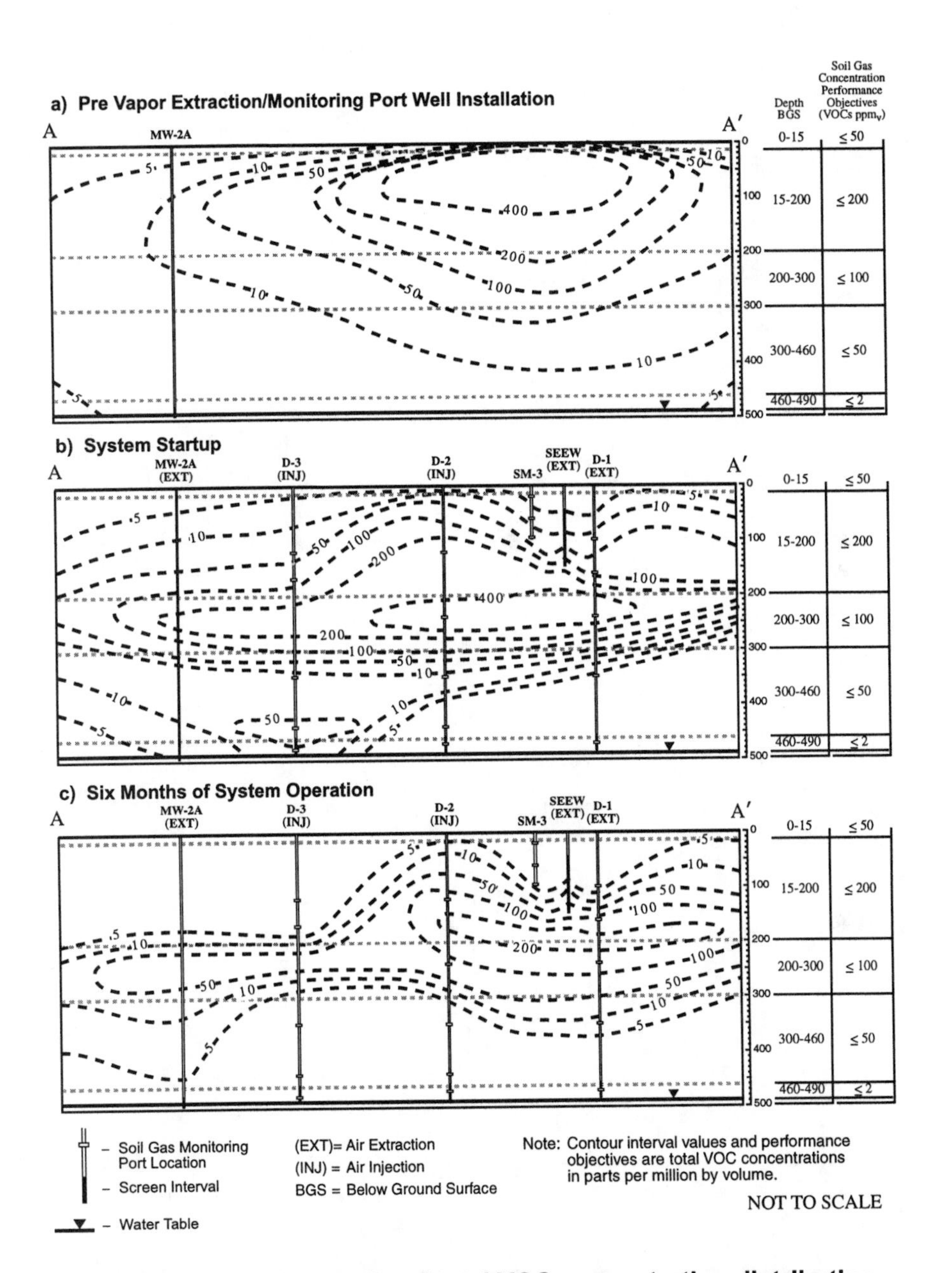

FIGURE 3. Vertical profile of total VOC concentration distribution along cross-section A-A′. a) pre VE VCM (1993), b) during VE VCM startup (May 1997), and c) at six-month performance sampling event (November 1997).

APPLICATION OF ISOTOPIC FINGER-PRINTING FOR BIODEGRADATION STUDIES OF CHLORINATED SOLVENTS IN GROUNDWATER

Ramon Aravena, Kevin Beneteau, Shaun Frape, Barbara Butler, University of Waterloo, Waterloo, Ont, Canada; Teofilo Abrajano, Memorial University of Newfoundland, St John's, Nf, Canada; David Major and Evan Cox, Beak International, Guelph, Ont, Canada

ABSTRACT: This study evaluates the chemical and isotopic patterns of inorganic and organic compounds in a site characterized by active biodegradation of PCE. These patterns on sulfate and dissolved inorganic carbon show the existence of anaerobic conditions (sulfate reducing conditions) in uncontaminated areas.. The persistence of sulfate reduction and methanogenesis is observed in the contaminated areas. ^{13}C data obtained on PCE and its degradation products show no significant isotopic difference between PCE, TCE and 1,2 DCE in the high concentration areas. A trend toward a more enriched ^{13}C content is observed in the TCE after more than 90 % of this compound has been degraded. A significant isotopic fractionation is observed in the step DCE-VC. The remaining DCE becomes more and more enriched in ^{13}C as the biodegradation reaction proceeds. No ^{13}C data is available yet for ethene, but the ^{13}C-VC data suggest the step VC-ethene is also affected by significant isotopic fractionation. Preliminary data obtained in microcosm experiments using soil samples from the contaminated site show similar trends to that of the field data. These results have significant relevance for the use of isotopes as a fingerprinting tool and for the identification of sources of chlorinated solvents in groundwater.

INTRODUCTION

The widespread occurrence of groundwater contaminated with chlorinated solvents has promoted significant research efforts in the understanding of the fate and behaviour of these compounds in the environment (e.g. Cherry, 1992). Several laboratory and field studies have shown that biodegradation under anaerobic conditions is a viable option for remediation of chlorinated solvents in groundwater (e.g. Major et al., 1991; 1994; Lorah et al., 1997). This process can occur under iron-, nitrate-, and sulfate reducing conditions and methanogenic conditions (e.g. Bagley and Gosset, 1990; Vogel et al., 1985). The characterization of biodegradation products and reaction rates in groundwater and the identification of sources of chlorinated plumes have significant relevance for remedial strategy for sites affected by chlorinated compounds. Our research aim is to develop new tools (isotopic finger-prints) that can be used to improve the characterization of sites contaminated with chlorinated solvents.

Environmental isotopes have been used extensively in geochemical and hydrogeological studies in groundwater systems (e.g. Clark and Fritz, 1997). Recent isotopic data reported for chlorinated solvents (van Warmerdam et al., 1995) show that specific compounds from different manufacturers seems to have a distinct isotopic fingerprint that can be linked to the manufacture process. New development in the area of compound-specific isotope analyses makes it possible to apply these techniques (isotopic finger- prints) in areas related to identification of sources and degradation of organic contaminants in groundwater (e.g., O'Malley et al., 1994). The work presented at this conference is part of a major research program aimed at evaluating the potential of environmental isotopes, mainly ^{13}C and ^{37}Cl, as tracers to provide information about sources and transformation of organic compounds in groundwater systems.

Study Site: The study site is a 4.5-acre chemical transfer facility located in Weston, Ontario. Organic solvents are delivered to the site by rail and are transferred by below and above-ground transfer pipes to an above-ground tank farm and eventually to tanker trucks. Releases of PCE during the coupling and uncoupling of railway tanker cars caused the contamination. Historically, the solvents stored and transferred at the site have been methanol, methyl ethyl ketone, vinyl and ethyl acetate, and butyl acrylate.. The previous owner of the facility had stored and transferred PCE at the site. The presence of TCE, 1, 2 DCE, VC and ethene indicated that active biodegradation of PCE is occurring at the study site (Major et al., 1991).

The subsurface geology is composed of clay till (upper clay) underlain in order by a sand-silt unit (upper sand/silt), a clay unit (lower clay), a layered sand/silt unit (lower sand/silt), and bedrock. The contaminant plume is mainly located in the upper clay and the upper sand /silt unit. The upper clay unit is weathered and fractured and contains numerous sand and silt lenses and stringers of uncertain lateral extent. The lower clay is very stiff and seems to be a competent aquitard.

MATERIALS AND METHODS

The site is well-instrumented with wells screening the four main geologic units. These wells are installed in uncontaminated areas, source areas and downgradient from the potential contaminant sources. This instrumentation was used to characterize the groundwater flow system and for groundwater sampling. An extensive sampling was carried out at the site by Beak International to characterize the contaminant plume (Major et al., 1991). Selected wells were used for the isotope study (Table 1). Groundwater samples were collected for chemical and isotope analyses. The inorganic analyses consisted of main cations including iron and main anions. Organic analyses were conducted for samples selected for ^{13}C analysis. Isotope analysis was performed on sulfate (^{34}S), dissolved inorganic carbon (^{13}C) and chloride (^{37}Cl). These analyses were carried out at the Environmental Isotopes Laboratory, University of Waterloo. The compound-specific ^{13}C analysis on the chlorinated compounds was performed at the Isotope Biogeochemistry Facility of Memorial University of Newfoundland.

RESULTS AND DISCUSSION

The groundwater in the clay units upgradient and downgradient of the contaminant plume are characterized by relative high concentration of bicarbonate (500 to 780 mg/L) and the presence of sulfate (120 to 470 mg/L). The dissolved inorganic carbon, which is equivalent to bicarbonate, show very depleted $\delta\ ^{13}C$ values (-15 to -18.6 ‰). These data and the ^{34}S data (~ +5 ‰) in sulfate suggest that sulfate reduction is occurring in the clay units. An increase in bicarbonate (2740 mg/L) and in its ^{13}C content was measured in the clay close to the areas of high concentration of chlorinated solvents (source areas). No sulfate is present in these areas. The positive $\delta\ ^{13}C$ values (+ 4.2 ‰) in these areas is a clear indication of methanogenic conditions. The groundwater in the upper sand/silt aquifer upgradient of the contaminant plume is characterized by lower bicarbonate concentration (160 mg/l) and enriched $\delta\ ^{13}C$ values (-7.2 ‰) compared to the clay units. No sulfate is present in the groundwater in these areas.

The bicarbonate concentration in the areas of active biodegradation range between 600 and 1130 mg/L. The $\delta\ ^{13}C$ values range between -14.3 and -9.0 ‰, indicating that biodegradation is occurring under sulfate reducing condition. Evidence of methanogenic condition (HCO_3^- ~ 600 mg/L; $\delta\ ^{13}C$ = +7.7 ‰) was also measured in groundwater collected from a former pumping well screening the clay and upper sand/silt aquifer in the source areas.

Comparison of ^{37}Cl data on the inorganic chloride obtained from uncontaminated and contaminated areas show noticeable difference that could be related to the input of Cl from dechlorination of PCE. However, the input of Cl from road salt at the study site complicates the ^{37}Cl interpretation.

The PCE in the two areas of high concentration is characterized by $\delta\ ^{13}C$ values close to -31 ‰. This indicates that the PCE is of similar origin, probably indicating a common source. No significant $\delta\ ^{13}C$ difference is observed between PCE and TCE in the groundwater with high concentration of both compounds. Slight $\delta\ ^{13}C$ enrichment is observed in the 1,2 DCE in the same groundwater (Table 1). A significant $\delta^{13}C$ shift toward more enriched values on TCE is only observed at very low TCE concentration, probably representing less than 5 % of the TCE produced from the dechlorination of PCE. The most appreciable isotopic fractionation occurred during the step DCE-VC (Table 1). The isotopic trend is toward more enriched ^{13}C in the remaining DCE. The bacteria prefer to dechlorinate the isotopically lighter organic molecules leaving the residual enriched in the heavy isotope (^{13}C). This is the general pattern that occurs in redox processes mediated by bacteria.

Preliminary data obtained in microcosm experiments using the same aquifer materials tend to support the isotopic pattern observed in the field site in both the direction and magnitude of carbon isotopic fractionation. Both field and laboratory observations confirm that anaerobic microbial dechlorination of chlorinated ethenes lead to the production of ^{13}C-enriched residue and ^{13}C-depleted product or intermediate. These results have significant implications for the use of isotopes as fingerprinting tools and for the identification of sources of chlorinated solvents in groundwater.

TABLE 1. Isotopic composition and concentration of chlorinated compounds in groundwater

Well Name	Depth Well Screen (m)	PCE $\delta^{13}C$ ‰	PCE CONC µg/l	TCE $\delta^{13}C$ ‰	TCE CONC µg/l	cis 1,2 DCE $\delta^{13}C$ ‰	cis 1,2 DCE CONC µg/l	VC $\delta^{13}C$ ‰	VC CONC µg/l
CEL 2[+]	2.0-4.5	-31.9	18870	-33.3	2927	-27.9	41210		6853
		-31.1		-32.1		-28.4			
CEL 1[++]	7.6-9.1	-29.6	1643.5	-32.8	829		933		71
PW-1[+++]	2.8-12.4		22	-26.0	128	-22.3	45125	-36.8	13980
CEL 3[+]	1.8-4.3	-31.7	43240	-30.3	6730	-29.8	24280		2653
		-31.8		-31.8		-29.5			
ISPR 5-3[+]	2.9-4.3		ND	-27.7	93		ND		51
				-27.1					
CEL 10[++]	5.8-7.2		6.3	-24.6	68	-10.3	52150	-27.2	33660
				-26.6		-11.4			
ISRP 5-1[++]	7.6-9.0		ND	-25.7	33	+13.3	10685	-6.4	20920
				-26.2		+11.1		-8.3	
ISRP 4-1[++]	7.7-9.1		7.9	-26.2	32	-1.8	18870		
				-25.2		-0.1		-26.6	29750

[+] wells in clay aquitard
[++] wells in sandy silt aquifer
[+++] former pumping well

REFERENCES

Clark, I., and P. Fritz (Eds.). 1997. Environmental Isotopes in Hydrogeology. Lewis Publishers.

Cherry, J. A. 1992. "Chlorinated Solvents in Groundwater: Field Experimental Studies of Behaviour and Remediation". Journal of Hazardous Materials, 32: 275-278.

Major, D.W., E.H. Hodgins, and B.H. Butler. 1991. "Field and Laboratory Evidence of in Situ Biotransformation of Tetrachlroethene to Ethene and Ethane in a Chemical Transfer Facility in North Toronto". In R.E. Hinchee & R.F. Olfenbuttel (Eds.). In Situ and On Site Bioreclamation, Buttersworth-Heineman, Stoneham, MA.

Major, D.W., E.E. Cox, E.Edwards, and P.W. Hare. 1994. "The Complete Dechlorination of Trichloroethene to Ethene Under Natural Conditions in a Shallow bedrock Aquifer Located in the New York State". In Symposium on Instrinsic Bioremediation of Groundwater, United States Environmental Protection Agency EPA/540/R-94/515, August 1994, Denver. CO.

Lorah, M.M., L.D. Olsen, B. L. Smith, M.A. Johnson, and W.B. Fleck. 1997. "Natural Attenuation of Chlorinated Volatile Organic Compounds in a Freshwater Tidal Wetland, Aberdeen Proving Ground, Maryland". US. Geological Survey, Water Resources Investigations Report 97-4171. 95 p.

O'Malley, V., T. Abrajano, and J.Hellou. 1994. "Determination of $^{13}C/^{12}C$ ratio of Individual PAH from Environmental Samples: Can PAH be Source-Apportioned?". Organic Geochemistry, 21, 809-822.

Van Warmendam, E.L., S.K. Frape, R. Aravena, R.J. Drimmie, H. Flat, and J. Cherry. 1995. "Stable Chlorine and Carbon Isotope Measurements of Selected Organic Solvents". Applied Geochemistry, 10: 547-552.

Vogel, T.M., and P.L. McCarty. 1985. "Biotransformation of Tetrachloroethylene to Trichloroehylene, Dichloroethylene, Vinyl chloride and Carbon Dioxide Under Methanogenesis Conditions". Applied. Environmental Microbiology, 49: 1080-1083.

RBCA COMPLIANCE STATISTICS FOR SMALL DATA SETS

Melinda W. Hahn (ENVIRON International, Houston TX)
Angela E. Sevcik (The ERM Group, Chicago IL)
Roy O. Ball (ENVIRON International, Chicago IL)

ABSTRACT: In Risk-Based Corrective Action (RBCA), it is necessary to calculate a statistical estimate of the spatially averaged concentration (the 'statistic' that represents average dose) for comparison with risk-based remediation objectives. This is especially important for recalcitrant compounds which are expected to persist over the default exposure duration (30 years) used in RBCA calculations for carcinogenic compounds. The statistic for environmental data that is compared to the remediation objective to determine compliance (i.e., the RBCA statistic) is typically calculated by 1) assuming that the data set can be represented by a distribution such as the normal or the lognormal and estimating the upper confidence limit of the mean based on the mathematics of the assumed distribution (the Risk Assessment Guidelines for Superfund [RAGS] method, or 2) transforming the data such that the transformed values are well represented by the normal distribution, calculating the upper confidence limit of the mean in the transformed regime, then transforming the statistic back to the original regime (the RCRA or SW-846 method). The "goodness of fit" of these distributions is tested for data selected with an exponential generating function (similar to a transport equation), and the implications of selecting a statistical distribution are explored for the calculation of the RBCA statistic. Although environmental data are generally well represented by the lognormal distribution, the skewness of this distribution to the right is more pronounced than that seen in environmental data sets, and it biases the estimate of the RBCA statistic to the right. To avoid the pitfalls of assuming the perfect fit of a mathematical distribution, an alternative is to use non-parametric methods to calculate the statistic, or to rely on statistics that are not biased by the extreme upper tail of the lognormal distribution.

INTRODUCTION

Environmental risk is calculated as the product of toxicology factors and dosage. Dosage is generally expressed as the mass of contaminant ingested, inhaled or otherwise absorbed, divided by the receptor mass per unit time. The concentration term is directly proportional to the dosage or exposure, and, therefore, to the corresponding risk. There are significant economic disadvantages of excessive conservatism in calculating the site concentration term for risk assessment, without any corresponding benefit to public health. Therefore, calculating a reliable estimate for the true average exposure concentration is important.

According to the May, 1992 Supplemental Guidance to RAGS, "... the concentration term (C) in the intake equation is an estimate of the ... average concentration for a contaminant based on a set of site sampling results." An estimate of average concentration is used because it "...is most representative of the concentration that would be contacted at a site over time...based on lifetime average exposures..."and "...the spatially averaged soil concentration can be used to estimate the true average concentration contacted over time." Therefore, the problem of reliably determining the environmental risk due to site contamination is directly related to, *inter alia*, estimating the spatially averaged soil concentration, generally using small data sets. 'Small data sets' are not precisely defined in the literature, but generally are taken to mean less than 30 samples per exposure (or source) area.[1]

Soil and ground water contamination are usually due to an accidental spill or localized release. Such generating events are thought to create a lognormal or near-lognormal distribution of contamination (Ott 1990), as the contaminant pattern in space and time is caused by successive dilution. Even if there is doubt about the underlying distribution of the contamination, Berthouex and Brown (1994) conclude that, "... making the lognormal transformation is usually beneficial, or at worst harmless."[2] This also avoids the problem discussed by Shumway *et al.* (1989), "...for most environmental data, the assumption that the underlying distribution is normal will not be appropriate, so the usual sample mean" (*i.e.*, the arithmetic mean) "will not be a good estimator of the population mean."

WHICH DISTRIBUTION IS BEST FOR ENVIRONMENTAL DATA?

To evaluate the accuracy of the selection of a particular distribution, a generating function was used to simulate environmental data. The distribution of contamination from the source is defined as:

$$C_{rz} \equiv C_o e^{-(z/\vartheta_z + r/\vartheta_r)} \tag{1}$$

where z is the depth from the ground surface, r is the radial distance from the release centroid and the release attenuation parameters (θ_r and θ_z) define the characteristic distance of attenuation along each axis. This exponential decay equation provides a realistic distribution of contamination. In actuality, such distributions do have a random component due to variations of soil properties and the rate and conditions of contaminant release. For this idealized example however, θ_r is taken as 8 feet and θ_z is taken as 5 feet, and our domain is defined by r ranging from 0 to 32 feet and z ranging from 0 to 20 feet. Ball and Hahn (1998) showed that this function yields a spatial average for C/Co which is equivalent to the geomean (rather than the expected value of the lognormal distribution which appears to fit the data) of data drawn randomly from this distribution.

---

[1] For example, the RAGS Supplemental Guidance of May, 1992, states that, "...fewer than 10 samples ... provide poor estimates of the mean concentration ... while data sets with 20 to 30 samples provide ... consistent estimates of the mean"

[2] Many other authors concur with Berthouex and Brown *(e.g.*, Newman *et al.* 1989, etc.). Stoline (1990), however, cautions against, "...the automatic use of the lognormal model..." and recommends, "...checking the adequacy of the lognormal model prior to use."

To test the hypothesis that data drawn from this distribution are drawn from a lognormal distribution, the Shapiro-Wilks test for normality was performed on the log-transform of 30 "data points" drawn from Equation 1. With a resulting W value of approximately 0.95, the hypothesis was accepted for $\alpha = 0.05$.

To further test the appropriateness of the lognormal distribution for data drawn from Equation 1, five hundred additional "data points" were drawn randomly from it. The sample mean and standard deviation were calculated (0.067 and 0.125, respectively). The sample mean and standard deviation were also calculated for the natural log transformed data (-3.93 and 1.62, respectively). Five hundred "data points" were then drawn from Normal(0.067, 0.125) and Lognormal(-3.93, 1.62) for comparison. The frequency plots for the original data and the fitted distributions are shown in Figure 1. Clearly, the normal distribution is a poor fit because it predicts negative values of C/Co, and because the distribution is symmetric, not skewed to the right. The lognormal, on the other hand, is a much better fit, especially for low values of C/Co. A critical disadvantage of the lognormal distribution, however, is that it can return values of C/Co above one – clearly a physical impossibility.

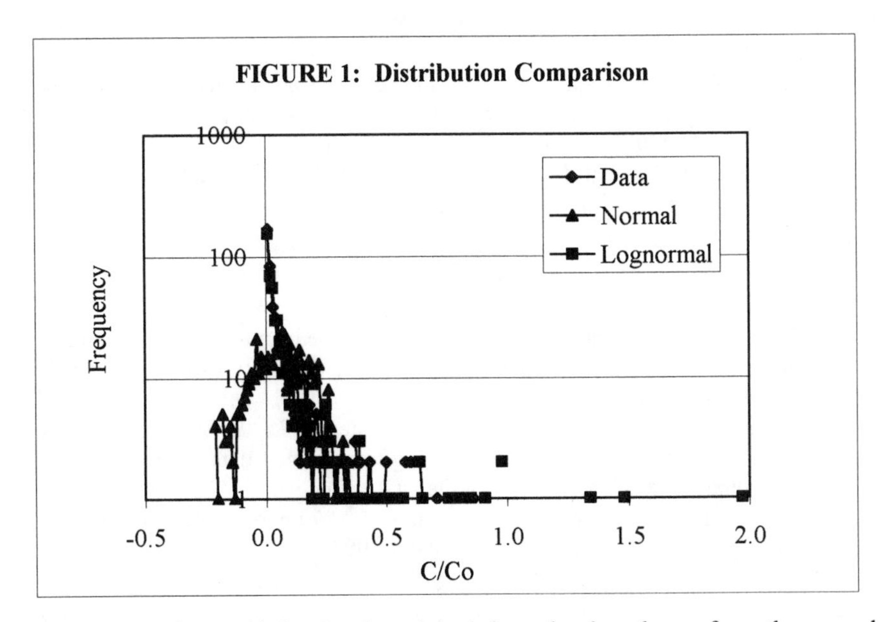

Summary statistics for the original data, the data drawn from the normal distribution, and the data drawn from the lognormal distribution are given in Table 1. Again, the normal distribution is clearly flawed due to the negative values. The lognormal distribution, in addition to returning values for C/Co greater than one, generates data with a larger sample average and a significantly larger sample standard deviation than the original data. This will significantly bias the estimate of the upper confidence limit of the mean, as shown below.

	Data	Fitted Normal	Fitted Lognormal
Average	0.067	0.062	0.083
Standard Deviation	0.125	0.132	0.273
Minimum	0.000	-0.388	0.000
Maximum	0.945	0.443	3.978

TABLE 1

THE IMPLICATIONS OF ASSUMING A DISTRIBUTION

To test the impact of the selection of a distribution on the calculation of the RBCA statistic, 500 random sets of 30 data points were selected from the generating function and the RBCA statistic was calculated using the following methods

1. Assume a normal distribution – calculate the 95th % upper confidence limit (UCL) of the arithmetic mean

$$\mathrm{UCL} = \bar{x} + t\left(s/\sqrt{n}\right) \qquad (2)$$

where $\bar{x}$ is the sample mean, s is the sample standard deviation, n is the number of data points, and t is the Student t statistic (e.g., from table published in Gilbert, 1987),

2. Assume a lognormal distribution – calculate the 95th % upper confidence limit of the arithmetic mean

$$\mathrm{UCL} = \exp\left(\bar{x} + 0.5s^2 + sH/\sqrt{n-1}\right) \qquad (3)$$

Where $\bar{x}$ is the mean of the log transformed data, s is the standard deviation of the transformed data, and H is the H-statistic (as given in Gilbert, 1987),

3. Assume a lognormal distribution – calculate the 95th % upper confidence limit of the arithmetic mean of the log-transformed data (using Equation 2) and back-transform the result (the UCL geomean)
4. Assume no distribution – calculate the 95th % upper confidence limit of the 50th percentile (the median) (Gilbert, 1993)

$$k = p(n+1) + Z_{0.95}[np(1-p)]^{1/2} \qquad (4)$$

where p is the percentile desired (in our case, 0.50) and Z is the Z-statistic given in Gilbert, 1987. The UCL of the median is then the kth value in the data set if it is ordered from smallest to largest.

For each of the 500 "trials", the average, standard deviation, minimum and maximum values were calculated for the RBCA statistic generated by each method. The results of this simulation are presented in normalized concentration

(C/Co) in Table 2. Despite the prediction of negative concentration, the UCL of the mean assuming a normal distribution overestimates the true spatial average (0.012 as described in Ball and Hahn, 1998[3]) by several times. Similarly, the UCL of the mean assuming a lognormal distribution overestimates the true spatial average by, on average, an order of magnitude. It is noteworthy that the maximum UCL mean using this method is almost as large as the average maximum value of the original data set. Clearly, for the lognormal distribution, the standard deviation is exaggerated by large, physically impossible values (C/Co > 1), and the UCL mean is biased high because it is very sensitive to large values of the standard deviation.

	Sample Statistics				
	Maximum	Average	Standard Deviation	Median	UCL Median
Average	0.444	0.060	0.102	0.019	0.044
Standard Deviation	0.187	0.020	0.039	0.008	0.020
Minimum	0.097	0.021	0.029	0.005	0.011
Maximum	0.927	0.132	0.210	0.053	0.148

	Normal Distribution		Lognormal Distribution			
	Mean	UCL Mean	Mean	UCL Mean	Geomean	UCL geomean
Average	0.060	0.092	0.019	0.129	0.019	0.032
Standard Deviation	0.020	0.031	0.006	0.059	0.006	0.010
Minimum	0.021	0.031	0.008	0.034	0.008	0.012
Maximum	0.132	0.189	0.041	0.418	0.041	0.069

TABLE 2

It is also noteworthy that, for our simulated data set, both the geomean (assuming a lognormal distribution) and the sample median provide an accurate estimate of the true spatial average (0.012). The UCL geomean (lognormal) and the UCL median (non-parametric) are accurate predictors of the upper bound of the true spatial average. The UCL geomean is the most stable predictor of the RBCA statistic as is evidenced by the ratio of the standard deviation to the average and by the relatively narrow range defined by the minimum and maximum predicted values. Clearly the geomean is less biased by the extended right tail of the fitted lognormal distribution than the UCL mean and is a superior predictor of the central tendency of the distribution.

CONCLUSION

As shown in Ball and Hahn (1998) and again here, the UCL geomean (assuming a lognormal distribution) is the most accurate and stable predictor of the true exposure concentration when an idealized exponential generating function is used to simulate the data. Environmental data, however, seldom yield a perfect fit to any statistical distribution. The Shapiro-Wilks test for normality was applied to the log-transformed data from a site in Pensylvania with

[3] Calculated by numerically integrating Equation 1 over the domain.

perchloroethene contamination in soil, and the hypothesis that the data are lognormally distributed was rejected for $\alpha = 0.05$. We have shown that applying even the lognormal distribution -- arguably the most appropriate for environmental data --can be harmful in some cases. Resorting to non-parametric tests is a sound alternative when "goodness-of-fit" tests indicate that data is neither normal nor lognormal. The non-parametric UCL of the median is an appropriate and unbiased statistical test for compliance with RBCA remediation objectives when the assumption of a statistical distribution is either unwarranted or undesired. The use of the non-parametric UCL of the median also is consistent with the letter and the intent of SW-846.

REFERENCES

ASTM, 1995, *Guide for Risk-Based Corrective Action Applied at Petroleum Release Sites*, ASTM E 1739-95.

Berthouex, P.M., and Brown, L.C., 1994, *Statistics for Environmental Engineers*, CRC Press, Boca Raton, Florida.

Gilbert, R.O., 1987, *Statistical Methods for Environmental Pollution Monitoring*, Van Nostrand Reinhold, New York.

Gilbert, R.O., 1993, *Statistical Evaluation of Cleanup: How Should It Be Done?*, Waste Management February 1993 Symposia, Tucson Arizona

Newman, M.C., Dixon, P.M., Looney, B.M., and Pinder, J.E. III, "Estimating Mean and Variance for Environmental Samples with Below Detection Limit Observations", *Water Resources Bulletin*, Vol. 25, No. 4, pp. 905-916.

Ott, W., 1990, "A Physical Explanation of the Lognormality of Pollutant Concentrations", *Journal of the Air and Waste Management Association*, Vol. 40, pp. 1378-1383.

Shumway, R.H., Azari, A.S., and Johnson, P., 1989, "Estimating Mean Concentrations Under Transformation for Environmental Data with Detection Limits", *Technometrics*, Vol. 31, No. 3, pp. 347-356.

Stoline, M.R., 1991, "An Estimation of the Lognormal and Box and Cox Family of Transformations in Fitting Environmental Data", *Environmetrics*, Vol. 2, No. 1, pp. 85-106.

U.S. EPA, May 1992, *Supplemental Guidance to RAGS: Calculating the Concentration Term*, Publication 9285.7-081.

U.S. EPA, 1986, *Test Methods for Evaluating Solid Waste, Field Methods*, USEPA SW-846.

GEOSTATISTICAL CHARACTERIZATION OF MICROBIOLOGICAL AND PHYSICAL PROPERTIES FOR BIOREMEDIATION MODELING

Christopher J. Murray, Timothy D. Scheibe, Fred J. Brockman, Gary P. Streile, and Ashok Chilakapati, Pacific Northwest National Laboratory, Richland, Washington, USA

ABSTRACT: Geostatistical methods were used to quantify the spatial heterogeneity in the distribution of microbiological activity and physical properties in subsurface sediments at two field sites, the White Bluffs and Oyster/Abbott's sites. Variograms were calculated for the permeability and aerobic microbial activity at the White Bluffs site. Sequential Gaussian simulation was used to generate a 2-D realization of permeability in the sand and silt facies at the White Bluffs. Indicator methods were then used to simulate the distribution of microbial activity. Using a calibration between aerobic microbial activity data and phospholipid data, the activity simulation was transformed into a simulation of microbial biomass at the site. A combination of indicator and Gaussian geostatistical methods was also used to generate realizations of the physical and microbiological properties at the Oyster site. The simulations of permeability and microbial biomass are being used as inputs for reactive flow and transport models to study the effects of microbial heterogeneity on intrinsic bioremediation of trichloroethylene (TCE) in the presence of phenol and oxygen.

INTRODUCTION

Subsurface heterogeneity has a major impact on reactive transport processes in the subsurface, including those associated with bioremediation. Physical heterogeneity, principally the spatial variability in pore and pore throat sizes that control hydraulic conductivity, affects our ability to predict the movement of fluids through the subsurface (e.g., Dagan, 1984). Preliminary studies suggest that geochemical heterogeneity impacts the migration of reactive fluids through subsurface aquifers and complicates the prediction of transport and degradation of contaminants in geochemically heterogeneous aquifers (e.g., Tompson et al., 1996). Recent field studies have identified spatial heterogeneity in microbial activity in subsurface sediments that appears to be linked to variations in geological properties (Brockman and Murray, 1997). Preliminary modeling studies suggest that interaction between physical and microbiological heterogeneity would impact the intrinsic bioremediation of chlorinated hydrocarbons (Murray et al., 1997).

This paper concentrates on the use of geostatistical techniques to evaluate field data from two sites and generate numerical grids of microbiological and hydrogeological properties. The grids are being used in reactive flow and transport models to estimate the effect that microbiological heterogeneity would have on intrinsic bioremediation of chlorinated hydrocarbons.

SITE DESCRIPTIONS AND METHODS

Site Descriptions. Field data were taken at two areas, the Oyster/Abbott's (O/A) sites in Virginia and the White Bluffs (WB) site in south-central Washington. The sites are not contaminated, but were chosen for the modeling study because high-resolution vertical and horizontal data on microbiological and hydrogeological heterogeneity were available. The two areas occupy very different climate zones. The O/A sites are located in a high precipitation area with high groundwater recharge rates and the WB site is in the semi-arid interior West. The difference in recharge rates indicates a much higher flux of moisture, and presumably accompanying nutrients, through subsurface sediments at the O/A sites than at the WB site.

The Oyster, Virginia, site has been used for bacterial transport experiments (DeFlaun et al., 1997), and the nearby Abbott's site was used in microbial heterogeneity studies (Zhang et al., 1998). Characterization of O/A sediments was performed with a spatial resolution of 10-20 cm using boreholes, geophysics, and nearby exposures in sand and gravel pits. Hydrogeologic facies identified in marginal marine sediments at the sites include a horizontally bedded sand facies, a cross-bedded sand facies, and a cross-bedded shelly, gravelly sand facies (DeFlaun et al., 1997).

The WB site contains lacustrine fine sand and silt beds that were sampled using several vertical and horizontal boreholes designed to resolve the spatial heterogeneity. Sampling within the boreholes was on the scale of 5-30 cm. Bedding at the site is horizontal, with only minor variations in bed thickness over distances of several hundred meters.

Physical and Microbiological Measurement Methods. The distribution of sedimentary facies at the O/A site was determined from facies identifications in boreholes. Grain size, porosity and falling head conductivity was measured on more than 100 samples from 19 boreholes (DeFlaun et al., 1997) at the Oyster site. Microbial properties were measured on samples from boreholes drilled at the Abbott's site, using special coring and sampling techniques to prevent contamination of the samples (Zhang et al., 1998). Grain size was measured on those samples and used to classify the sedimentary facies at Abbott's using the classification rules developed at Oyster. Aerobic mineralization potential (AMP) of subsurface microorganisms was measured by incubating core samples with a ^{14}C-labeled glucose/acetate mixture and measuring the evolution of $^{14}CO_2$ (Zhang et al., 1998). Viable microbial biomass was determined in a subset of samples by measuring the phospholipid fatty acid (PLFA) content.

The thickness of lacustrine sand and silt beds at the WB site was measured in vertical boreholes. Over 200 samples from that site were analyzed for permeability and AMP. Permeability was measured by air injection on core plugs. AMP was measured using the radiolabel method employed at the O/A site. PLFA measurements were made on a subset of samples to estimate viable microbial biomass.

Geostatistical Methods. Spatial variability of the data was determined by variogram analysis (Brockman and Murray, 1997; Deutsch and Journel, 1992). At the O/A site, the spatial distribution of sedimentary facies was modeled using categorical indicator variograms (Deutsch and Journel, 1992). At both sites, variograms were calculated and modeled for normal score transforms of permeability data, and were analyzed separately for each sedimentary facies. Variograms of AMP rate data from each WB facies were calculated and modeled for indicator transforms of the data.

Generation of the numerical grids of permeability and microbiology was a two-stage process. First, the distribution of sedimentary facies was generated, because sedimentary facies are often the primary control on spatial variability for both physical and microbiological variables (Brockman and Murray, 1997). For the O/A site, the spatial distribution of sedimentary facies was generated using sequential stochastic indicator simulation (program SISIMPDF from the GSLIB geostatistical subroutine library [Deutsch and Journel, 1992]). At the WB site, with it's much simpler facies distribution, a layered system was developed with average bed thickness determined from vertical boreholes at the site.

Once the facies were generated, the permeability and AMP rates within each facies were simulated. Sequential Gaussian simulation was used at both sites for simulation of permeability, employing SGSIM (Deutsch and Journel, 1992). A similar process was followed for the AMP rate grids at the WB, except that the sequential indicator simulation program SISIM (Deutsch and Journel, 1992) was used rather than sequential Gaussian simulation because of the highly non-Gaussian nature of the AMP data. There was insufficient data at the O/A sites to allow definition of the horizontal variogram model. The grid of AMP rates for that site was generated by drawing a random value from the univariate distribution of the AMP rate corresponding to the facies present at a grid node. For both the O/A and WB sites, the grids of AMP rates were transformed into grids of microbial biomass using conversion factors determined from the PLFA data.

RESULTS AND DISCUSSION

Physical and Microbiological Measurements. All 3 sedimentary facies at the O/A site are dominated by sandy sediments with an average permeability of 35,000 mD (milliDarcies). The contrast between permeability in the three facies is less than an order of magnitude. Examination of grain size data suggests that the same sedimentary facies exist at both O/A sites, allowing combination of physical and microbiological data from the two sites. Microbial AMP rates measured at O/A varied considerably between the 3 sedimentary facies, with microbial activity concentrated in more permeable coarse-grained facies.

The permeability of WB sediments is 2-3 orders of magnitude lower than at the O/A site, with average permeability of 40-50 mD in silt beds and 850-950 mD in sand beds. Although average permeability is lower at the WB site, the contrast between different sediment types is much greater than it is at O/A. Microbial AMP rates are lower at WB, and no activity was detected in 38% of the samples (even after long incubations with added water and inorganic nutrients). Microbial AMP at WB tends to be concentrated in finer-grained silt layers.

Variogram analysis. For the O/A site, variogram models were fit to categorical indicators for the 3 sedimentary facies. Vertical variogram models were fit to facies data available from boreholes. Horizontal variogram models were constrained by the continuity of facies seen in outcrop exposures. The maximum horizontal correlation range was estimated to be 20 m, with vertical correlation ranges of the facies between 0.1 and 1 m. Correlation ranges for permeability were about half that of sedimentary facies.

At the WB, models were fit to horizontal and vertical variograms calculated on normal score transforms of permeability. This was performed separately for the sand and silt facies. In both cases there was a pronounced anisotropy, with a stronger anisotropy in the more finely layered silt beds than in the sand beds (25:1 vs. 2.8:1). The horizontal variogram models possessed longer ranges (250 cm for silt and 110 cm for sand) than the vertical variograms. Sills of horizontal permeability variograms were lower than those of vertical variograms, so that zonal as well as geometric anisotropy was present at the WB (see Deutsch and Journel [1992] for a discussion of the properties of anisotropic variogram models). This was expected given the highly ordered nature of the layering in the WB sediments.

Variograms for WB AMP rates were fit using indicator transforms of the data for each facies type. For the sand facies, the few locations with activity appeared to be random, so the indicator variogram model used for microbial activity in that sediment type was a pure nugget effect. The variogram of AMP rate for the silt facies displayed geometric anisotropy, with greater continuity in the horizontal. That variogram was fit by a model with a 10:1 anisotropy and a horizontal variogram range of 150 cm.

Permeability and biomass grids. Simulations of permeability and AMP rate at the two sites were generated using the geostatistical procedures outlined in the Methods section. The AMP grids were then transformed to biomass grids using calibrations between AMP rates and viable biomass. The resulting simulations of permeability and biomass provide grids that reproduce the important heterogeneities of properties at the two sites (Figure 1). The simulations in Figure 1 represent 2-D vertical slices at each site. At WB, the grids portray horizontally layered beds with low permeability but with a high permeability

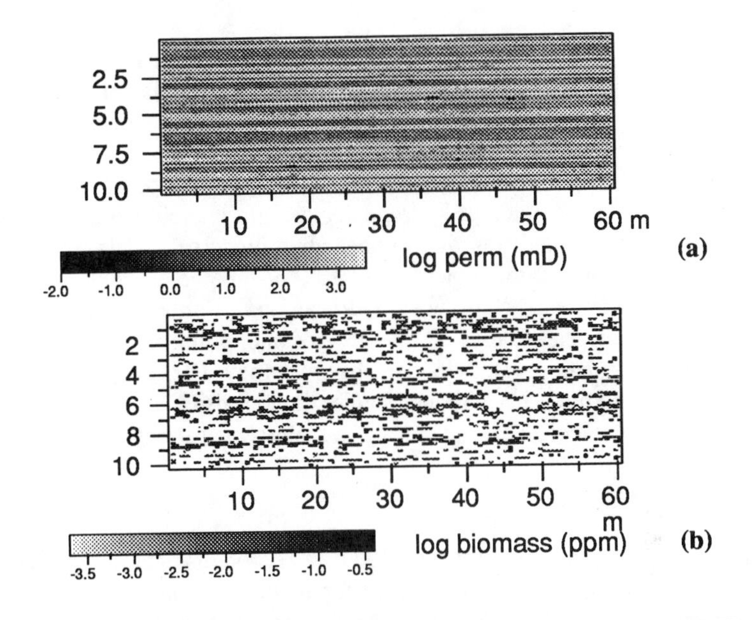

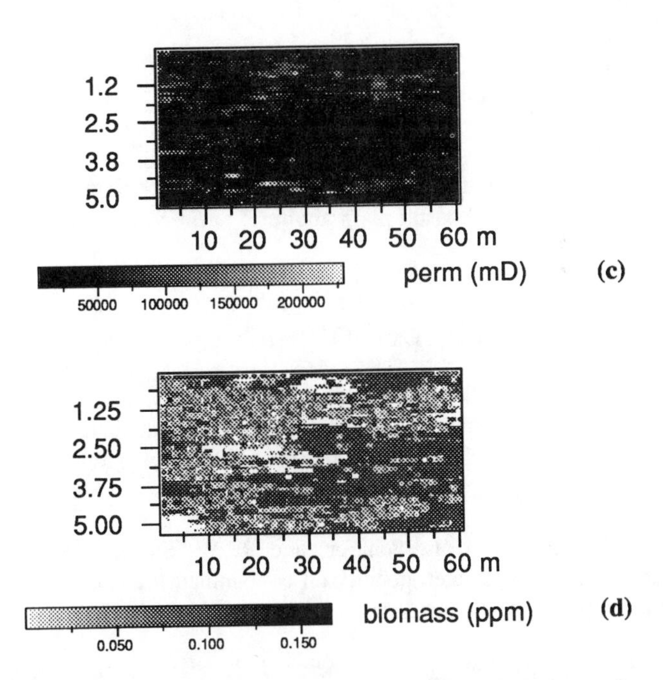

Figure 2. Permeability and biomass grids. (a) Permeability grid for the White Bluffs Site. (b) Biomass grid for the White Bluffs site. (c) Permeability grid for the Oyster site. (d) Biomass grid for the Oyster site.

contrast between the sand and silt beds (Figure 1). Low biomass is concentrated in high permeability zones at the WB. At O/A, the layering is not as structured, there is a lower permeability contrast between the facies, and high microbial biomass is concentrated in high permeability zones (Figure 1).

The results indicate that geostatistical methods provide a useful method of producing numerical grids of hydrogeological and microbiological properties suitable for computer modeling of bioremediation processes. This could lead to better understanding of the potential impact of microbial heterogeneity on intrinsic bioremediation of chlorinated hydrocarbons. Further research needs to be done to extend the studies to the use of more direct measurements of contaminant degradation rates, rather than a proxy (i.e., degradation of a glucose/acetate mixture). Additional studies should also be performed at sites having different geological, geochemical, and microbial heterogeneity patterns.

REFERENCES

Brockman, F. J., and C. J. Murray. 1997. "Microbiological heterogeneity in the terrestrial subsurface and approaches for its description." In P. Amy and D. Haldeman (Eds.), *Microbiology of the Terrestrial Deep Subsurface*, pp. 75-102, CRC Lewis Press, Boca Raton, FL.

Dagan, G. 1984. "Solute transport in heterogeneous porous formations." *Journal of Fluid Mechanics*, *145*: 151-177.

DeFlaun, M. F., C. J. Murray, W. Holben, T. Scheibe, A. Mills, T. Ginn, T. Griffin, E. Majer, and J. L. Wilson. 1997. "Preliminary observations on bacterial transport in a coastal plain aquifer." *FEMS Microbiology Reviews*, *20*: 473-487.

Deutsch, C.V., and A.G. Journel. 1992. *GSLIB: Geostatistical Software Library and User's Guide*. Oxford University Press, New York.

Murray, C., G. Streile, A. Chilakapati, F. Brockman, and T. Scheibe. 1997. "Potential effects of microbial heterogeneity on bioremediation in subsurface sediments." *Geological Society of America Abstracts with Programs*, *29*(6): 149.

Tompson, A. F. B., A. L. Schafer, and R. W. Smith. 1996. "Impacts of physical and chemical heterogeneity on contaminant transport in a sandy porous medium." *Water Resources Research*, *32* (4), 801-818.

Zhang, C., A. V. Palumbo, T. J. Phelps, J. J. Beauchamp, F. J. Brockman, C. J. Murray, B. S. Parsons, and D. J. P. Swift. In press. "Grain-size and depth constraints on microbial variability in coastal plain subsurface sediments." *Geomicrobiology*.

CONTAMINANT PLUME AND CHARACTERIZATION USING 3D GEOSTATISTICS

R.O. Ball, Ph.D., P.E. (ENVIRON, Buffalo Grove, IL)
Mindy Hahn, Ph.D. (ENVIRON, Buffalo Grove, Ilinois)
Angela E. Sevcik, (ERM Group, Vernon Hills, Illinois)

ABSTRACT: An historical remediation project at a site which was characterized by extensive data collection is revisited and analyzed with 3-D kriging techniques in an effort to test the hypothesis that the use of geostatistics can save significant effort and cost in the process of plume delineation. A site in Pennsylvania was found to have soil and ground water contaminated with chlorinated organic compounds (VOX) and benzene, toluene, ethylbenzene, and xylenes (BTEX) due to leaking underground storage tanks.

The Pennsylvania Department of Environmental Protection (PaDEP) required a remediation objective for soil of 1 part per million (mg/kg) total VOX. Because removal of the source of contamination was required, the soil source area was characterized by approximately 150 discrete samples representing three depths on a regular grid. This produced an unusually dense data set ideal for evaluating sampling strategies for VOX contaminated soil.

The historical data set was used as a basis for evaluating the utility of 3-D kriging techniques for identifying areas of uncertainty that require sampling and for plume delineation. Small data sets (e.g., 15 data points) were sampled randomly from the historic set and kriged in three dimensions. Areas of high variance, or high "uncertainty" were identified and "re-sampled" by selecting from the historic set. The results show that the plume could have been fully delineated to the required remediation objective within a reasonable confidence limit with as few as 30 samples (15% of the original data set). This use of 3-D geostatistics to optimize sampling for definition of soil source areas above a regulatory concentration isolevel can save up to 85% of investigative costs when compared to traditional RCRA grid sampling procedures described in SW-846, with equivalent confidence.

INTRODUCTION

An industrial site had been used for manufacturing of three piece aluminum steel cans. In support of this manufacturing, the site used various chlorinated solvents including tetrachloroethene (PCE) for cleaning selected cans and a solvent for ink materials used in can decoration. The PCE was stored in quantity in an above ground storage tank. During a refilling event the truck driver allowed the tank to overfill which forced hundreds of gallons of solvent out the tank vent onto a roof which then drained via roof drains to a grassy area near the loading dock. This created a distinct area of soil contamination which required remediation.

In order to reliably characterize the contamination, which was shown by the field organic vapor meters to cover an area of approximately 200 ft. by 150 ft. with a maximum penetration into the ground of approximately 6 ft., fifty soil borings were installed and samples were taken at depths of 1, 3 and 5 ft. for a total of 148 soil samples.

This unusually dense database for a distinct contaminant PCE provides an excellent opportunity for testing the use of 3-D kriging and visualization software. Among the potential benefits of 3 dimensional geostatistical kriging and the subsequent visualization of the kriged images minimizing or limiting the cost of sampling without jeopardizing the objective of obtaining a reliable characterization in the soil is one of the strongest advantages. The 3-D geostatistical software used is Environmental Visualization System™ by C Tech Development Corporation of Huntington Beach, CA. The 148 samples were numbered and data subsets were drawn from the 148 samples at random. Data sets consisting of 15, 20, 25, 30, 40, 50 and 100 samples were drawn from the data set and in addition the full data set was available. One of the features of the EVS software is the ability to calculate a confidence interval on the soil volume containing contamination above a given concentration, or isolevel. In all cases, an isolevel of 100 parts per million was used and the +/- 60% confidence level of volume was calculated.

The results of this exercise are shown in Table 1 in which the average minimum and maximum volumes (with minimum and maximum being the 60th percentile confidence limit from the mean) are displayed. It is assumed in this case that the average, minimum and maximum for the full 148 samples represented the "true" volume of soil to be remediated. It could be seen that on the average the volumes ranged from approximately 2/3 of the true volume to approximately 1 and ½ times the volume. These data are shown graphically in Figure 1 which indicates that, except for the 15 sample case and the 30 sample case, the minimum to maximum volume range for all cases encompassed the true volume.

TABLE 1. Volumes with random selection of samples.

Number of Samples	Average Volume (yd^3)	Minimum Volume (yd^3)	Maximum Volume (yd^3)
15	727.1	433.8	1064
20	1565	950	2382
25	1302	986.9	1684
30	1977	1603	2287
40	1164	833.6	1632
50	1372	1118	1806
100	1491	1189	1943
148	1263	1066	1483

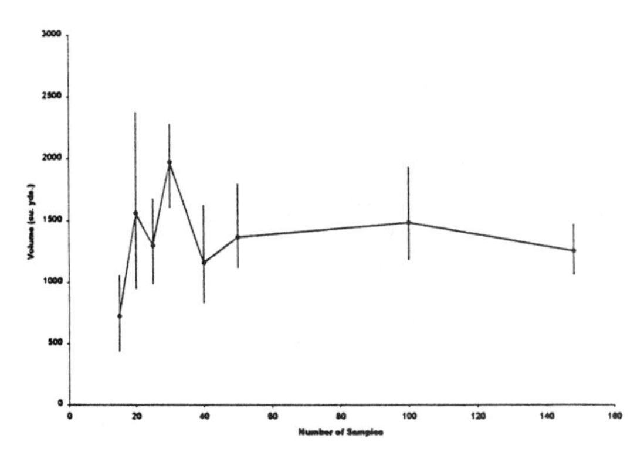

FIGURE 1. Random selection of samples.

The EVS program provides measurement of a parameter defined as uncertainty. Because the program calculates a regular grid of values based on the available data using kriging the value at any given data point also has a calculated confidence limit. The confidence should reasonably be weighted by the importance of the measurement, and therefore, uncertainty is defined as the variance at a point multiplied by the calculated value. This gives areas of higher concentration greater significance in terms of uncertainty as opposed to confidence limits. A feature of the program known as Drill Guide then suggests additional sample variations based on this uncertainty analysis. For each initial number of samples, Drill Guide was used to improve the estimate and to see at what additional number of samples the precision of the estimate approached the precision obtained by taking all 148 samples. The improvement in the estimates after taking 15 samples are shown in Table 2 and the trajectory of the estimate in shown in Figure 2. It is clear that with as few as five additional samples the precision of the estimate was greatly improved and after 10 initial samples little more improvement occurred in the estimate.

TABLE 2. Volumes with random selection of 15 samples and drill guide selection of additional samples.

Number of Samples	Average Volume (yd^3)	Minimum Volume (yd^3)	Maximum Volume (yd^3)
15	727.1	433.8	1064
20	1504	1180	1830
25	1383	1065	1682
30	1328	984	1651
35	1183	896.3	1559
40	1150	889.3	1476
148	1263	1066	1483

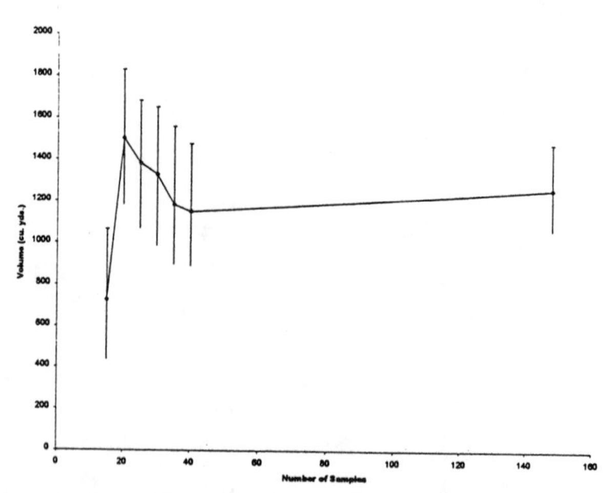

FIGURE 2. Random selection of 15 samples and drill guide selection of additional samples.

A similar process was followed for 20 samples as shown in Table 3 and correspondingly in Figure 3. Here again approximately 10 additional samples created a stable improvement in prediction. Finally, Table 4 and Figure 4 show this process with an initial start of 25 samples. Here an additional 5 samples at most, were needed to arrive at stable and reliable estimate.

TABLE 3. Volumes with random selection of 20 samples and drill guide selection of additional samples.

Number of Samples	Average Volume (yd^3)	Minimum Volume (yd^3)	Maximum Volume (yd^3)
20	1565	950	2382
25	1800	1026	3276
30	1326	902.3	2356
35	1194	700.9	1876
40	1219	757.3	1776
148	1263	1066	1483

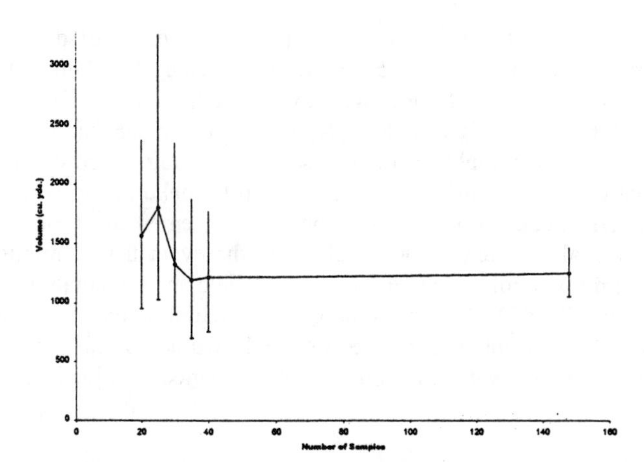

FIGURE 3. Random selection of 20 samples and drill guide selection of additional samples.

TABLE 4. Volumes with random selection of 25 samples and drill guide selection of additional samples.

Number of Samples	Average Volume (yd^3)	Minimum Volume (yd^3)	Maximum Volume (yd^3)
25	1302	986.9	1684
30	1066	808.8	1461
35	1082	837.7	1395
40	1093	874.2	1369
148	1263	1066	1483

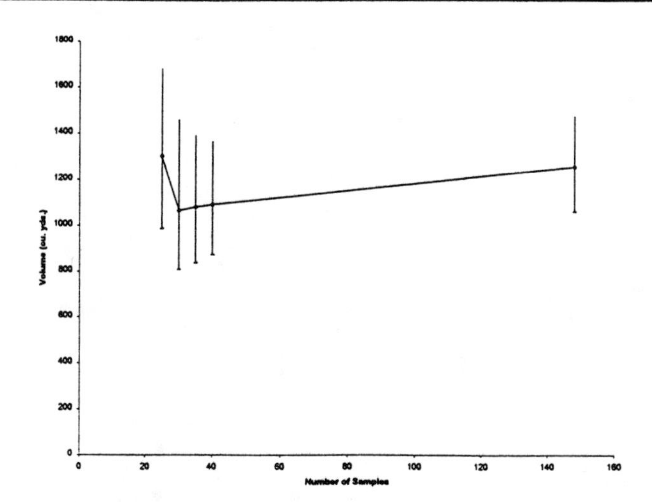

FIGURE 4. Random selection of 25 samples and drill guide selection of additional samples.

It is noteworthy that because of the effectiveness of drill guide, it was equally productive to start with 15 samples followed by 10 additional for a total of 25 versus starting with 20 followed by 10 additional for 30 or 25 for five additional for a total of 30 samples. Regardless, it is clear that only 15 to 20% of the total number of samples were necessary to realiably determine the volume of contamination if 3-dimensional kriging is used to make the volume estimate, and if the initial round of data collection is supplemented through the use of undertainty analysis such as Drill Guide. In the event that sampling was taking place utilizing geoprobe-type equipment and field instrumentation, it is obvious that use of EVS and Drill Guide could greatly streamline the investigative process as well significantly enhance the precision and accuracy of the estimate of volume of soil and/or ground water contaminated above remedial objectives.

RANK KRIGING FOR SUPERFUND SITE CHARACTERIZATION

A.K. Singh (Department of Mathematical Sciences, University of Nevada, Las Vegas, Nevada)
M.M.A. Ananda (Department of Mathematical Sciences, University of Nevada, Las Vegas, Nevada)
Anita Singh (Lockheed Martin Environmental System & Technologies Company, Las Vegas, Nevada)

ABSTRACT: Kriging is one of the most commonly used methods of spatial interpolation of contamination concentration data from Superfund sites. Kriging is performed in two steps: 1) the contaminant concentration data collected from the site is used to model the spatial continuity in data (through variogram or covariance matrix), and 2) the estimated variogram or covariance matrix is then used to compute the best linear unbiased estimates (BLUE) of either the average contaminant concentration over a user-defined block of the site, or the contaminant concentration at an unsampled location. In case the observed histogram is positively skewed, it is common practice to use log-normal kriging, in which case kriging is performed on log-transformed concentration data. The estimate and its kriging variance is then back-transformed to the original scale. It has been the experience of the authors that, for some Superfund sites, the method of log-normal kriging gives very unsatisfactory results. In this paper, we propose an alternative kriging method based on the rank transformation of the data.

INTRODUCTION

Geostatistics has been defined as a branch of statistics dealing with spatial phenomena (Journel, 1986). The term 'kriging' refers to a range of methods for estimating point or block values at a given location from a given set of spatial data. In the parametric setup, kriging yields an estimator which is a best linear unbiased estimator (BLUE) under certain assumptions on the joint distribution of the set of observations. Some of these assumptions are not easy to verify. In practice, these assumptions may not be met, and the optimality of the estimators obtained by kriging on the data values may then be in question. Non-parametric procedures are statistical procedures which have certain desirable properties under a relatively mild set of assumptions. There are two types of non-parametric geostatistical methods in the literature. Henley(1981) uses a 'weighted median' or a 'weighted varying quantile' for point estimation. This approach is quite different from the methodology of kriging. Journel(1988) uses kriging as a tool to develop the indicator kriging and probability kriging algorithms; probability kriging utilizes the rank-transformed values in co-kriging. In standard statistical literature, the rank transformation is viewed as a tool to develop non-parametric procedures to solve new problems (Conover and Iman ,1981). Singh, Ananda and Sparks (1993) developed a non-parametric variogram model, and used the resulting estimate to krige concentration data. This method was succesfully used for site characterization of the East Fork Poplar Creek Superfund Site in Oakridge, Tennessee (Gerlach *et al.*, 1995). The advantages of this approach are obvious: the procedures developed in this manner would be less sensitive to departures from the underlying assumptions of the parametric procedures, and current software packages could easily incorporate their implementation. In this paper, we propose the following method of kriging for Superfund site characterization: use the estimated variogram of ranks to krige rank-transformed concentration data, and then use linear interpolation to back-transform the estimates. An approximation for back-transforming the kriging variance is also be given. Data from a Superfund site is used in one of the examples to illustrate our procedure.

RANK KRIGING

Let $\{Z(\underline{s}): \underline{s} \in D \subset R^p\}$ be a real-valued stochastic process defined on a domain D of the p-fold real line. One of the assumptions of ordinary kriging (see, for example, Cressie, 1989) is that the process being measured is *intrinsically stationary*, i.e., it satisfies

$$E[Z(\underline{s})] = \mu, \ \underline{s} \in D \tag{1}$$

$$Var[Z(\underline{s}+\underline{h}) - Z(\underline{s})] = 2\gamma(\underline{h}), \ \underline{s}, \underline{s}+\underline{h} \in D \tag{2}$$

This paper will be concerned with the case of *isotropic variogram* in which $\gamma(\underline{h}) = \gamma(||\underline{h}||)$, where $||\underline{h}||$ = length of the vector $\underline{h}$.

Let $\{Z(\underline{s}_i) : i = 1, 2, ..., n\}$ be n observations at n distinct locations in the set D. The problem is to predict $Z(\underline{s}_0)$ at some unsampled location $\underline{s}_0$. The method of kriging finds the linear unbiased estimator with minimum variance. The method of Lagrangian multiplier yields an (n+1) x (n+1) system of linear equations

$$\begin{aligned}
&\Gamma\underline{\alpha}_0 = \underline{\gamma}_0, \text{ where} \\
&\Gamma_{ij} = \gamma(\underline{s}_i - \underline{s}_j), \ i, j = 1, \ldots, n \\
&\quad = 1, \ i = n+1, \ j = 1, \ldots, n \\
&\quad = 0, \ i = n+1, \ j = n+1 \\
&\underline{\alpha}_0 = (a_1, \ldots, a_n, m)^T \\
&\underline{\gamma}_0 = (\gamma(0,1), \ldots, \gamma(0,n), 1)^T
\end{aligned} \tag{3}$$

The BLUE is given by

$$\hat{Z}(\underline{s}_o) = \sum_{1}^{n} \alpha_i Z(\underline{s}_i) \tag{4}$$

and its variance, called kriging variance, is given by

$$E(Z(\underline{s}_0) - \hat{Z}(\underline{s}_0))^2 = \sum_{1}^{n} \alpha_i \gamma\left(\underline{s}_i - \underline{s}_0\right) + m \qquad (5)$$

The proposed method of rank kriging can now be described as follows:

Step 1 : Rank the data set $\{Z(\underline{s}_i), i = 1, ..., n\}$ in increasing order, and let $R_i = \text{rank}(Z(\underline{s}_i)), i = 1, ..., n$. Use mid-ranks if ties are present.

Step 2 : Replacing $Z(\underline{s}_i)$ by R_i, estimate the semivariogram, and then solve the linear system (3); also compute the predicted rank R_0 of the unknown value $Z(\underline{s}_0)$ from (4), and its kriging variance using (5).

Step 3 : If required, estimate the unknown value $Z(\underline{s}_0)$ by

$$\begin{aligned} \hat{Z}_0 &= Z(\underline{s}_j) + (R_0 - R_j)(Z(\underline{s}_{j+1}) - Z(\underline{s}_j)), \text{ if } 1 \le R_j \le R_o \le R_{j+1} \le n \\ &= Z(\underline{s}_1), \text{ if } R_0 < 1 \\ &= Z(\underline{s}_n), \text{ if } R_0 > n \end{aligned} \qquad (6)$$

where the index j is given by the inequalities

$$Z(\underline{s}_j) \le Z(\underline{s}_0) \le Z(\underline{s}_{j+1}).$$

Step 3 is the inverse linear interpolation commonly used in the nonparametric methods for monotone regression (Conover(1980)). A closed form expression for the kriging variance of the estimator (6) is not available. We have been able to derive an upper bound for the kriging variance as

$$E[\hat{Z}(\underline{s}_0) - Z(\underline{s}_0)]^2 \le \gamma(j,0) + \gamma(j+1,j) + 2|\gamma(0.j+1) - \gamma(j,j+1) - \gamma(0,j)|$$

EXAMPLES

EXAMPLE 1:The data for this example is taken from the data file EXAMPLE.DAT, that is included in the geostatistical software package GEO-EAS (Englund and Sparks, 1988). We present kriging results for Arsenic concentrations; Figure 1 shows the estimated variogram for Arsenic concentrations, and Figure shows that same estimated from ranks of Arsenic concentrations. Figure 3 shows the estimated Arsenic concentrations obtained from kriging the raw data, and Figure 4 shows the estimated ranks of Arsenic concentrations. The results of back-transformation of estimated ranks is not included.

EXAMPLE 2: The data from Cal West Metals Superfund Site is used to illustrate the method of rank kriging. Cal West Metals Site is located nine miles north of Socorro, New Mexico. This facility was operated as a battery recycler between 1979 and 1981. In 1991, measurements of lead concentrations in soils were taken, and ordinary kriging was used to estimate lead concentrations in soils at the Site (Cole *et al.,* 1995). In this paper, we show the results of rank kriging on observed lead concentartions (Figures 5 and 7). The results of kriging raw lead concentrations are also included (Figures 6 and 8). We have not included the results of back-transformation of estimated ranks.

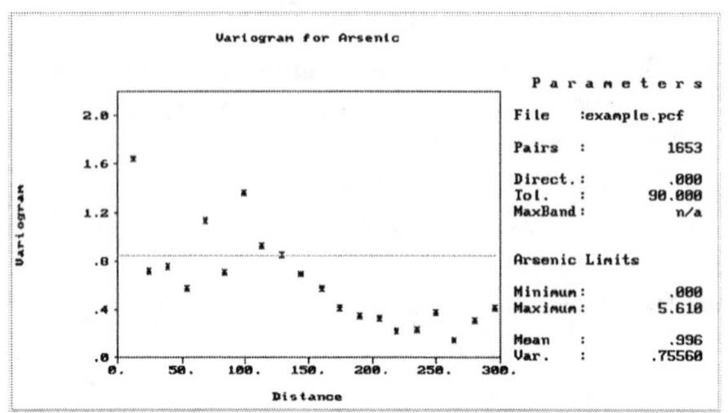

Figure 1: variogram for Arsenic concentrations in data file Example.dat of GEO-EAS (PURE NUGGET MODEL, nugget = 0.85)

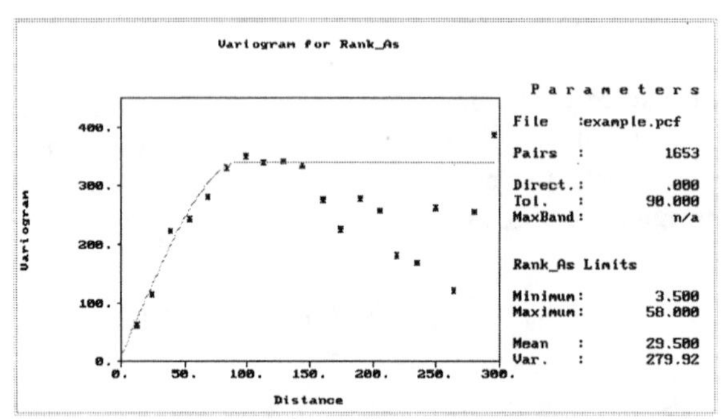

Figure 2: variogram for ranks of Arsenic concentrations in data file Example.dat of GEO-EAS (SPHERICAL, nugget=10, sill=330, range=95)

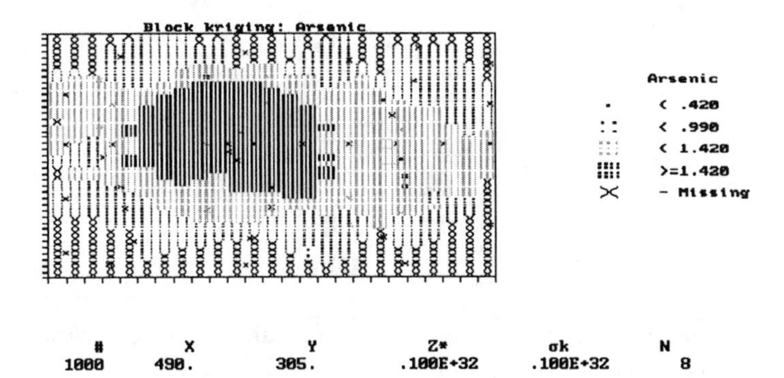

Figure 3: results of kriging Arsenic concentrations

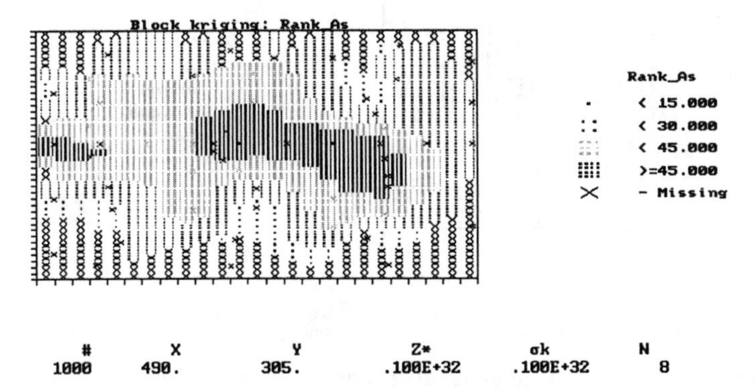

Figure 4: results of kriging ranks of Arsenic concentrations

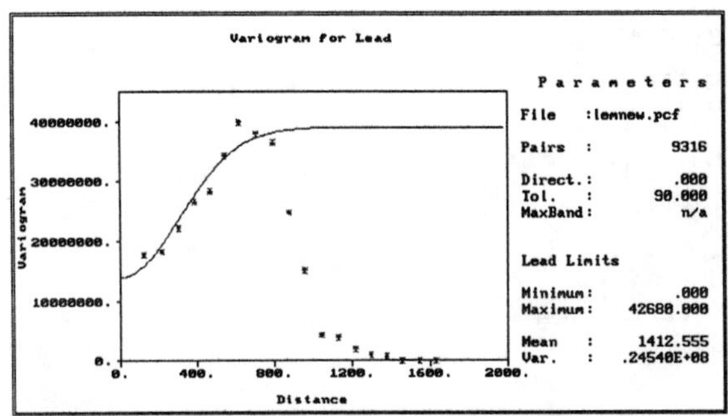

Figure 5: variogram for observed lead concentrations at the Cal West Metals Superfund Site (GAUSSIAN, nugget=14000000, sill=25000000, range=750)

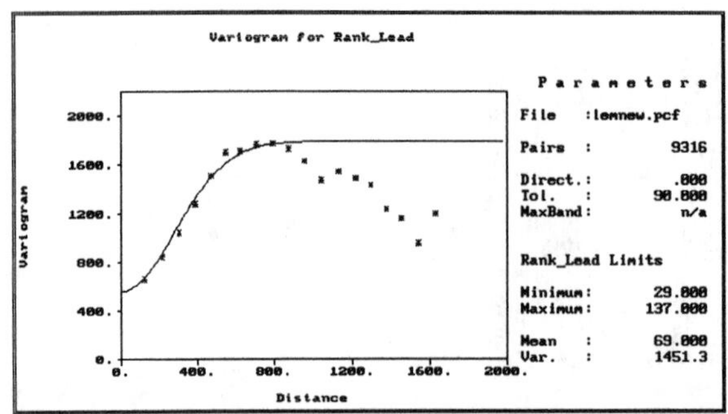

Figure 6: variogram for ranks of observed lead concentrations at the Cal West Metals Superfund Site (GAUSSIAN, nugget=560, sill=1230, range=680)

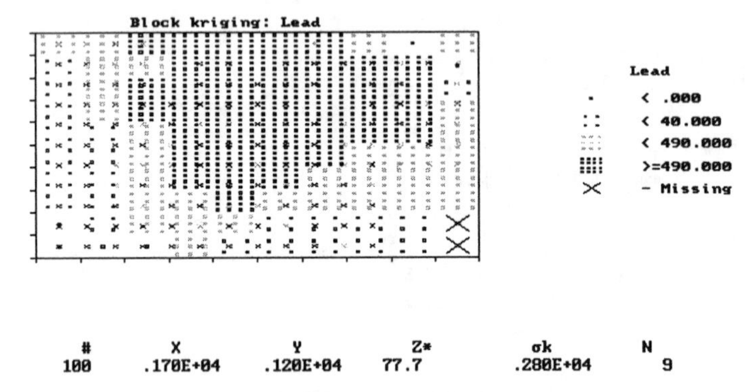

Figure 7: results of kriging lead concentrations

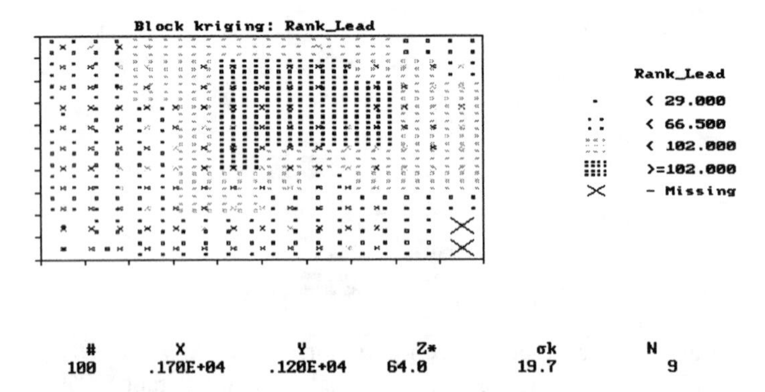

Figure 7: results of kriging ranks of lead concentrations

REFERENCES

Cole, W.H., Gonzales, C., Kuharic, and Singh, A.K. (1991), X-Ray Fluorescence Site Screening and Geostatistical Analysis of Soil Lead Data from cal West Metals NPL Site, EPA\EMSL-LV TSC-12.

Conover, W. J., 1980, Practical Nonparametric Statistics, 2nd ed: John Wiley & Sons, New York, p. 272-276.

Conover, W. J. and Iman, R. L., 1981, Rank Transformation as a Bridge between Parametric and Nonparametric Statistics, The American Statistician, v. 35, p. 124-129.

Cressie, N., 1989, Geostatistics, The American Statistician, v. 43, p. 197-202. Henley, S., 1981, Nonparametric Geostatistics: Elsevier Applied Science Publishers, London, p. 66-104.

Englund, E. and Sparks, A., 1988, Geo-EAS User's Guide, EPA/600/4-88/033.

Gerlach, C., Dobb, D., Miller, E., Cardenas, D., Singh, A.K., Page, D., Combs, D., Heithmar, E.M. (1995), Characterization of Mercury Contamonation at the East Fork Poplar Creek Site, Oak Ridge, Tennessee: A case Study, EPA\600\R-95\110.

Isaaks, E. H. and Srivastava, R. M., 1989, An Introduction to Applied Geostatistics: Oxford University Press, New York, p. 4-9.

Journel, A.G., 1986, Geostatistics:Models and Tools for the Earth Sciences, Mathematical Geology, v. 18, 119-140.

Journel, A.G., 1988, Nonparametric Geostatistics for Risk and Additional Sampling Assessment, in L. H. Keith, ed., Principles of Environmental Sampling, ACS Professional Reference Book, American Chemical Society, p. 45-72.

Singh, A.K., Ananda, M.M.A., Sparks, A.R., 1993, Superfund Site Characterization Using Non-parametric Variogram Modeling, Analytica Chimica Acta, 255-266.

Weber, D. and Englund, E., 1990, Comparison of Estimation Methods by using a Factorial Design Sampling, in preparation.

HANDLING CHEMICAL DATA BELOW DETECTION LIMITS FOR MULTIVARIATE ANALYSIS OF GROUNDWATER

Irene Farnham and Amy Smiecinski (Harry Reid Center for Environmental Studies, Las Vegas, Nevada)
A.K. Singh (UNLV, Las Vegas, Nevada)

ABSTRACT: Principal components analysis (PCA) and formal clustering techniques are used to evaluate similarities in the trace element chemistry of groundwaters from southern Nevada. Many of the trace elements, however, occur at concentrations below the detection limits (DL) which presents problems for statistical analyses. Two approaches to prepare these data sets for statistical analysis are addressed: (1) replace the values '<DL' with a constant (0, DL, or DL/2), and (2) use all measured values including those below the DL. A comparison to determine the best constant to substitute for the '<DL' values was performed. The results showed that substitution with DL/2 gave superior results compared to substitution with DL or 0. The results also suggested that the use of uncensored data is preferred when a large number of '<DL' values and multiple detection limits are present.

INTRODUCTION

The trace element chemistry of groundwater may be used to assign a fingerprint with which water from similar flow paths can be identified. The trace element chemistry is measured for waters collected from many springs and wells in southern Nevada. Principal components analysis (PCA) is then used to reduce the data and informal clustering of springs and wells, based on these results, is then used to evaluate chemical similarities between the waters. Similarities are also examined using formal clustering techniques.

Difficulties arise in the application of PCA and cluster analysis when the concentration of an element is below the detection limit (DL). It is often standard laboratory practice to report data that is below the method detection limits simply as '<DL'. This 'censoring' of data then complicates all subsequent statistical analyses.

The procedures for handling censored data depends on the technical application involved. The best method to use generally depends on the amount of data below the detection limit and also the size of the data set. When the number of '<DL' is small, replacing them with a constant (ie. DL/2) is generally satisfactory (EPA, 1989). The values that are commonly used to replace the '<DL' are 0, detection limit (DL), or DL/2. Distributional methods such as the marginal maximum likelihood estimation (Chung, 1993) or more robust techniques (e.g., Helsel, 1990) are often required when a large number of '<DL' are present. Substitution of a constant may still be preferred over the more complicated methods such as maximum likelihood techniques and regression order statistics when the data sets are very small

(n < 10). This is due to the inability to accurately infer the distributional properties from small data sets which is required for these methods (Clarke, 1998). Substitution of censored data with a constant value has also been shown to give preferred results over substitution of missing data using PCA (Aruga, 1997).

Another approach to deal with data below the detection limit is to use the measured value. Numerical values are often produced from the analytical measurement but are not reported by the chemist because the value is judged "unreliable"(Hinton, 1993). Censoring of data prevents the misuse of low quality data and can be appropriate in some cases (Keith, 1994), however, it commonly leads to a loss of important information. Recording only the DL can create an uncorrectable bias in determining long term trends (Chambless, et al. 1992). In fact, trends are more effectively detected in uncensored data than censored data even when the censored data are highly unreliable (Gilliom, et al., 1984). Samples taken over time may be censored at different levels as changes in analytical technology alter the precision of a method. This often leads to very complicated patterns of censoring (Porter et al., 1988). Censoring data is therefore not recommended in most cases. Instead, it is better to report the measured value along with an estimate of measurement precision (Porter et al., 1988; Chambless et al., 1992).

The goal of this study was to determine the best constant (0, DL, or DL/2) to substitute for values that are below the detection limit. PCA results were compared for groundwater trace element data with no values below the detection limit to those after imposing artificial detection limits and then substituting the "<DL" with the three constants. The use of constants for substituting values below detection limit was further examined by comparing multiple measurements of samples collected from the same site at different times. Different detection limits are associated with each set of samples and consequently different DL/2 substitutions as well as differing degrees of censoring are observed. These data were compared using PCA and cluster analysis. Results using uncensored data were also evaluated.

METHOD

PCA was applied to five data sets that contained no '<DL'. Statistica for Windows 5.1, 1997, (StatSoft Inc) was used for all statistical analyses. PCA is performed on the correlation matrix and cluster analysis is performed on standardized data. The number of samples in each set varied from 21 to 25 and the number of variables varied from 13 to 17. The PC scores from these original data sets are considered the 'true' values. Values below the detection limit were then generated by imposing an artificial DL. Detection limits were selected so that the '<DL' made up 5% - 10% of the total data. All results below this DL were then substituted with: a) 0, b) DL, and c) the DL / 2. PCA was re-applied to the new data sets and the results (factor scores) from each of these substitutions were compared to the 'true' values. This was repeated on the same five data sets after selecting higher detection limits that resulted in approximately 25% values '<DL'. This experiment was then repeated after applying separate detection limits to different portions of the data. The detection limits used previously that resulted in 5% - 10% '<DL' values were applied to half of the data and the detection limits that resulted in 25% '<DL' values were applied to the second

half. The values '<DL' were again substituted with 0, DL, and DL/2. PCA was applied to these new data sets and the results compared to the 'true' values.

RESULTS AND DISCUSSION

The Euclidean distances between the 'true' PC scores and those calculated from the data with the artificially imposed detection limits substituted with 0, DL, and DL/2 are listed in Table 1. The distances are listed for the first three principal components (PC) calculated for each of the five data sets. In some cases, Euclidean distances were quite large due to an inversion of the signs of all PC scores within a single component. These distances were significantly lowered (see values in parentheses) after inverting the signs of all scores within the component. Substitution with the DL or DL/2 is preferred over substitution with 0. Overall, DL/2 gave results most similar to the original data set. It is also important to note that the deviations from the 'true' values greatly increased with the increasing number of non-detects.

The ICP-MS detection limits for the trace elements vary over time and therefore different sets of data have different detection limits associated with them. To test which substitution value is preferred in this case, the experiment was repeated after applying two sets of detection limits to different portions of the data. The results are listed in Table 1. Again, substituting the DL or DL/2 is superior to substituting with 0.

TABLE 1. The Euclidean distances between 'True' PC scores and those calculated for censored data substituted with 0, DL, and DL/2

5 - 10 % Values Below the Detection Limit

	Substitute ND w/0			***Substitute ND w/ DL***			***Substitute ND w/ DL/2***		
Set	PC1	PC2	PC3	PC1	PC2	PC3	PC1	PC2	PC3
1	8.9 (1.1)	1.2	0.9	0.3	0.7	0.9	0.4	0.4	0.3
2	0.1	0.4	9.6 (0.5)	0.2	0.7	0.7	0.1	0.2	0.2
3	0.9	0.8	2.1	0.2	0.3	0.6	0.4	0.5	1.2
4	0.1	0.0	0.1	0.0	0.0	0.0	0.0	0.0	0.0
5	0.3	0.3	0.2	0.2	0.2	0.4	0.0	0.1	0.2
Sum	**10.2(2.5)**	**2.7**	**12.9(3.8)**	**0.9**	**2.1**	**2.6**	**1.0**	**1.2**	**1.9**

25 - 30 % Values Below the Detection Limit

	Substitute ND w/0			***Substitute ND w/ DL***			***Substitute ND w/ DL/2***		
Set	PC1	PC2	PC3	PC1	PC2	PC3	PC1	PC2	PC3
1	1.1	2.6	2.6	8.9(1.2)	8.4(3.1)	8.4(3.1)	0.4	2.2	2.3
2	0.5	0.7	0.7	0.8	2.3	9.3(2.4)	0.4	1.0	0.9
3	1.1	1.5	8.8(3.1)	0.5	0.9	7.4(5.9)	0.6	1.0	8.5(4.0)
4	0.5	0.5	9.6(0.4)	0.8	0.8	0.5	0.1	0.1	0.2
5	0.3	0.5	0.7	0.4	0.7	9.8(1.1)	0.1	0.2	0.9
Sum	**3.5**	**5.8**	**22.4(7.5)**	**11.4(3.7)**	**13.0(7.7)**	**35.3(13.0)**	**1.7**	**4.5**	**12.8(8.3)**

Multiple Detection Limits

	Substitute ND w/0			***Substitute ND w/ DL***			***Substitute ND w/ DL/2***		
Set	PC1	PC2	PC3	PC1	PC2	PC3	PC1	PC2	PC3
1	0.8	1.1	8.9(0.9)	0.3	0.7	0.9	0.4	0.7	0.5
2	0.1	0.5	0.5	0.2	0.7	0.7	0.1	0.4	0.4
3	1.1	1.3	2.3	0.4	0.4	1.3	0.6	0.8	1.0
4	0.1	0.2	0.2	0.5	0.5	0.0	0.2	0.2	0.1
5	0.3	9.8(0.5)	8.9(3.9)	0.3	0.7	9.8(1.1)	0.1	9.8(0.3)	2.7
Sum	**2.4**	**13.0(3.7)**	**20.7(7.8)**	**1.7**	**3.0**	**12.6(3.9)**	**1.4**	**11.9(2.4)**	**3.0**

The use of DL/2 was further examined using a data set containing multiple measurements of the trace element chemistry from several springs in Death Valley. These samples were collected in November 1993, March 1994, and November 1994. In order to make an accurate comparison between waters from different locations, it must first be shown that waters from a single location can be measured reproducibly. Approximately 20% of these data were listed as "<DL" and were substituted with DL/2. A higher detection limit was associated with the third sample set. The trace element concentrations were averaged and PCA was applied to the averages as well as the individual samples. A plot of the scores for the first two principal components is shown in Fig. 1. This figure clearly illustrates the excellent reproducibility of the multiple measurements for each site. With the exception of Tex / TrvA /TrvB and Scot / Surp, distinct clusters were formed that contain each of the multiple measurements and corresponding average for a given spring. The waters from Texas, TrvA and TrvB cannot be distinguished from each other using these PCA results because the chemical composition of these springs is essentially identical. The same can be said for Scotty's and Surprise Springs. (Kreamer et al., 1996). These results show that substitution of '<DL' values with DL/2 gives highly reproducible PCA results while maintaining the ability to distinguish waters from different sources.

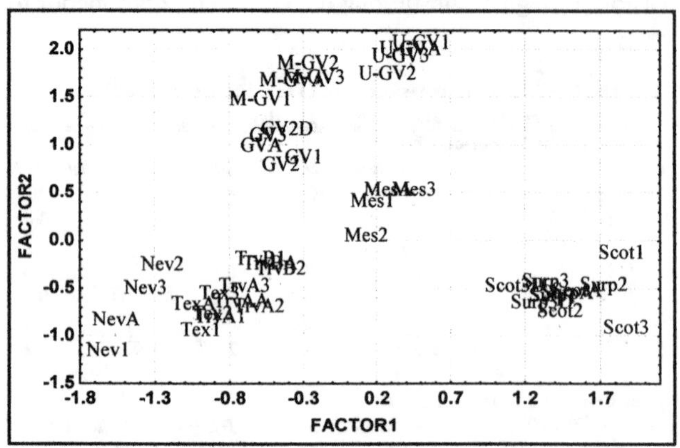

FIGURE 1 PCA score plot for Death Valley trace element data (Values <DL were Substituted with DL/2)

A sample set that contained a higher percentage of values below the detection limit was examined next. This sample set, consisting of the trace element chemistry for groundwater from the Oasis Valley area, contains up to 50% values below the detection limit. The values below the detection limit were again substituted with DL/2 and analyzed using PCA. Most of the multiple measurements for the same site cluster together with the exception of one sample. These results were further examined using cluster analysis of the standardized concentrations. Vertical icicle plots are shown in Fig. 2. Again, distinct clusters were formed between the multiple measurements with the exception of one sample (Windmill1) which instead clusters with Coffer1. Coffer1 and Windmill1 were collected and analyzed at the same time and therefore have the same detection limits associated with them. The detection limit was higher for many of the elements and therefore more values below the detection limit were reported for these samples. Because the large number of "<DL" values are replaced by the same values for both samples, the ability to distinguish between them is lost.

The data set was then regenerated by reporting all values that were greater than zero. A significant portion of data was deleted because it was below the DL although it was detected with a %RSD (n=3) of <25%. The concentrations were standardized and cluster analysis re-applied. The vertical icicle plots are shown in Fig. 3. The samples from the same site all clustered together when the measured values are used.

K-Mean cluster analysis was then applied to a data set that contained ground waters collected from several areas within Southern Nevada. Samples were collected

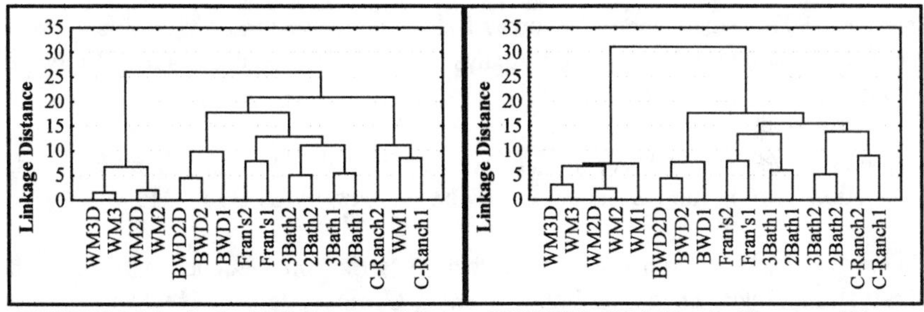

FIGURE 2 Icicle plot for DL/2 Substituted Data **FIGURE 3 Icicle Plot for Uncensored Data**

from Oasis Valley (OV), Ash Meadows (AM), Spring Mountains (SM), Furnace Creek (FC), Pahranagat Valley (PV), Furnace Creek, Death Valley (FC), Air Force Range (AFR), Northern Death Valley (DV) and the Nevada Test Site (NTS). The uncensored data set was used even though relatively low uncertainty was associated with several of the measurements. The results, in Table 2, show that the uncensored data can be used to distinguish waters from different locations.

TABLE 2. Results of K-Means Clustering of Uncensored Data

Cluster 1	Cluster 2	Cluster 3	Cluster 4	Cluster 5	Cluster 6	Cluster 7
Big-AM	Coffer-OV	Cold-SM	Nev-FC	Crys-PV	Scot-DV	WM-OV
Crystal-AM	Bath2-OV	Deer-SM	Tex-FC	Hiko-PV	Surprise-DV	BWD-OV
Kings-AM	Bath3-OV	GV-SM	TrvA-FC	Ash-PV		GldFlat-AFR
Scruggs-AM	Fran's-OV	WilCrk-SM	TrvB-FC			Tolicha -AFR
Corn-SM						ER30-NTS
Indian-SM						J13-NTS
						WC-TRI

CONCLUSIONS

This study showed that substitution of non-detects with DL and DL/2 is preferred over 0, and that substitution with DL/2 gives results most similar to the 'true' values. These results were consistent when two sets of detection limits were applied to the data. It was also shown that the deviation from the 'truth' increased with the increasing number of values requiring substitution. Inaccurate clustering can result due to the presence of different detection limits when a large number of values below the detection limit are observed. The results were improved when uncensored data was used. Groundwaters from springs/wells in close proximity to each other were distinguishable from those from other locations based on the uncensored trace element chemistry. The use of uncensored data is therefore preferred.

Interpretation of statistical results obtained from data sets with a large number of non-detects must be approached with caution. A large uncertainty is present whenever a large number of values below the detection limit are reported regardless of the approach used to pretreat the data. The results of the statistical analyses must be examined to insure that data with high uncertainty is not driving the results.

REFERENCES

Aruga, R. 1997. "Treatment of Responses Below the Detection Limit: Some Current Techniques Compared by Factor Analysis on Environmental Data". *Analytica Chimica Acta*. 354: 255-262.

Chambless, D.A., S.S. Dubose, and E.L. Sensintaffar.1992.Detection Limit Concepts Foundations, Myths, and Utilization." *Health Physics*. 63(3): 338 - 340.

Chung, C.F. 1993. "Estimation of Covariance Matrix from Geochemical Data with Observations Below Detection Limits." *Math., Geology*. 25(7): 851 - 865.

Clarke, J.U. 1998. "Evaluation of Censored Data Methods to Allow Statistical Comparisons Among Very Small Samples with Below Detection Limit Observations." *Environ. Sci. Technol*. 32(1):177 - 183.

Gilliom, R.J., R.M. Hirsch, E.J. Gilroy. 1984. "Effect of Censoring Trace-Level Water-Quality Data on Trend-Detection Capability." *Environ. Sci. Technol*. 18(7): 530-535.

Helsel, D.R. 1990. "Less than Obvious; Statistical Treatment of Data Below the Detection Limit." *Environ. Sci. Technol*. 24(12): 1767 - 1773.

Hinton, S.W. 1993."Δ Log-Normal Statistical Methodology Performance." *Environ. Sci. Technol*. 27(10): 2247 - 2249.

Keith, L.H. 1994. "Throwaway Data." *Environ. Sci. Technol*. 28:8: 389A-390A.

Porter, P.S., R.C. Ward, and H.F. Bell. (1988). "The Detection Limit, Water Quality Monitoring Data are plagued with Levels of Chemicals that are too Low to be Measured Precisely." *Environ. Sci. Technol*. 22(8): 856 - 861.

Statistical Analysis of Ground-Water Monitoring Data at RCRA Facilities.1989. U.S. Environmental Protection Agency Technical Report, EPA530-SW-89-026, Office of Solid Waste, Washington, D.C.

SOIL REMEDIAL CONCENTRATION FOR METHYL-TERTIARY-BUTYL-ETHER BY MONTE CARLO SIMULATION MODELING

Yue Rong, Ph.D.
Environmental Specialist
California Regional Water Quality Control Board - Los Angeles Region
101 Centre Plaza Drive, Monterey Park, California, U.S.A

INTRODUCTION

Methyl tertiary butyl ether (MTBE) is a gasoline oxygenating additive that is used to enhance automobile combustion and consequently reduce air pollution. However, due to its high solubility and mobility in the environment, MTBE has become a significant groundwater contaminant that is difficult to cleanup once it arrives in groundwater.

Modeling groundwater flow and contaminant transport in the subsurface has been increasingly used in the field of environmental site assessment, remediation, and risk assessment. The proper application of a model and selection of model input parameters rely on many quality assurance procedures throughout the modeling process. One of the most important quality assurance procedures is to conduct parameter uncertainty analysis.

This paper uses Monte Carlo Simulation (MCS) for analysis of parameter uncertainties associated with a commonly used groundwater mixing model in a soil remediation case in Los Angeles, California, U.S.A. The mixing model is used in conjunction with a one-dimensional vadose zone transport model, VLEACH, to predict the impact of residual soil concentrations of MTBE on groundwater quality. Results of MCS can produce a probability distribution of model outputs. Such a statistical characterization would assist in decision making on soil remediation numerical criteria. The results of MCS demonstrate that model simulation based on a probabilistic approach could provide more information to decision makers than based on the conventional approach using single point average input values.

METHODOLOGY

This paper first applies a one-dimensional vadose zone transport model, VLEACH, to predict downward transport of residual MTBE left in soil after remediation at an industrial site. Then a groundwater mixing model is used to predict the dilution of MTBE in groundwater from the point of entering aquifer underneath the source area to the point of compliance, which is some distance downgradient from the source.

VLEACH is developed to simulate the mobilization and downward movement of volatile organic compounds (VOCs) in the vadose zone, including the processes of liquid phase advection, vapor phase diffusion, and adsorption (Rosenbloom et al. 1993; Ravi and Johnson 1994; Rong 1995). The model output produces a calculated mass loading onto the underlying groundwater in grams per year (g/yr). This output is then used as input for the groundwater mixing model (USEPA 1996). Result of the mixing model is groundwater concentration at the point of compliance, and can be used to evaluate potential impacts of the residual soil concentration on groundwater quality. Finally, this paper uses MCS to conduct uncertainty analysis for the groundwater mixing model. MCS is a methodology to analyze and quantify uncertainties in model outputs propagated from the uncertainties associated with input parameters.

GROUNDWATER MIXING MODEL

Groundwater mixing model, presented in USEPA (1996), is a type of "continuous stirred tank reactor" model. Although being overly simplified, it still attracts certain groups of users; and there is a need for its evaluation (Brusseau 1996). The mixing model calculates a steady-state groundwater concentration at the point of compliance based on the assumption that the total mass discharged from the vadose zone is completely mixed within the total volume of the entering mass impacted zone in the aquifer. The conceptual model is presented in Figure 1, and the mathematical expression in the following equations:

$$C_w = M_T / (Q_P + Q_A) \quad (1)$$

Where C_w is groundwater concentration at the point of compliance (μg/l); M_T is total mass released to groundwater during time T (μg/yr); Q_P is the vertical percolation flow rate (l/yr); and Q_A is the longitudinal groundwater flow rate (l/yr).

Here,

$$Q_P = q \cdot A \quad (2)$$

and

$$Q_A = W \cdot d \cdot K \cdot (dh/dL) \quad (3)$$

Where q is the average infiltration rate (m/yr); A is area of the source (m^2); W is the width of the source area perpendicular to the direction of groundwater flow (m); d is depth of mixing zone in aquifer (m); K is hydraulic conductivity in aquifer (m/yr); and dh/dL is hydraulic gradient (m/m).

The depth of mixing zone, d, is calculated by equation (4):

$$d = (2 \cdot \alpha_v \cdot L)^{1/2} + Z_a \cdot [1 - \exp(-(L \cdot q)/(Z_a \cdot v_l \cdot \phi))] \quad (4)$$

Where α_v is the vertical dispersivity (m); L is distance from the source to the point of compliance (m); Z_a is thickness of aquifer (m); v_l is the longitudinal seepage velocity (m/yr) = $K \cdot (dh/dL)/\phi$; and ϕ is effective porosity (dimensionless).

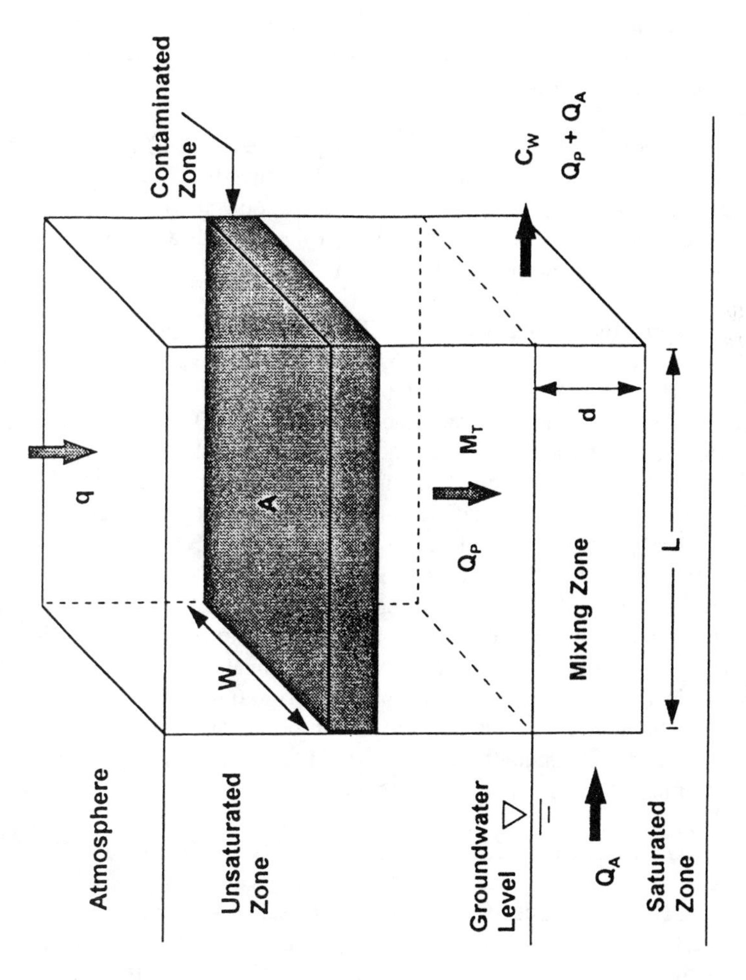

Figure 1: Conceptual Model

The first term in equation (4), $(2 \cdot \alpha_v \cdot L)^{1/2}$, is an estimate of the mixing depth due to vertical dispersivity related to the length of groundwater travel (L). The α_v is usually estimated using the following empirical equation:

$$\alpha_v = \alpha \cdot L \tag{5}$$

The α is a scaling factor (dimensionless), which we call it coefficient of dispersivity in this paper. Its values can range from 0.00002 to 0.0025 based on field studies (Gelhar et al. 1992).

The second term in equation (4), $Z_a \cdot [1 - \exp(-(L \cdot q)/(Z_a \cdot v_l \cdot \phi))]$, is the estimate of mixing depth due to downward velocity of infiltrating water. The mixing depth relates positively to the infiltration rate q, and inversely to longitudinal velocity v_l. Similarly, the longer the groundwater travel distance L, the longer mixing time, and the larger the mixing depth.

INPUT PARAMETERS FOR VLEACH

Values of bulk density ($\rho_b = 1.67 g/cm^3$), porosity ($\phi = 0.23(-)$), soil water content ($\theta = 0.14(-)$), and soil organic carbon content ($f_{oc} = 0.0011(-)$) are calculated based on the average values of field data from 27 soil samples collected at different locations and depths at the site (McLaren 1996). The number of simulated soil column is limited to one, only the source area, where residual MTBE exists. The source area is estimated to be 12 meters by 12 meters (144 m^2) (1600 ft^2). Depth to groundwater is approximately 23 meters (75 feet) based on the groundwater monitoring data at the site. The initial input MTBE concentration varies to fit groundwater concentration less than 35 $\mu g/L$ at the point of compliance. The concentration is assumed to be uniformly distributed throughout a volume of soil defined by the contamination source area multiplied by the vertical interval from 3.35 meters (11 feet) to 6 meters (20 feet) below the surface. Initial concentrations at other places in the soil column are assumed to be zero. The model simulation time is chosen to be long enough to produce the maximum mass loading at the water table.

INPUT PARAMETERS FOR GROUNDWATER MIXING MODEL

The total mass released to groundwater (M_T) can be directly obtained from the VLEACH output. The average infiltration rate (q=189mm/yr), source area (A), and effective porosity (ϕ) are the same as VLEACH input parameters above. Width of the source perpendicular to the flow is calculated as the square root of the VLEACH polygon area, i.e., $W = A^{1/2}$. The thickness of aquifer is determined to be the uppermost regional unconfined aquifer composed of mainly alluvial deposits extending to a depth of approximately 30 meters below the water table (McLaren 1996). Hydraulic conductivity is based on the results of the slug test and aquifer pumping test conducted at the site (McLaren 1996). Hydraulic gradient is determined based on groundwater monitoring data obtained from 1994 to 1996. The coefficient of dispersivity is equal to 0.0056 from Gelhar and Axness (1981). The longitudinal seepage velocity (v_l) is calculated based on the

Darcy's law as $v_l = K \cdot (dh/dL)/\phi$. The vertical dispersivity is calculated by equation (5). Here we choose L to be 30 meters downgradient from the source area.

MONTE CARLO SIMULATION

A sensitivity analysis identifies that model output is most sensitive to parameter hydraulic conductivity (K). Based on this result, MCS is conducted by varying hydraulic conductivity (K), coefficient of dispersivity (α), and the hydraulic gradient (i=dh/dL). All other input parameters utilize the average values as presented above. The ranges and distributions for the above identified MCS parameters are based on field data and listed in Table 1.

TABLE 1. Parameter Range and Distribution

Parameter	Range	Distribution	Reference
α (--)	0.0001 - 0.01	Uniform	Gelhar et al. 1992
dh/dL (m/m)	0.0058-0.0135	Normal	McLaren 1996
K (m/yr)	0.07-5,075	Log-normal	McLaren 1996

RESULTS OF MCS

Results of 1,000 MCS runs show a log-normal distribution of model outputs, based on which we calculate that C_w(50%ile)=6, C_w(75%ile)=32, and C_w(95%ile)=144 (μg/l). These results relate to soil concentrations as presented in Table 2.

Table 2. Soil Concentration vs. Probability Confidence Level

MTBE Soil Concentration (μg/kg)	Pr ($C_w \leq 35$ μg/l)
1,020	50%
190	75%
43	90%

CONCLUSIONS

Mathematical transport model in conjunction with Monte Carlo simulation (MCS) can be used to calculate soil remedial concentrations. Results of MCS can produce a probability distribution of model outputs. Such a statistical characterization would assist greatly in decision making on soil remediation numerical criteria. This paper provides calculation of MTBE soil remedial concentration ranging from 43 μg/kg in a 90% confidence level to 1,020 μg/kg in a 50% confidence level in terms of protecting groundwater quality. The results of MCS indicate that the probabilistic characterization of model outputs provides much more information to decision makers than the results from a conventional model using single-point average input parameters.

REFERENCES

Brusseau, M.L. 1996. "Evaluation of simple methods for estimating contaminant removal by flushing." *Ground Water* 34(1):19-22.

Gelhar, L.W. and Axness, C.J. 1981. "Stochastic analysis of macro-dispersion in three-dimensionally heterogeneous aquifers." Report No. H-8. Hydraulic Research Program. New Mexico Institute of Mining and Technology, Socorro, NM.

Gelhar, L.W., Welty, C., and Rehfeldt, K.R. 1992. "A critical review of data on field-scale dispersion in aquifers." *Water Resources Research*, 28(7):1955-1974.

McLaren/Hart 1996. "Results of groundwater characterization of the oil field reclamation project study area (OFRP), Santa Fe Springs, California (July 31)." Submitted to California Regional Water Quality Control Board - Los Angeles Region, Monterey Park, CA 91754,

Ravi, V. and Johnson, J.A. 1994. *VLEACH (Version 2.1)* Center for Subsurface Modeling Support, Robert Kerr Environmental Research Laboratory, P.O.Box 1198, Ada, OK 74820.

Rong, Y. 1995. "Uncertainty analyses of a vadose zone transport model for soil remediation criteria of volatile organic compounds." Ph.D. dissertation, UCLA School of Public Health, Los Angeles, CA 90095.

Rosenbloom, J., Mock, P., Lawson, P., Brown, J., and Turin, H.J. 1993. "Application of VLEACH to vadose zone transport of VOCs at an Arizona Superfund site." *Groundwater Monitoring and Remediation* 13(3):159-169.

USEPA 1996. *Soil Screening Guidance: Technical Background Document (May)*. EPA/540/R-95/128, PB96-963502.

MODELING SOIL-WATER PHASE TRANSFER PROCESSES OF ORGANIC BASES

Chad T. Jafvert , Jose R. Fabrega, Linda S. Lee and Hui Li
(Purdue University, West Lafayette, Indiana)

ABSTRACT: A distributed parameter function describing the association of an organic cation to soil cation exchange sites is invoked within a larger speciation model to describe mass transfer of organic and inorganic bases between saturated soil and the aqueous phase at equilibrium. The distributed parameter function is a Gaussian distribution of logarithmic association constants (log K_L^{BH}) with mode = log μ, and standard deviation = σ. Reactions considered in the overall model are: a) Acid dissociation of the protonated organic base, b) sorption of the neutral organic species to soil organic carbon with partition coefficient K_{oc}, and c) ion-exchange on the soil between the protonated organic base (BH^+) and inorganic divalent cations (D^{2+}). Mathematically, the last reaction was expressed as separate association reactions for each type of cation to unoccupied sites. The material balance equations considered were those on the organic base, the cation-exchange sites, and the sum of Mg^{2+} and Ca^{2+}. Values for K_{oc}, μ and σ were calculated for aniline and α-naphthylamine by minimizing residuals between experimental and calculated aqueous phase base concentrations. The best fit values for K_{oc}, log μ, and σ were 16.5 L/kg, 23.7 and 1.66; and 100 L/kg, 25.1 and 2.04 for aniline and α-naphthylamine, respectively.

INTRODUCTION

Organic bases such as aniline and aniline derivatives are important environmental contaminants because of the combined effects of their high potential toxicity and carcinogenicity and the large mass produced each year.

The environmental transport and fate of these chemicals is partially determined by soil-water mass transfer processes. Under high pH conditions, many organic bases exist predominately in their neutral form, and mass transfer occurs through hydrophobic interactions. Greater sorption of organic bases to soils generally is observed at low pH values, with cation exchange being the dominant mechanism under these conditions. Due to the wide variety of functional groups present as ligands on soil, a range in soil affinities for protons and other cations (i.e., organic amines) has been reported (see for example Perdue et al., 1984).

Discrete and continuous frequencies are often employed to represent the frequency of sites possessing specific binding constants in soils. In this study, we have utilize within an overall speciation model, a continuous frequency distribution (i.e., distributed parameter) to describe the transfer of monovalent organic cations to water-saturated soils, and compare calculated results to those of a simple model.

MODEL DEVELOPMENT

In this work, two models were developed and compared to experimental data and to each other. These models are: (*i*) a general mass action model which does not consider site heterogeneity of cation exchange sites (i.e., a two-site model), and (*ii*), a distributed parameter model that assumes a range of affinities of cation exchange sites for BH^+.

Two Site Model. The 'two-site' (TS) mass action model considers th e following mass transfer processes,

$$K_a = \frac{[B][H^+]}{[BH^+]} \tag{1}$$

$$K_d = f_{oc}K_{oc} = \frac{[B_{org}]}{[B]} \tag{2}$$

$$K_G = \frac{[BHS][D^{2+}]^{0.5}}{[BH^+][D_{0.5}S]} \tag{3}$$

where K_a is the acid dissociation constant of the conjugate acid (mol / L), B is the neutral aqueous organic base (mol / L), K_{oc} is the partition coefficient to soil organic carbon (L / kg), f_{oc} is the fraction of organic carbon, B_{org} is the concentration of B associated with the soil (mol / kg), K_G is the selectivity coefficient ($M^{-0.5}$) (Gapon, 1933), D^{2+} is the sum of divalent inorganic cations (Ca^{2+} + Mg^{2+}) (mol / L), BHS and $D_{0.5}S$ are the organic and inorganic cations, respectively, that are attached to cation exchange sites on the soil (S^-). The corresponding material balance equations are,

$$S_T = BHS + D_{0.5}S \tag{4}$$

$$D_T = D^{2+} + 0.5\frac{m}{v}D_{0.5}S \tag{5}$$

$$B_T = B + BH^+ + \frac{m}{v}\left(B_{org} + BHS\right) \tag{6}$$

where S_T (mol / kg) is the total concentration of negatively charged cation exchange sites equal to the pH-specific cation exchange capacity (CEC), D_T (mol / L) is the total concentration of divalent inorganic cations expressed as mol / L, and B_T (mol / L) is the total mass of the organic base normalized to the aqueous phase volume. This model was solved by introducing eqs 1-3 into eqs 4-6, evaluating the values of BH^+, D^{2+} and $D_{0.5}S$ by Newton-Raphson iterations with estimates on K_{oc} and K_G.

Distributed Parameter Model. In this model, S^- replaces $D_{0.5}S$ as a component for the system, modeling cation exchange as two independent half reactions (Griffioen, 1993) involving vacant sites (S^-),

$$K_{BH} = \frac{[\mathrm{BHS}]}{[\mathrm{BH}^+][\mathrm{S}^-]} \tag{7}$$

$$K_D = \frac{[\mathrm{D}_{0.5}\mathrm{S}]}{[\mathrm{S}^-][\mathrm{D}^{2+}]^{0.5}} \tag{8}$$

A normal probability distribution function on log $K_{BH,i}$ is employed to model the frequency of sites,

$$f(X) = \frac{1}{\sigma\sqrt{2\pi}}\left(e^{-\frac{1}{2\sigma^2}(X-\mu)^2}\right) \tag{9}$$

where μ and σ represents the mode and standard deviation, respectively; X = log $K_{BH,i}$, where the subscript i denotes the specific site or group of sites having a specific binding coefficient; and $f(X)$ is the frequency of sites. This function is represented with a discrete number of sites, with groups of sites possessing the same value of $K_{BH,i}$. With each group having the same number of sites ($S_{T,i}$), discrete expressions for BHS and $D_{0.5}S$ are obtained,

$$\mathrm{BHS} = \sum_{i=1}^{i=n} \mathrm{BHS}_i = [\mathrm{BH}^+]\sum_{i=1}^{i=n} K_{BH,i}[\mathrm{S}^-]_i \tag{10}$$

$$\mathrm{D}_{0.5}\mathrm{S} = \sum_{i=1}^{i=n} \mathrm{D}_{0.5}\mathrm{S}_i = K_D[\mathrm{D}^{2+}]^{0.5}\sum_{i=1}^{i=n}[\mathrm{S}^-]_i \tag{11}$$

where the total number of sites in each compartment is defined by,

$$\mathrm{S}_{T,i} = [\mathrm{S}^-]_i + K_{BH,i}[\mathrm{BH}^+][S^-]_i + K_D[\mathrm{D}^{2+}]^{0.5}[\mathrm{S}^-]_i = \frac{CEC}{n} \tag{12}$$

In eqs 10-12, $[S^-]_i$ is the concentration of free sites in compartment i, and $K_{BH,i}$ is the associated cation exchange constant corresponding to that specific group. A summation on all groups of $[S^-]_i$ can be made,

$$[S^-]=\sum_{i=1}^{i=n}[S^-]_i=\frac{CEC}{n}\sum_{i=1}^{i=n}\frac{1}{\{1+K_{BH,i}[BH^+]+K_D[D^{2+}]^{0.5}\}} \tag{13}$$

Upon introducing eqs 1, 2, 7, 8, and 13 into the mass balance equations for B_T and D_T (eqs 5 and 6), the following system (*i.e.*, pair) of nonlinear equations with two unknowns (BH^+ and D^{2+}) is obtained,

$$B_T=[BH^+](1+\frac{K_a}{H^+}+\frac{m}{v}\frac{K_{oc}f_{oc}K_a}{H^+})+[BH^+]\frac{CEC}{n}\frac{m}{v}\sum_{i=1}^{i=n}\left(\frac{K_{BH,i}}{1+K_{BH,i}[BH^+]+K_D[D^{2+}]^{0.5}}\right) \tag{14}$$

$$D_T=[D^{2+}]+0.5K_D[D^{2+}]^{0.5}\frac{CEC}{n}\frac{m}{v}\sum_{i=1}^{i=n}\left(\frac{1}{1+K_{BH,i}[BH^+]+K_D[D^{2+}]^{0.5}}\right) \tag{15}$$

Because only the relative affinities of D^{2+} and BH^+ for sites are discernible from the experimental data it is reasonable to make K_D constant over all sites. Equations 14-15 are referred to as a distributed parameter (DP) model and are easily solved by the Newton-Raphson iterative method.

Parameter Estimation. Singular values for K_{oc} and K_G (TS model) or K_{oc}, μ, and σ (DP model) were calculated by adjusting these values until the global minimum was found for the following objective function,

$$SSR=\sum_{j=1}^{j=n}(Ce_{\exp}-Ce_{pred})^2 \tag{16}$$

where C_e is the total concentration of organic base in the aqueous phase subscripts *exp* and *pred* denote experimental and model predicted values, respectively, SSR is the sum of squared residuals between experimental and model predictions, and n is the number of datum points. The optimum fitting parameters occur at the minimum of this function.

RESULTS AND DISCUSSIONS

Both models were evaluated by employing aniline and α-naphthylamine data obtained from Lee *et al.* (1997) on three different soils, at 3 pH values, at three electrolyte concentrations. The fitting parameters obtained for both models are shown in Table 1.

Two-Site Model Results. Figure 1 shows typical aniline experimental data and TS and DP model isotherms calculated with the optimum fitting parameters for each model. Experiments with Chalmers-4 soil (i.e., pH ≅ 4.5) resulted in lower aqueous phase recoveries of aniline than experiments with the near-neutral pH

TABLE 1. TS and DP Models Best Fitting Parameters (*with log* (K_D)*= 25.00*)

TS Model	**Aniline**[a]	**α-Naphthylamine**[b]
Log K_G	0.76	1.94
K_{oc} (L / kg)	30.0	245.0
DP Model		
Log μ	23.66	25.14
σ	1.66	2.04
K_{oc} (L / kg)	16.50	100.00

[a,b] K_{oc} values calculated from the equation presented by Sabljic *et al.* (*1995*), are 25.7 and 174 L/kg for aniline and α-naphthylamine respectively.

soils (Chalmers-6 soil) as expected due to the importance of ion-exchange at low pH. The TS model captured the magnitude, as well as direction, of this trend. The linearity of the isotherms suggests that the relative affinities of the cation exchange sites for D^{2+} (Ca^{2+} or Mg^{2+}) divided by the affinities for the protonated form of aniline is nearly constant. Examples of experimental and TS model isotherms for α-naphthylamine are provided in Figure 2. In this case, the experimental isotherm is nonlinear, whereas the TS model predicted isotherm is nearly linear.

Distributed Parameter Model Results. Figures 1 and 2 also show the results obtained for the DP model. Again, the optimum values of K_{oc}, log μ, and σ - those used to construct the model isotherms - are provided in Table 1. Recall that the magnitude of log μ is relative to the value assigned to log K_D, and can be compared to log K_G by dividing log μ by log K_D. For all isotherms, including those displayed in Figures 1 and 2, the DP model provided better or similar results as those predicted by the TS model, based on the SSR values. The parameter σ is larger for α-naphthylamine than for aniline as expected as this parameter is a good indicator of isotherm nonlinearity. The larger value of σ for α-naphthylamine results in a wider distribution of K_{BH} values.

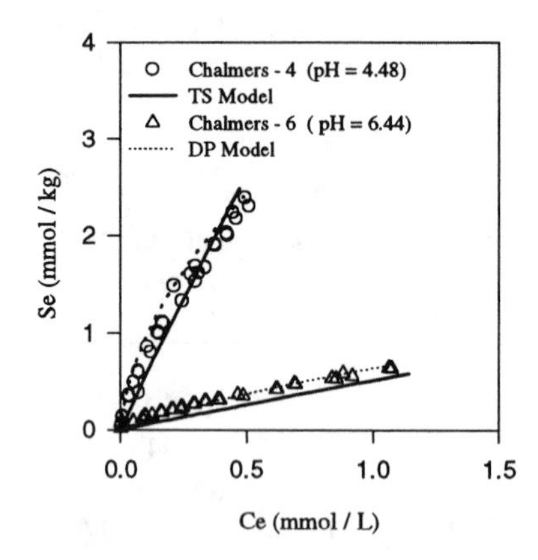

FIGURE 1. Sorption of aniline to Chalmers-4 and Chalmers-6 soils from 5 mM $CaCl_2$ solutions.

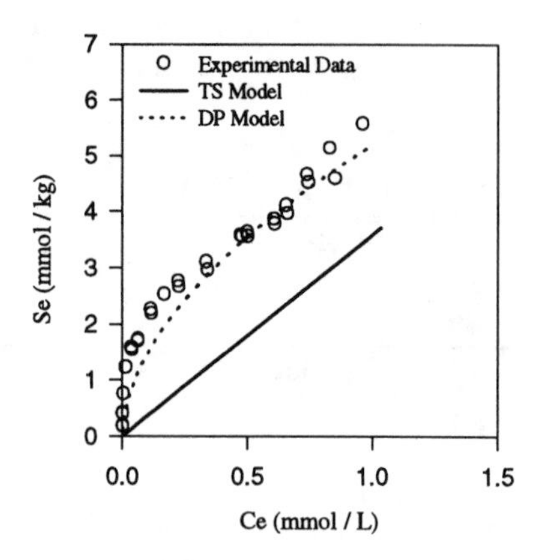

FIGURE 2. Sorption of α-naphthylamine to Chalmers-6 soil from 0.5 mM $CaCl_2$ solution.

REFERENCES

Gapon, E. N. 1933. "On the theory of exchange adsorption in soils." *Zh. Olschei. Khim.*, 3: 144-152.

Griffioen, J. 1993. "Multicomponent cation exchange including alkalinization/acidification following flow through sandy sediment." *Water Resour. Res.* 29(9): 3005-3019.

Lee, L. S., A. V. Nyman, H. Li, M. C. Nyman, and C. T. Jafvert . 1997. "Initial Sorption of Aromatic Amines to Surface Soils." *Environ. Toxicol. Chem.* 16(8): 1575-1582.

Perdue, E. M., J. H. Reuter, and R. S. Parrish. 1984. "A statistical model of proton binding by humus." *Geochim. Cosmochim. Acta.* 48: 1257 - 1263.

Sabljic, A., H. Gusten, H. Verhaar, and J. Humens. 1995. "QSAR Modeling of Soil Sorption. Improvement and Systematics of log K_{oc} vs log K_{ow} Correlations." *Chemosphere*. 31: 4489-4514.

DETERMINING RISK BASED REMEDIATION REQUIREMENTS USING RAPID FLUX CHAMBER TECHNOLOGY

William A. Frez and James N. Tolbert (Earth Tech, Grand Rapids, MI, USA),
Blayne Hartman (TEG, Solana Beach, CA, USA), and
Travis R. Kline (TechLaw, Boston, MA. USA)

ABSTRACT: Risk assessments for industrial facility closures frequently eliminate exposure pathways from further evaluation. However, uncertainties often require confirmatory field analyses to determine air pathway completeness. Available models requiring contaminant concentration, effective diffusivity, biodegradation effect, etc., have high uncertainty and yield unreliable air concentrations. As part of an RCRA Facility Investigation at a former chemical manufacturing facility, surface flux-chambers were used to measure vapor phase fluxes of chemicals associated with subsurface contamination. This approach mitigates uncertainties introduced by existing volatilization models, simplifying and increasing exposure assessment accuracy. Surface fluxes of hexachlorobenzene (C66), hexachloro-cyclopentadiene (C56), hexachloro-1,3-butadiene (C46), chloroform, vinyl chloride (VC), and tetrachloroethene (PCE) were measured to evaluate potential risks and hazards associated with the air pathway. Trials show surface fluxes are related to areas of known subsurface impact, demonstrating the utility of this technology for defining risks and remediation requirements for the site. Based upon the success of the trial program, an extensive program consisting of 100 to 150 surface flux measurements will be conducted this summer with EPA approval.

INTRODUCTION

In 1990, the United States Environmental Protection Agency (USEPA) proposed guidelines for corrective action (CA) for solid waste management units (SWMU) at hazardous waste management facilities (USEPA 1990). While the guidelines have yet to be finalized, they have generally provided the basis for ensuring that releases from SWMUs do not result in adverse impacts to human health and the environment. In the published rule, several priorities were established for CA program management. These include, prompt remediation of all significant off-site contamination, streamlined facility investigations, expeditious remedy decisions, and an emphasis in flexibility whenever possible. A fundamental premise of these priorities is that remedial requirements are risk-based and may vary significantly according to differing site settings.

Risk based remediation is predicated on the presence of complete exposure pathways that in-turn require sources, transport mechanisms, exposure points and routes (USEPA 1989). In many instances, potential exposure to soils and groundwater can readily be determined by comparisons to published preliminary remediation goals (PRG) such as those set forth by Region 9 of the USEPA (USEPA 1997). However, estimation of exposures to air secondarily impacted from intermedia transport is complicated not only by lack of specific knowledge of receptor characteristics but also by assumptions required to estimate volatilization to the overlying atmosphere.

Objective. In a 1997 collaborative study (approved by Region 5 of the USEPA), a surface flux chamber was designed and tested with the objective of determining its utility in Phase II investigations, and ultimately for setting remediation requirements for chlorinated hydrocarbons (CHC) at a former chemical manufacturing facility. Prior investigations at the facility delineated subsurface contamination of chlorinated hydrocarbons (hexachlorobenzene (C66), hexachlorocyclopentadiene (C56), hexachloro-1,3-butadiene (C46), chloroform, vinyl chloride (VC), and tetrachloroethene (PCE) in both soils and groundwater, including the presence of free product. While complete characterization of the soil and groundwater pathways continues, the volatility of these chemicals suggested that the upward vapor migration pathway could be substantial. After considering various alternatives to estimate and determine the risk, including the use of volatilization models, it was proposed that direct measurement of surface fluxes at potential source areas would give the most representative values of the upward vapor fluxes, and therefore best define remedial requirements at the site. To test this premise, a trial program consisting of ten (10) flux chamber measurements was conducted in October 1997 at the subject site.

METHODOLOGY

The methodology consisted of essentially two steps. Flux data were collected and analyzed using cost effective field methods. Flux-based risk criteria were then calculated in order to determine the relevancy of flux results to investigation and remediation requirements. Each of these steps is described below.

Field Methods. Cylindrical surface flux chambers were placed at eight locations at the site, including locations with known soil CHC contamination and a background location. The chambers were constructed of galvanized steel, measuring approximately 6 inches (15 cm) in diameter and 6 inches (15 cm) high, and equipped with sampling and purge ports. Their simple design resulted in a low per unit cost (<$10 per unit) which enabled ten to be deployed simultaneously. The deployment of a multitude of chambers allows much greater coverage of the site and yields a much more representative data base, features that are typically cost-prohibitive with more complex and expensive chamber designs.

Chambers were buried approximately 1 inch into the ground leaving an inner chamber height of approximately 5 inches (12 cm). A solar shade was placed over the chamber to reduce radiant heat effects. The chambers were left in-place for 10 to 11 days.

At the end of the deployment period, a gas tight syringe was inserted into the sampling port of the chamber and approximately 20 cc of air was withdrawn and immediately analyzed on-site by gas chromatography using a certified mobile laboratory. The on-site analysis was a unique feature of this program. Equipped with an Excel spreadsheet custom designed to calculate the fluxes from the measured data, samples were collected, measured for all the compounds of concern, and the upward vapor risk determined within one-hour of sample collection. This capability enabled real-time decisions regarding the need to leave the chambers in place for additional periods and the need to deploy additional chambers. In addition, the on-site analysis eliminated potential complications and errors introduced when larger volume samples are collected and analyzed off-site.

Determination of Flux-Based Remediation Goals. Published preliminary remediation goals (PRG) such as those set forth by Region 9 of the USEPA (USEPA 1997) are in terms of concentration units. In order to use flux chamber data, it is necessary to relate the concentrations measured in a flux chamber to the published concentrations.

The flux of a compound into an enclosed space (flux chamber or room) is defined as:

$$\text{Flux (mass/area*time)} = (\Delta C * V)/(\text{Area} * \Delta T) = \Delta C * H/\Delta T \quad (1)$$

Where ΔC is the change in concentration in the enclosed space over time
V is the volume of the enclosed space
Area is the contact area with the ground surface
ΔT is the deployment time.
H is the inner height of the enclosed space.

Using the preceding expression, the flux of a compound into a surface flux chamber (F_c) and an overlying room (F_r) is given by:

Flux chamber: $\text{Flux}_c = \Delta C_c * H_c/\Delta T$ (2)

Overlying Room: $\text{Flux}_r = \Delta C_r * H_r/\Delta T$ (3)

Because flux is in terms of unit area, the fluxes into a surface chamber or room will be identical at the same location, so that:

$$\text{Flux}_c = \text{Flux}_r = \Delta C_c * H_c/\Delta T = \Delta C_r * H_r/\Delta T \quad (4)$$

ΔT divides out, leaving:

$$\Delta C_c = \Delta C_r * H_r/H_c \quad (5)$$

Equation (5) indicates that although the flux into different-sized enclosed spaces are the same, the resulting concentration in the space depends upon the ratio of the volume to area. For a room 8 feet high, and a flux chamber 5" high, $H_r/H_c = (8/0.4) = 20$ and:

$$\Delta C_c = \Delta C_r * 20 \quad (6)$$

In other words, for the same flux, the concentration in a 5 inch high chamber will build up twenty times faster than the concentration in an 8 foot high room. If the allowable concentration in the room (PRG_r) is set to the EPA Region 9 residential remediation goals in ambient air, then the acceptable concentration for the chamber will be:

$$PRG_c = PRG_r * 20 \quad (7)$$

Finally, a correction needs to be made for the fact that room air exchanges during the course of a day, but the air in these chambers did not. If room air only flushed once per day, then the allowable concentration buildup in the chamber each day would be equal to the value in equation (7). However, air exchanges much more rapidly in residential and industrial buildings, typically on the order of a complete room volume every 1 to 2 hours. Thus, the allowable concentration buildup rate in the chamber is equal to:

$$\Delta C_c/\Delta T = PRG_c/E = PRG_r * 20/E \quad (8)$$

Where E is the air exchange rate in the room (exchanges/hr), and
ΔT is in hours.

For this study, a conservative value of one exchange every 2 hours (E=0.5), or 12 exchanges per day was used.

Equation 8 facilitates the interpretation of surface chamber data because it enables the measured concentration change in the chamber to be directly compared to published PRG values for overlying rooms.

Chamber Deployment Times: Once the required chamber concentrations are known as described in the preceding section, the required deployment time for the chambers can be calculated as the analytical detection limit divided by the amount of concentration buildup per hour. For the compounds of concern at this site, minimum deployment times ranging from 3 to 11 days were required depending upon the volume of vapor injected into the instrument. The longer deployment time was chosen since it was considered to yield a more representative flux, because it allowed short term fluctuations on the vapor flux, for example due to barometric pressure changes, to be averaged out.

RESULTS AND DISCUSSION

Table 1 shows results for the trial flux chamber measurements compared to permissible fluxes based upon ambient air concentrations protective of residential receptors established by USEPA Region 9 (USEPA 1997).

TABLE 1. Preliminary Flux Chamber Measurement Results (ug/L-h) for Selected Chlorinated Hydrocarbons at a RCRA Site.

Compound[a]	Area	PRG_r[b]	PRG_c[c]	$\Delta C/\Delta T_{PRG}$[d]	$\Delta C/\Delta T_m$[e]
C66	Impacted	4.2E-06	8.4E-05	4.2E-05	ND
	Background	--	--		ND
C56	Impacted	7.3E-05	1.5E-03	0.74E-03	5.6E-02
	Background	--	--	--	ND
C46	Impacted	8.7E-05	1.8E-03	0.88E-03	ND
	Background	--	--	--	ND
$CHCl_3$	Impacted	8.4E-05	1.7E-03	0.85E-03	ND
	Background	--	--	--	ND
VC	Impacted	2.2E-05	4.4E-04	2.2E-04	ND
	Background	--	--	--	ND
PCE	Impacted	3.3E-03	6.6E-02	3.3E-02	3.7E-02
	Background	--	--	--	ND

a-see text for definition.
b-in ug/L (USEPA 1997).
c-PRGr x 20 (in ug/L).
d-from Equation 8.
e-measured chamber concentration divided by deployment time (~260 hr).

Fluxes of two compounds, hexachlorocyclopentadiene (C56), and tetrachloroethylene (PCE), were detected. The measured C56 flux at the impacted area was nearly 80 times above the acceptable limit, strongly indicating that upward vapor migration is an exposure pathway of concern at this location. The measured PCE flux was within 10% of the acceptable limit, indicating a possible risk at this location.

Fluxes of the other volatile CHCs were below detection from both impacted and background areas, and did not indicate a potential risk from these contaminants at the 8 locations tested. To illustrate the effectiveness of the direct flux measurements versus results obtained from volatilization models, the upward flux and associated risk was calculated from the model described in the ASTM (1996) risk based corrective action (RBCA) guide, assuming free product of hexachlorobenzene (C66) at 20 feet bgs. The following default values were used in the model calculation:

Vapor pressure of C66: 1.089E-05 mm Hg
Total soil porosity: 0.3
Total air filled porosity: 0.15
Slab attenuation factor: 1 (no slab)
Room air exchange rate: 0.5 per hour (12 per day)

The computed flux would have been 1.9E-6 ug/hr-cm2 equivalent to a chamber concentration build-up of 1.6E-4 ug/l-hr, which would have over predicted flux and exceeded the risk criteria by 4 times.

The use of flux methodology to determine vapor emissions in general is not particularly new. For instance, trichloroethene fluxes from the unsaturated zone have been made to compare removal effectiveness to that of pump and treat systems at the Picatinny Arsenal (Smith et. al. 1996). Batterman et. al. (1992) used flux measurements to determine petroleum hydrocarbon emissions at waste sites endorsing the use of passive sampling as a means to conserve costs and improve logistics. Kim and Lindberg (1995) demonstrated the use of dynamic enclosure chambers to measure fluxes of mercury vapors from soils. However, all of the flux determinations discussed above depended on the collection of chemical vapors using absorbents that somewhat complicate handling and may introduce analytical error into the flux measurement.

The methodology used in this preliminary study was unique for several reasons:

1. The number of chambers deployed.
2. The addition of on-site analysis without preconcentration on adsorbants.
3. The on-site determination of risk.
4. The lengthy deployment time.

The preliminary results should not be used to conclude that risk to receptors is likely without consideration of additional site-specific factors such as mixing or dilution from ambient air, etc. Several such factors are being evaluated for incorporation into a much larger program consisting of 100 to 150 flux chambers, planned for the summer of 1998. Overall, however, the results of this trial program indicate that the methodology has significant potential to help determine investigation and remediation requirements at sites contaminated with volatile CHC, and has distinct advantages over using volatilization models.

REFERENCES

ASTM. 1996. Standard Guide for Risk-Based Corrective Action at Petroleum Release Sites. E-1739-96. American Society for Testing and Materials. West Coshohocton, PA.

Batterman, S.A., B.C. McQuown, P.N. Murthy, and A.R. McFarland. 1992. "Design and Evaluation of a Long-term Soil Gas flux Sampler." Environ. Sci. Technol. 26:709-714.

Kim, K. and S.E. Lindberg. 1995. "Design and Initial Tests of a Dynamic Enclosure Chamber for Measurements of Vapor-Phase Mercury Fluxes Over Soils." Water, Air and Soil Pollution. 80:1059-1068.

Smith, J.A., A.K. Tilsdale, and H.J. Cho. 1996. "Quantification of Natural Vapor Fluxes of Trichloroethene in the Unsaturated Zone at Picatinny Arsenal, New Jersey." Environ, Sci. Technol. 30: 2243-2250.

USEPA. 1989. "Risk Assessment Guidance for Superfund, Vol. I, Human Health Evaluation Manual (Part A)." United States Environmental Protection Agency. Office of emergency and Remedial Response. EPA/540/1-89/002. Washington, DC. 20460.

USEPA. 1990. "Corrective Action for Solid Waste Management Units at Hazardous Waste Management Facilities; Proposed Rule." Federal Register Vol. 55(145): 30798-30884.

USEPA. 1997. "Region 9 Preliminary Remediation Goals (PRGs) 1997." United States Environmental Protection Agency, Region 9. San Francisco, CA.

ENVIRONMENTAL STABILITY OF WINDROW COMPOSTING FOR EXPLOSIVES-CONTAMINATED SOILS

Mark Hampton, Wayne Sisk (U.S. Army Environmental Center, A.P.G., MD)

INTRODUCTION

The occurrence of explosives-contaminated soils presents an environmental cleanup challenge at many military installations across the country. The major explosives of concern include TNT, RDX, HMX, 2,4-DNT, 2,6-DNT, and Tetryl. Various biotreatment technologies have been evaluated and have proved to be successful in significant biodegradation of the explosive contaminants. One of the most successful demonstrations has been at the Umatilla Depot Activity, Oregon, where windrow composting has been shown to be 97-99% effective in the removal of TNT contamination from soils. The windrow composting project demonstrated that TNT and its major transformation products were reduced to approximately 1% of original concentrations within the first 20 days of composting, at up to half the cost of incineration.

RESULTS AND DISCUSSION

The success of the windrow composting demonstration generated further studies designed to determine the fate and mobility of the TNT and associated metabolites in the composting process. Early tracer studies using ^{14}C-labeled TNT showed that minimal $^{14}CO_2$ was released in the process, indicating that the degradation products were bound strongly to the treated compost. Additional studies using various solvent and aqueous leaching techniques, primarily conducted at Oak Ridge National Laboratory, confirmed the minimal release of CO_2, as a TNT transformation product, and showed that the treated compost was highly resistant to further breakdown or release. The Oak Ridge study also examined the effect of long-term exposure on compost stability, using the simulated 1,000-year acid rain leaching test (EPA 846 Method 1320). The results predicted that less than 10% of TNT transformation products would be released. Fate studies with ^{14}C-labeled TNT also found that a significant portion of the ^{14}C-labeled material remained bound in the organic components of the compost. Recent work conducted at the U.S. Army Waterways Experimental Station has corroborated those results and has provided an expanded analysis of the distribution of the ^{14}C-labeled material in the various compost components (i.e., primarily in the cellulose, humin, fulvic acid, and humic acid fractions). In summary, the combined results of these studies show that no significant TNT transformation products can be extracted or leached from the compost matrix.

In 1989, the U.S. Environmental Protection Agency (EPA) recommended that the toxicity of composted material be examined by using three separate methods: (1) direct ingestion of compost material by rats, (2) the Ames Assay of mutagenicity to solvent extracts as a measure of human health risk, and (3) the Ceriodaphnia dubia test using compost leachates to measure acute and chronic aquatic toxicity. All of the tests showed a 90 - 98% reduction in toxicity and mutagenicity as a result of the composting treatment.

The question of the bioavailability of TNT through inhalation of composted materials was studied in 1994 at the U.S. Army Center for Health Promotion and Preventive Medicine. Pure ^{14}C-labeled TNT, ^{14}C-labeled TNT-contaminated soil, and subsequent composted materials were installed directly into the lungs of rats. The ^{14}C signature showed a longer residence time in the organ tissues of the rats installed with the compost material than in those with the noncomposted materials. Research on the effects of dermal exposure to composted material, using in vitro testing and the pig skin model, has shown negligible dermal absorption. Although the rat inhalation studies raised questions on the stability of compost, the nature of the study itself raises some issues of concern. First, no discussion of the composition of the composted material that was installed in the lungs was provided. The study measures the concentration of ^{14}C in the various organ tissues and assumes that the ^{14}C signature resulted from TNT bound in the composted material, but no further attempt was made to examine the possible transfer of the ^{14}C signature to other organic fractions. Finally, no evidence was presented that suggested that the ^{14}C signature found in the organ tissues represents a hazard to human health.

The results of this study, combined with other research on composting, have raised several important issues. Care should be taken in drawing conclusions from laboratory-scale composting studies. Laboratory-scale, static pile composting does not follow the same process and may not result in the same products as windrow composting in the field.

One of the more important observations that arises from the wide variety of current research is the need to develop standardized treatability methods. Specifically, standardized methods for the preparation of laboratory-composted materials need to be developed to simulate different field composting technologies. If the use of any biotreatment method is proposed at a site, laboratory-scale treatability tests should be conducted to ensure that treatability tests accurately portray the site's capacity to support biotreatment and to show that the selected biotreatment method is compatible with site-specific conditions. Site managers should be provided with treatability test standards with procedures on preparing composting materials, selection of amendments, as well as sampling and analysis protocols. This standardization will allow the site managers to make more informed and reliable treatment decisions.

CONCLUSIONS

Even though these questions remain to be resolved, the application of composting for the treatment of explosives-contaminated soils continues, having been shown to meet the EPA's requirements for overall reduction in toxicity and mobility. Both laboratory research and field demonstration have provided evidence that the composted materials are not toxic and are highly resistant to leaching; and therefore, pose negligible risk to human health from dermal exposure, ingestion, or groundwater contamination.

ACKNOWLEDGMENTS

Dr. Cheryl Hastings of Argonne National Laboratory compiled the sources for this review.

REFERENCES

Hammell, Lowe, Marks, Myers, 1993, *Windrow Composting Demonstration for Explosives-Contaminated Soils at the Umatilla Depot Activity, Hermiston , Oregon*, U.S. Army Environmental Center report number CETHA-TS-CR-93043.

Griest, W.H., Stewart, A.J., Ho, C.-h, Tyndall, R.L., Vass, A.A., Caton, J.E., and Caldwell, W.M., 1994, *Characterization of Explosives Processing Waste Decomposition Due to Composting*, Prepared for the U.S. Army Environmental Center.

3D VOLUME MODELING FOR CONTAMINANT EXTENT AND SOURCE TERM DEFINITION

Jim Hicks, Tad Fox, Tom Naymik, and Neeraj Gupta; Battelle Memorial Institute, Columbus, Ohio, USA.

ABSTRACT: The distribution of contaminants in groundwater is commonly depicted using 2D contour maps and in some instances 3D images. The use of 3D volume modeling of contaminant distributions to characterize the extent and distribution of groundwater contamination and to develop source terms for input to solute transport modeling is presented. Solute transport modeling efforts generally require the definition of a source term describing the time, duration, location, and volume of contaminant released into the subsurface. In many instances, this information is unknown or can not be determined. In these cases, the current extent of the contamination can be used to define the source term representing the dissolved plume within the flow system. Solute transport modeling techniques can then be applied evaluate plume response to remedial alternatives and risk assessment calculations. A technique utilizing 3D volume modeling results to generate the source term for the dissolved plume is presented. Case studies at a DoD facility having large Trichloroethylene (TCE) and chromium (Cr) plumes and at a DOE facility having large high explosive (HMX and RDX) and Cr plumes are presented. At each site, the contaminants were released into the subsurface from multiple sources over long periods of time. The methodology permitted accurate representation of the contaminant distribution (as depicted by the 3D volume model) within the solute transport modeling framework. Groundwater flow is simulated using MODFLOW. Solute transport is simulated using Random Walk 3D (RWLK3D), a Battelle-developed contaminant transport and particle tracking code. 3D volume modeling of contaminant distribution is performed using EarthVision (DGI, 1996).

INTRODUCTION

The aerial extent of contaminants in the groundwater is often estimated through the use of 2D contour maps. It is assumed that there exists sufficient data coverage over the area of concern to define the contaminant plume. 2D contour maps are generally sufficient when examining a single aquifer or a discrete zone within an aquifer. However, when groundwater samples from varying depths within the same or different aquifer zones are analyzed, 3D grids and images of the plume can be generated using EarthVision 3D volume modeling software (DGI, 1996). The 3D images can then be used to evaluate plume extent and migration over time through the use of a series of images. Contaminated groundwater volumes can be estimated from the 3D image grid. This 3D grid can also be used to define a source term for the fate and transport modeling of the contaminants in the groundwater beneath the site.

A DoD facility was the focus of a study in which 3D volume modeling was used to define the extent of a large area of TCE and Cr contamination in the groundwater and to generate a source term component for these contaminants. The hydrogeology at this site consists of a shallow, locally perched water table aquifer overlying semi-regional and regional aquifers that supply groundwater to the surrounding municipalities. The aquifer consists of lithified interbedded sedimentary units with semiconfining units separating the aquifer zones and high vertical gradients between the different zones. Groundwater flow within the aquifers beneath the site were modeled (MODFLOW) as three primary hydrostratigraphic units called the upper saturated zone (USZ), lower saturated zone (LSZ), and production zone. The LSZ was further subdivided to examine vertical flow within this unit. Groundwater contamination occurs in the USZ and LSZ at different depths below ground surface. Multiple groundwater samples from each of the sample locations scattered across the site had been collected over time from the separate aquifer zones. Figures 1 and 2 show 2D contour maps of TCE from the USZ and the LSZ, respectively. The contaminant distribution data was combined with the site hydrostratigraphic data to generate a 3D volume model of the multiple layered TCE plume beneath the site as shown in Figure 3. A similar process was employed for the Cr data.

2D and 3D modeling at a DOE facility was used to estimate the extent and migration of Cr and a high explosive, RDX, within the groundwater beneath the site. Groundwater contamination exists in a thin unconfined perched aquifer overlying a regional sole-source aquifer. There is very little depth differentiation in the sample data due to the contamination being limited to a single aquifer zone. Therefore, in order to generate a 3D concentration grid, the perched aquifer was assumed to have uniform concentration vertically at each monitoring. The contaminant distribution data was again combined with the site hydrostratigraphic data to generate a 3D volume model of the contaminant plume beneath the site for both Cr and RDX.

OBJECTIVE

The objectives consist of (1) using 3D volume modeling to generate a 3D grid and image of the contaminant distribution within the hydrostratigraphic framework for a given site, (2) to evaluate plume migration over time through the comparison of 3D volume estimates from discrete sampling events, and (3) to estimate the mass of contaminant within a given hydrostratigraphic unit which is then converted into the input source term for RWLK3D solute transport modeling.

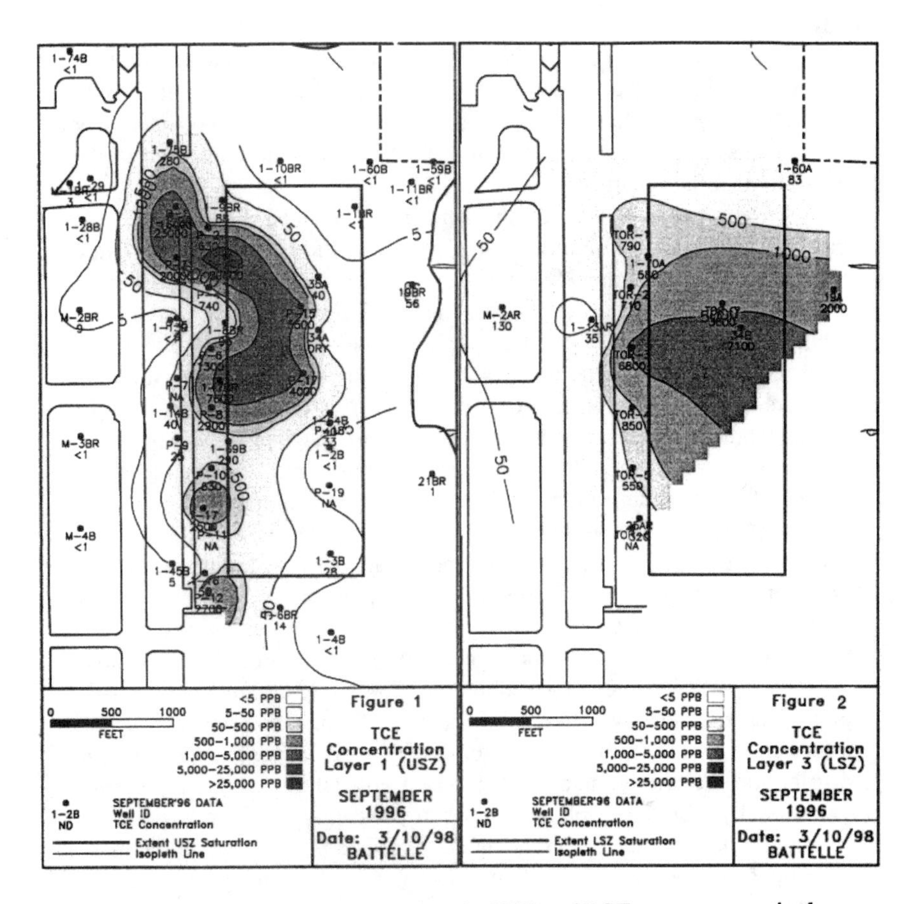

Figures 1 and 2. TCE concentrations in USZ and LSZ zones respectively.

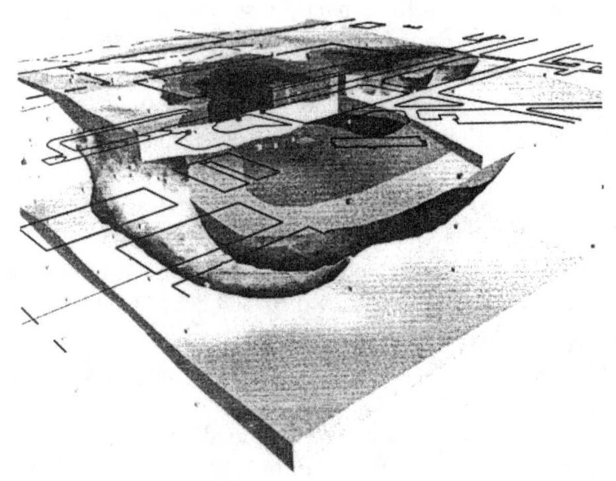

Figure 3. 3D image of TCE plume at DoD Facility.

METHODS

EarthVision was used to generate 3D volume models for data collected in September 1994 and September 1996 at the DoD facility to evaluate changes in the plumes over the 2-year period. Data collected over time at this facility were also examined statistically and compared to the 3D volume modeling results.

Concentration data for the contaminants of concern (COC) was used to make a 3D minimum tension grid with the EarthVision software program. Two separate 3D grids were created to represent the present TCE concentration levels and extent in the USZ and the LSZ at the DoD facility. The hydrogeologic stratigraphy at the site had been previously determined through lithologic data collected earlier during the site characterization. Two dimensional structure grids defining aquifer zones top and bottom surfaces were generated from the interpretation of the lithologic data. The USZ and LSZ are separated by a low permeability confining layer, the volume of which is not included in the total contaminant volume calculation. A total volume of 5.1×10^8 ft^3 was calculated for TCE contaminated groundwater and sediment using the volumetrics utility in EarthVision. If a porosity of 30% is assumed this means that approximately 1.5×10^8 ft^3 of groundwater is contaminated.

The September 1994 data were used to construct RWLK3D input source terms. Source terms for solute transport modeling generally consist of a mass released over time into the flow field. In many cases, this information is difficult at best to define. As a result, the current distribution of the contaminant is used to define the source term. Traditionally, for small plumes and simple geology, this was done by overlaying the plume map on the model grid, and approximating the mass from the concentration distribution. For complex 3D plumes, this method is cumbersome and results in numerous approximations. To overcome these limitations, a method for converting solute concentrations into mass was devised. The 3D gridded concentration distributions are used as the input concentration data. The X, Y, Z coordinates, and concentration data from the 3D grid are compared with the tops and bottoms for each model layer. Concentrations within each 3D cell are converted into the number of particles based on the concentration, cell volume, porosity, fluid density and the mass assigned to each particle. This process was employed for both the TCE and Cr data.

The DOE facility presented a challenge to 3D volume modeling in that the initial sample data was not distributed in a 3D manner. It was assumed that the contamination in the perched aquifer was homogeneously in the vertical dimension; therefore, two data points were added to each sample location to represent the top and bottom concentrations within the aquifer. The subsequent data set was gridded to create a 3D image to show the extent of Cr contamination in the groundwater. The volume of contaminated groundwater and sediment was 6.8×10^8 ft^3. The same Cr 3D

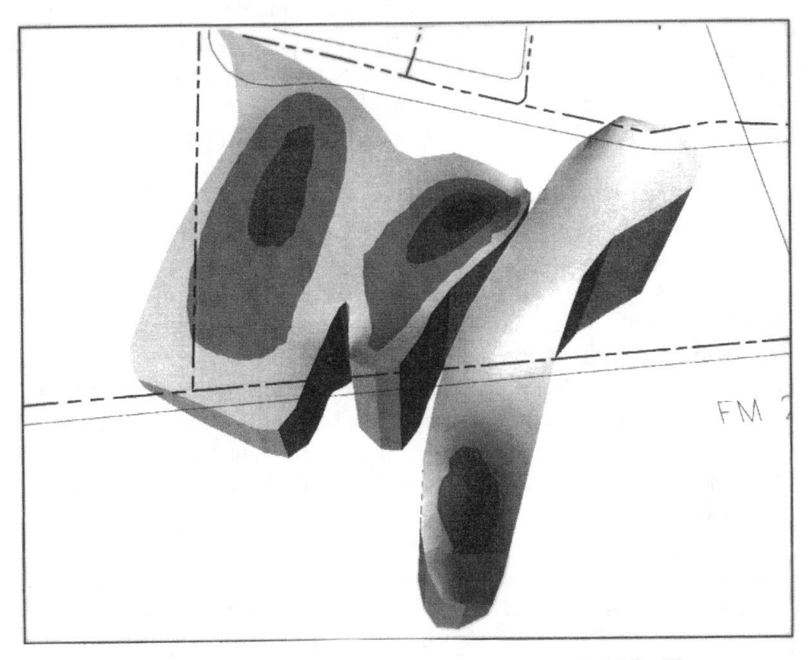

Figure 4. 3D image of Chromium plume at DoE facility.

Figure 5. Backward particle tracking scenario for six extraction wells at DoE facility.

p-133-145

grid (Fig the source term component for solute transport sport modeling was used to evaluate the move over time and to develop remedial alternativ vn in Figure 5 illustrates the capture zones of e the Cr and RDX plumes.

RESULTS AND DISCUSSION

The 3D volume modeling results for the September 1994 and 1996 TCE data collected at the DoD facility indicated a volume reduction of approximately 545 liters of TCE in the USZ. The decrease over this period likely results from a combination of extraction via pump-and-treat, natural attenuation, migration out of the USZ, and data variablility. This change in volume is less than the volume extracted by the treatment system and may be an indication of continued recharge through the source zones. The calculated volume for TCE in the LSZ increased from roughly 803 liters for the same period, which may indicate migration of TCE contaminated groundwater into the LSZ and data variability.

The method employed to generate input source terms for transport modeling provided an efficient and accurate approximation of the 3D contaminant distribution in groundwater. Initially, a particle mass of 0.02 kg for RDX at the DOE facility. This resulted in the generation of over 335,359 particles to represent the RDX plume. To ensure source terms were representative of existing conditions, a 1-day transport advective simulation was performed for RDX. Raw data used to develop the 3D RDX grid were then compared against the 1-day transport results. The R-squared value for RDX was 0.94 indicating a good match between the estimated current distribution and the input source term.

3D volume modeling is a valuable tool in determining groundwater contamination plume extent and volume. In addition, an accurate and interpretive 3D image meshed with site hydrogeology and basemaps can aid in interpretation of site data. However, as with any data interpretive model, the accuracy of the above modeling technologies depends on the accuracy and reliability of the input data sources.

REFERENCES

Battelle, 1995. *Solute Transport Code Verification Report for RWLK3D*, Internal Draft, Battelle, Columbus, Ohio, October 1995.

Dynamic Graphics, Inc. 1996. EarthVision, Version 3.1

McDonald, M.G. and A.W. Harbough, 1988. *A Modular Three-Dimensional Finite-Difference Ground-Water Flow Model*. Techniques of Water-Resources Investigations of the U.S. Geological Survey, Book 6, Chapter A1, 258pp.

ECOLOGICAL RISK ASSESSMENT UNDER THE MASSACHUSETTS CONTINGENCY PLAN

Jerome J. Cura (Menzie-Cura & Associates, Inc., Chelmsford, Massachusetts)

ABSTRACT: Ecological risk assessment under the Massachusetts Contingency Plan (MCP) occupies an integral position in determining Response Action Outcomes at state hazardous waste sites. The Massachusetts Department of Environmental Protection's (MADEP) uses an explicit tiered approach. There are two tiers: Stage I and Stage II, with three possible outcomes in Stage I (no significant risk, readily apparent harm, and potential risk). The concept of "readily apparent harm" and its clear definition within the regulations, allows site managers to avoid any further detailed assessment and proceed more directly to remedial solutions for sites exhibiting obvious risk. Additionally, a Stage I environmental screening eliminates pathways from Stage II consideration if the exposure pathway is incomplete; or the pathway is incomplete, but the exposure is so minimal that it clearly does not pose a significant risk. The process recognizes the use of local conditions at aquatic sites. When concentrations are consistent with local conditions, further assessment of the risk posed by that substance in that medium may not be required.

INTRODUCTION

This paper describes ecological risk assessment's central position in the remedial action decision making process under the Massachusetts Contingency Plan (MCP) at state hazardous waste sites. The Massachusetts Department of Environmental Protection's revised regulation and guidance (MADEP, 1996; 1997) use an explicit tiered approach incorporating water quality criteria, sediment criteria or effects levels, area of contamination, background, and the concept of "local conditions." The approach retains conservative assumptions, but is efficient in simultaneously moving sites to remedial solutions, called Response Action Outcomes (RAO). There are three general categories of RAOs which depend on a combination of factors including whether: the site poses no significant risk; all substantial hazards (short term significant risks) are eliminated; remedial actions have achieved a level of no significant risk; activity and use limitations are needed to maintain a level of no significant risk; concentrations exceed published upper concentration limits; actions have achieved background.

Clearly, risk assessment, and in particular achieving a level of no significant risk, is paramount in reaching decisions at contaminated sites. Given the importance of the term "condition of no significant risk" in this process, the regulations provide an explicit definition. Four criteria must be met to demonstrate that a condition of "no significant risk of harm to the environment" exists: (1) no physical evidence of a continuing release; (2) no evidence of biologically significant harm associated with current or foreseeable future conditions; (3) concentrations of oil and hazardous material at or from the disposal site do not exceed any applicable or suitably analogous formally promulgated environmental standards; (4) there is no indication of the potential for biologically significant harm to environmental receptors.

There are three important points here. (1) All four conditions must be met. (2) If the site surface waters influenced by the site exceed water quality criteria (a formally promulgated standard) for site contaminants, then a condition of no significant risk cannot be attained. (3) One must consider the indication of the potential for biologically significant harm. This last should be incorporated into the problem formulation by considering the possibility of future changes in contaminant distribution. The guidance also suggests that ecological clean-up goals may be identified by assessing risk over a gradient of concentrations at the site, and cautions that risk from harm to the environment must be balanced with potential harm resulting from habitat destruction during remediation.

THE TIERED APPROACH TO RISK ASSESSMENT

There are two tiers: Stage I and Stage II, with three possible outcomes in Stage I (no significant risk, readily apparent harm, and potential risk). The concept of "readily apparent harm" and its clear definition within the regulations, allows site managers to avoid any further detailed assessment and proceed more directly to remedial solutions for sites exhibiting obvious risk. Additionally, a Stage I environmental screening eliminates pathways from Stage II consideration if the exposure pathway is incomplete; or the pathway is incomplete, but the exposure is so minimal that it clearly does not pose a significant risk.

General Requirements of The Stage I Risk Assessment. Stage I is largely a pathway analysis. It attempts to analyze whether site contaminants in any media, soil, groundwater, sediments, air, or surface water can reach site biota or habitats. If the answer to this question is "yes", then Stage I tries to assess whether there is "readily apparent" environmental harm or a condition of "potentially significant exposure". If a Stage I reveals any potentially significant exposure pathways, then a Stage II Risk Characterization is required to assess the risk from those exposures. There are three possible conclusions from a Stage I Screening:

> A Stage II Environmental Risk Characterization is not required because there are no complete exposure pathways;
>
> A Stage II Environmental Risk Characterization is not required because, for each contaminated medium, harm is readily apparent, and therefore a condition of no significant risk of harm to the site biota and habitats clearly does not exist and a Stage II is redundant;
>
> A Stage II Environmental Risk Characterization is required because there is not enough information to determine whether a condition of no significant risk of harm exists, and therefore those media are assumed to pose a potentially significant pathway.

Stage I - Exposure Analysis. The first step in the Stage I is to evaluate whether there is current or potential future exposure to environmental receptors (plants, animals, or habitats) from site contaminants, based on historical records, site data, field observations, site history, or other information sources. Evidence of current or potential exposure includes:

> Current or past physical evidence (e.g. sheens, oil, tar, semi-solid hazardous waste) that oil or hazardous material (OHM) at or from the disposal site have come to be located in soil, surface water, sediment, or wetlands;
>
> Records or other evidence of past impacts on wildlife, fish, shellfish, or other aquatic biota;
>
> Analytical data indicating the presence of OHM in surface water, sediments or wetlands;
>
> Potential for transport of OHM in groundwater or surface water runoff to receptors such as surface water, sediments, or wetlands;
>
> Presence of OHM at the disposal site within two feet of ground surface and the potential for such contamination to result in exposure to wildlife.

Possible Outcomes of an Exposure Analysis. There are two possible outcomes from the exposure analysis: (1) If evaluation of the above lines of evidence do not indicate a current or potential exposure pathway, then a condition of "no significant risk of harm" to site biota or habitats exists and a Stage II is not required; (2) If the exposure evaluation indicates a current or potential future exposure, based on the above lines of evidence, then the risk assessor must evaluate site conditions to determine whether significant environmental harm is readily apparent.

Criteria for Determining "Readily Apparent Harm." The criteria for determining readily apparent harm are:

> visible evidence or stressed biota attributable to the disposal site;
>
> Surface water concentrations in excess of Massachusetts Surface Water Quality standards or USEPA Ambient Water Quality Standards;
>
> Visible presence of oil, tar, or other non-aqueous phase hazardous material in soil over and area equal to or greater than two acres or equal to or greater than 1,000 square feet in sediment;

Readily apparent harm has occurred if the disposal site exhibits any of the above three criteria. Under a conclusion of readily apparent harm, the MCP specifies that a condition of "no significant risk" does not exist, and a Stage II is not required.

Evaluation of Potentially Significant Exposures. If the result of the above analysis for potential exposure pathways is a conclusion of "no readily apparent harm", then the risk assessor must decide whether the potential exposure pathway results in a "potentially significant exposure". This decision is based on an "effects based screening approach" which varies among aquatic, terrestrial and wetland environments.

STAGE I ASSESSMENT IN AQUATIC ENVIRONMENTS

There are two "pre-Stage I steps for an aquatic assessment: (1) comparison of sediment and surface water to background; and (2) comparison of sediment and surface water to local conditions. Under the MCP, background concentrations are those which are ubiquitous and consistently present near the disposal site, **and** are attributable to geologic or ecological conditions, atmospheric deposition of industrial processes or engine emissions, fill materials containing wood or coal ash, releases to groundwater from a public water supply system, and/or petroleum residues that are incidental to the normal operation of motor vehicles. Compounds which are consistent with background are not assessed further.

Comparison to local conditions is another step which the guidance recommends **only for aquatic environments,** (this does not include wetlands which are discussed separately) specifically sediment and surface water. Local conditions are levels of OHM present consistently and uniformly throughout the surface water body, or throughout a larger section of a river that contains the area potentially affected by contamination at or from the site. Hot spots and localized contamination are not considered local conditions. Like background, local conditions may be assessed on a chemical specific basis. When concentrations are consistent with local conditions, further assessment of the risk posed by that substance in that medium may not be required.

In aquatic environments, the detection of elevated levels of contamination in sediment or surface water, or the potential for elevated levels to occur in the future, constitutes identification of a complete exposure pathway. For any complete pathway, an effects based screening is necessary in Stage I. For effects based screening levels, the guidance recommends:

NOAA ER-Ls for marine and estuarine sediment;

Ontario Ministry of the Environment Guidelines for freshwater sediment;

AWQC, LOELs published with the AWQC, or LOELs for the literature (in that order) for surface water.

Note that the effects screening criteria are only for ruling out pathways, not individual chemicals. **If a pathway is not ruled out, risks from all chemicals that result in exposure by that pathway should be evaluated in Stage II, even if those substances are present at levels below their screening criteria.**

STAGE I ASSESSMENT IN TERRESTRIAL HABITATS

If soil concentrations are consistent with background, those contaminants are not carried through the risk assessment. The terrestrial Stage I depends on "an evaluation of habitat quality" which incorporates the size of the site and the nature of the surrounding area. There are four habitat evaluation steps.

Step 1. If exposure of a state listed rare or endangered species or other species of special concern is possible, a Stage II risk characterization should be conducted.

Step 2. If contaminant transport from surface soil to a state designated Area of Critical Environmental concern is possible, a Stage II Risk Characterization should be conducted.

Step 3. If the undeveloped portion of the affected area is less than two acres and neither of the two preceding criteria apply, no further assessment is necessary.

Step 4. If the undeveloped portion of the affected area is greater than six acres, the assessment should proceed to an effects based screening or a Stage II (note - the practical result is a Stage II, because the guidance does not recognize any terrestrial screening levels). If the area is less than six but greater than two the assessor may proceed to an effects based screening (not possible at this time) or a Stage II or conduct a further evaluation to determine the significance of exposure pathways, based on four questions:

> Is total contiguous area six acres or more (counting adjacent connected parcels)?
>
> Is there a unique or unusual niche for valued species?
>
> Is there a vernal pool within 150 meters?
>
> Does the affected area contain habitat for species under restoration by he state?

If the answer is yes to any of these questions, proceed to an effects based screening step. If all answers are no, then no further assessment is required. There are no effects based screening levels for soil, therefore, if the habitat evaluation indicates the need for further assessment, the risk assessor should proceed directly to a Stage II.

STAGE I ASSESSMENT IN WETLANDS

The Stage I should first identify complete exposure pathways, and then proceed to an effects based screening using:

> NOAA ER-Ls for sediment regardless of whether the wetland is marine or freshwater (not this is different than for aquatic environments;
>
> AWQC or LOAELs in the AWQC document.

The guidance notes that many of these values are based on toxicity to species that do not inhabit wetland areas, and they should not necessarily be used as benchmarks in quantitative wetland assessments. However, the values are sufficient for wetland screening criteria.

DECIDING ON POTENTIALLY SIGNIFICANT PATHWAYS

If the application of the above criteria indicate that the measured concentrations in a medium (e.g. soil, sediment, surface water) are less than the effect level, then the risk assessor can rule out that medium as a "potentially significant exposure". A medium so characterized does not require a Stage II to determine whether a condition of no significant risk exists. If a screening analysis cannot rule out a medium as a potentially significant exposure, then a Stage II must be conducted to determine if a "condition of no significant risk of harm exists."

REFERENCES

MADEP, 1996. Guidance for Disposal Site Risk Characterization, Section 9, Method 3 – Environmental Risk Characterization. Massachusetts Department of Environmental Protection, Boston, MA.

MADEP, 1997. The Massachusetts Contingency Plan, 310 CMR 40.00, Massachusetts Department of Environmental Protection, Boston, Massachusetts.

ECORISK OF PAHs IN SUBSURFACE SOILS: ESTIMATING TRANSPORT

Jerry M. Neff (Battelle, Duxbury, MA)

ABSTRACT: The ecological risk of subsurface polycyclic aromatic hydrocarbons (PAHs) to surface receptors is dependent on the exposure concentration at the point of contact between dissolved PAHs and the receptors. Exposure concentrations can be estimated by estimating transport of the PAHs from a subsurface non-aqueous phase liquid (NAPL) to ground water and with the ground water to the surface expression of the PAH-contaminated plume. Partitioning of PAHs from a NAPL into ground water can be estimated by a NAPL/water partition coefficient which is approximated by the octanol/water partition coefficient (K_{ow}) for the PAH. Actual concentrations of PAHs in water in contact with a NAPL are lower than concentrations estimated this way, so the estimates are conservative. Transport of PAHs in ground water is controlled by partitioning between the aqueous phase and soil organic matter, which can be approximated by the soil organic matter/water partition coefficient (K_{oc}). Migration of PAHs is strongly retarded by sorption to soils containing more than about 0.1 percent organic matter. Higher molecular weight PAHs usually do not migrate in ground water and, therefore, do not pose a risk to surface receptors.

INTRODUCTION

Subsurface soils at many contaminated sites contain elevated concentrations of PAHs. The main sources of PAHs in soils are petroleum (petrogenic) and combustion (pyrogenic) products (Neff, 1979). Coal tar from gas manufacture and coke production and creosote contain both petrogenic and pyrogenic PAHs. PAH-contaminated NAPLs from different sources vary widely in physical/chemical properties, affecting release of the PAHs from the NAPL phase into ground water. Once dissolved in ground water, PAHs from different sources can be expected to behave the same.

The most difficult step in estimating the ecological risk of PAHs in subsurface soils and ground water to various aquatic and terrestrial receptors is estimating the exposure concentration at the point of contact with valued ecosystem components. While in the subsurface, PAHs and other contaminants usually do not pose a direct health risk to aquatic and terrestrial receptors, unless animals can reach the subsurface deposit by excavation or burrowing. Therefore, the potential for exposure depends on the transport of the PAHs to a surface expression of the PAH-contaminated plume.

Complex PAH mixtures may be present in soils as part of a nonaqueous phase liquid (NAPL), in solution in ground water, or tightly sorbed to a solid phase, such as soot, ash, coal dust, or asphalt. PAH assemblages from pyrogenic sources in soils usually are tightly bound to a solid phase (soot or ash). However, PAHs associated with a NAPL phase or dissolved in ground water may be mobile and, if they reach the surface, may be bioaccumulated by terrestrial and aquatic organisms. The objective of this paper is to review approaches that can be taken to estimate migration of PAHs in soils and ground water as part of ecological risk assessments.

SUBSURFACE MIGRATION OF NAPL

Crude and refined petroleum, coal tars, and creosote are extremely complex liquid mixtures of hundreds or thousands of organic chemicals, mostly hydrophobic, poorly water-soluble hydrocarbons. When released to subsurface soils, they tend to retain their identity as a separate oil phase, the NAPL. The rate and extent of migration of a NAPL into and though soils and ground water depends on the viscosity, density, and interfacial tension of the oil, and the permeability and porosity of the soil (Sale et al., 1992). Petroleum products, coal tars, and creosote vary widely in density, interfacial tension, and viscosity. These physical properties also change as the NAPL weathers in place, primarily through evaporation, dissolution, and microbial degradation. Thus, rates of migration of NAPL vary widely in soils of different texture and porosity, and with time as the NAPL weathers.

Crude oils, heavy petroleum distillate fractions, and some coal tars are sufficiently viscous that they do not penetrate soils rapidly (Neff et al., 1994). Soils composed of fine-grained silt or clay, as well as sandy soils containing more than a few percent silt/clay, have very low permeabilities (Scheidegger, 1957); migration of all but the lowest viscosity oils in such soils is slow (Mott and Weber, 1991). The interfacial tension between oil and water inhibits dispersal of oil into waterlogged soils. Because of the hydrophobicity of petroleum products and coal tars, water-wet, saturated soils are not easily wetted or penetrated by oils. NAPLs less dense than the water in the saturated zone of the soil (LNAPL) tend to perch on the upper boundary of the saturated zone. NAPLs more dense than the soil water (DNAPL) may migrate down through water-saturated soil if the capillary pressure exceeds a threshold value called the entry pressure; the entry pressure is directly proportional to the interfacial tension of the NAPL and inversely proportional to the size of the pore spaces (Sale et al., 1992). If the thickness and density of the DNAPL is great enough, it will sink through the saturated zone and accumulate on the impermeable stratum at the bottom of the aquifer.

In the vadose zone, the NAPL gradient is predominantly vertical, producing a downward flow of the NAPL. The NAPL continues to migrate downward until the mass of NAPL is depleted due to the residual saturation left behind, the soil permeability declines or the viscosity of the oil increases to the point where gravitational force can not overcome capillary force, or the NAPL reaches the top of the saturated zone. Most oil and coal tar NAPLs are less dense than water and, therefore, float on the ground water surface.

A floating NAPL often accumulates as a mound on the ground water surface and diffuses outward in all directions due to gravitational force. However, the major direction of movement is down-gradient in the direction of net ground water flow. The rate of horizontal NAPL migration is dependent on the physical/chemical properties of the oil and the permeability of the soil (Sale et al., 1992). The rate of NAPL migration usually is a small fraction of the rate of flow of the ground water.

PAH PARTITIONING FROM NAPL INTO GROUND WATER

PAHs are in solution in a petroleum or coal tar NAPL and so can be considered to behave ideally in the complex mixture; that is, the behavior of each component in the mixture is determined by its physical/chemical properties and its concentration in the mixture (Lee et al., 1992b; Peters et al., 1997). Under these conditions, the concentration of each PAH in an aqueous phase in equilibrium with the NAPL phase is proportional to the mole fraction of the chemical in the organic phase in agreement with Raoult's Law (Shiu et al., 1990; Lee et al., 1992a,b; Lane and Loehr, 1995).

Because both the NAPL/water partition coefficient (K_o) and the octanol/water partition coefficient (K_{ow}) bear a similar log/log relationship to aqueous solubility of the PAH, there is a log/log relationship between K_o and K_{ow} for the PAHs in the NAPL (Lee et al., 1992a). For many of the PAHs in petroleum and coal tar NAPLs, the difference between log K_o and log K_{ow} is small and so K_{ow} is a reasonable surrogate for K_o in predicting aqueous phase concentrations of individual PAHs based on their NAPL concentrations. For example, Shiu et al. (1990) estimated log K_o for naphthalene in three crude oils ranging from 3.28 to 3.32, slightly lower than the log K_{ow} for naphthalene (3.37). Log K_{ow} values have been published for a large number of PAHs (Mackay et al., 1992). Thus, K_{ow} is a reasonable surrogate for estimated NAPL/water partition coefficients for PAHs. The concentration of a PAH in water in equilibrium with a petroleum product or coal tar NAPL in soil or ground water can be estimated as the ratio of the concentration of the PAH in the NAPL to its K_{ow}.

Actual partitioning behavior between the NAPL and aqueous phases depends on the relative surface area of NAPL in direct contact with the ground water and the rate of flow of water past the NAPL phase, both of which are dependent on the porosity of the soil. Weathering of the NAPL may result in production of an interfacial film or phase separation in the NAPL, resulting in a decrease in the partitioning of PAHs from the residual NAPL (Luthy et al., 1993; Peters et al., 1997). As the NAPL weathers, it may become more viscous. Mixing is inefficient in a viscous NAPL, and partitioning of PAHs out of the NAPL decreases as the NAPL/water interface is depleted of PAHs. In a highly viscous NAPL, partitioning of PAHs to the aqueous phase is limited by the low diffusivity of PAHs in the viscous NAPL/soil matrix (Yeom et al., 1996; Peters et al., 1997). Estimated aqueous phase concentrations of PAHs near a NAPL will always be lower than actual concentrations in solution.

SOIL-WATER PARTITIONING OF PAHs

As water migrates away from the NAPL/ground water interface, it carries any PAHs dissolved from the NAPL with it. These dissolved PAHs partition between the aqueous phase and the soil organic phases according to their relative solubilities in the different phases. This partitioning is described by the soil organic carbon/water partition coefficient (K_{oc}) (Karickhoff, 1981). Several formulas have been published for estimating the K_{oc} of PAHs, including (Means et al., 1980):

$$\log K_{oc} = \log K_{ow} - 0.317 \quad (1)$$

The actual value of K_{oc} for different PAHs varies slightly, depending on the texture of the soil and nature of the organic carbon in the soil. In soils containing less than about 0.1 percent organic carbon, sorption is primarily to inorganic phases, particularly surfaces of clay particles, and the K_{oc}/K_{ow} relationship is not valid.

Both particulate and dissolved organic matter in soils compete for sorption of nonpolar organic chemicals such as PAHs. However, sorption of PAHs to dissolved organic matter in soils may facilitate transport through the soil, whereas sorption to particulate organic matter will retard migration (Tan and Tomson, 1990; Sheunert et al., 1992). The concentration of dissolved organic matter usually is not high enough to affect PAH transport.

In fact, the main effect of soil organic carbon/water partitioning on PAH transport in ground water is retardation of PAH transport rates relative to ground water flow rates. This relationship can be represented by a retardation factor, R, that is mathematically related to the soil particle/water partition coefficient, K_d, by the equation:

$$R = \rho_b K_d / n \quad (2)$$

where ρ_b is the bulk or mass density of the porous medium and n is the porosity of the medium (Neff et al., 1994). Retardation of PAH migration in ground water increases with increasing soil organic matter concentration and PAH K_{oc} as shown in the examples in Table 1.

Retardation increases dramatically with increasing PAH K_{oc}, particularly in soils with high organic carbon concentrations. The higher molecular weight PAHs that also have very low aqueous solubilities are essentially completely retarded and do not migrate at all in ground water of soils containing more than about 0.1 percent organic carbon.

In summary, PAHs associated with a subsurface oil or coal tar NAPL in soil have little potential to migrate long distances in the NAPL itself or in association with a ground water aquifer. This is particularly the case for the higher molecular weight (>3-ring) PAHs. Partitioning from a liquid NAPL into ground water always yields aqueous concentrations of PAHs that are well below their single-phase aqueous phase solubilities. Transport of the dissolved PAHs with ground water is retarded by sorption to soil organic matter. Thus, the risk of exposure of surface receptors to potentially toxic concentrations of PAHs from subsurface NAPLs is low.

Table 1. K_d and retardation factors (R) for several PAHs in soils containing 0.1 and 0.5 percent organic carbon and the same porosity. Benzene and *m*-xylene are included for comparison. From Neff et al. (1994).

Chemical	K_d (0.1%OC)	R (0.1%OC)	K_d (5% OC)	R (5% OC)
Benzene	0.08	1	4	26
m-Xylene	0.98	7	49	296
Naphthalene	1.30	9	65	391
1-Methylphenanthrene	85.11	512	4,256	25,535
5-Methylchrysene	1,622	9,730	81,090	486,544
Benz(a)anthracene	1,380	8,280	69,000	414,000
Benzo(a)pyrene	5,500	33,000	275,000	1,650,000

REFERENCES

Karickhoff, S.W. 1981. "Semi-empirical estimation of sorption of hydrophobic pollutants on natural sediments." Chemosphere *10*: 833-846.

Lane, W.F. and R.C. Loehr. 1995. "Predicting aqueous concentrations of polynuclear aromatic hydrocarbons in complex mixtures." Water Environmental Research *67*: 169-173.

Lee, L.S., M. Hagwall, J.J. Delfino, and P.S.C. Rao. 1992a. "Partitioning of polycyclic aromatic hydrocarbons from diesel fuel into water." Environmental Science and Technology *26*: 2104-2110.

Lee, L.S., P.S.C. Rao, and I. Okuda. 1992b. "Equilibrium partitioning of polycyclic aromatic hydrocarbons from coal tar into water." Environmental Science and Technology *26*: 2110-2115.

Luthy, R.G., A. Ramaswami, S. Ghoshal, and W. Merkel. 1993. "Interfacial films in coal tar nonaqueous-phase liquid-water systems." Environmental Science and Technology. *27*: 2914-2918.

Mackay, D., W.Y. Shiu, and K.C. Ma. 1992. *Illustrated Handbook of Physical-Chemical Properties and Environmental Fate for Organic Chemicals. Vol. II. Polynuclear Aromatic Hydrocarbons, Polychlorinated Dioxins, and Dibenzofurans.* Lewis Publishers, Chelsea, MI.

HOLISTIC RISK MANAGEMENT OF MTBE IN GROUND

Brendan P. Dooher (University of California, Los Ang
Anne M. Happel (Lawrence Livermore National Laboratory, I
William E. Kastenberg (University of California, Berkeley, CA)

ABSTRACT: Methyl tertiary butyl ether (MTBE) is currently the preferred oxygenate to meet the requirement of California's Reformulated Gasoline Program. Its occurrence in public and private drinking water wells has raised concerns about its potential to affect groundwater supplies. In a recent study by Lawrence Livermore National Laboratory (LLNL), MTBE was detected at approximately 80% of California leaking underground fuel tank (LUFT) sites examined. Due to the ubiquitous nature of LUFT sites, the cumulative impact to aquifers and the resultant impact to drinking water supplies from a recalcitrant constituent such as MTBE requires a holistic approach to its analysis. A comprehensive groundwater management system is necessary to evaluate this problem in the most economical manner possible, in order to preserve the integrity of groundwater resources. To properly assess this threat, groundwater basins should be examined as whole entities. Because of the particular geographical makeup of many groundwater basins and accompanying watersheds, they may often represent a logical partition for dealing with population growth patterns as well.

The following presents a study performed using the Napa Valley watershed. This valley was chosen because of its mixture of rural and suburban settings, its hydrogeological and geographic isolation, and its proximity to the rapidly growing population center of the San Francisco Bay Area. Regional hydrogeological parameters are used as *a priori*, non site-specific inputs for the initial evaluation of the risk. The large data sets involved are represented and analyzed using a combination of analytical tools, probabilistic forecasting methods, and Geographical Information System (GIS) presentation software. Risk managers and water system planners may use the results in order to assess MTBE impacts to groundwater resources and human receptors both now and in the future. With some modification, this methodology may be extended for use with other constituents of concern that have different source points or different risk probability domains than MTBE.

INTRODUCTION

MTBE is a recalcitrant, highly mobile compound that has been used for the past six to eight years as an oxygenate in reformulated gasoline at levels between eleven and fourteen percent by volume. Prior to this time, likely from the mid 1970's, it was used as an octane enhancer at approximately two percent by volume. Unlike other constituents of gasoline, such as the benzene, toluene, ethylbenzene, and the xylenes (BTEX) which are degrading and are thus self-limiting in their volumetric zone of influence, conservative species, such as MTBE, have the capacity to "go far". That is, MTBE studies to date suggest a lack of intrinsic biodegradation. Thus biodegradation, which for other constituents of gasoline actually removes mass from an aquifer system, has no such affect on MTBE. Therefore, there is the potential for accumulation of MTBE in aquifer systems that can lead, over time, to a degradation of the entire system. This can occur even when MTBE has attenuated by advection and dispersion beyond detection limits.

MTBE, from a health perspective, is considered to be a carcinogen only at chronic concentration levels (well into the ppm range) and is not considered to cause cancers by genotoxic effects (Mennear, 1997). It is not considered harmful at levels that would be tolerable to humans. Human taste and odor thresholds range from as low as 5 ppb to a level of 200 ppb, where the majority of the population can detect it in un acclimatized conditions. Where there has been prior exposure to gasoline constituents, these levels may be much higher.

MTBE is very soluble in water, up to 30 times more soluble than benzene, but has very little capacity to sorb to soils. Due to its relatively large proportion in gasoline, releases from many sites may have a cumulative affect. There are an estimated 40,000 Underground Storage Tank (UST) sites in California, 150 of which are known to have affected wells. There are approximately 100 public wells that are known to have been affected by benzene, the majority of which are likely to have been affected by LUFT sites. Since MTBE has been sampled in public well systems, 18 public wells have reported MTBE impacts out of a sample of 1628 wells (Happel et al., 1998). The "Failure Rate" associated with MTBE then is 1.1%, though MTBE has only been used at ~11% mixtures for around the last 6 years. This is approximately the same failure rate as is associated with benzene, which has been in use as a major constituent for a much longer duration.

Objective. Due to the large number of UST sites, and the potentially large number of associated LUFT hydrocarbon releases, there is a need for better prioritization of risk management decisions associated with these sites. The way in which this can be accomplished is to look at the risk paradigm of *source*, *pathway*, and *receptor*. To do this, we need to look at the macro-scale problem, disassociated from the individual LUFT sites and placed at the regional level. Then an appropriate model can connect the source and the receptor, using what is known of the region's hydrogeologic conditions. This model is one in which a plume domain of the chemical release can be established, along with associated probabilities of extent and growth. When this is accomplished, the vulnerability of drinking water well populations from the hazard can be estimated. Finally, the need for mass based estimates of releases are stressed in order to quantify the potential for degradation of the aquifer system.

The region in which the initial investigation occurs is the Napa Valley watershed. Information presented here come from a series of sources. Large data sets are represented using a combination of analytical tools and Geographic Information System (GIS) presentation software. Risk managers and water system planners may use the results in order to properly deal with FHC impacts to groundwater resources. With some modification, this methodology may be extended for use with constituents of concern (COCs) that have larger areal zones of influence than FHC constituents. Figure 1(a) shows a coarse threat estimate for the State of California while Figure 1(b) shows the overall estimate of threat to private wells for the next five years time due to MTBE releases in Napa County, CA.

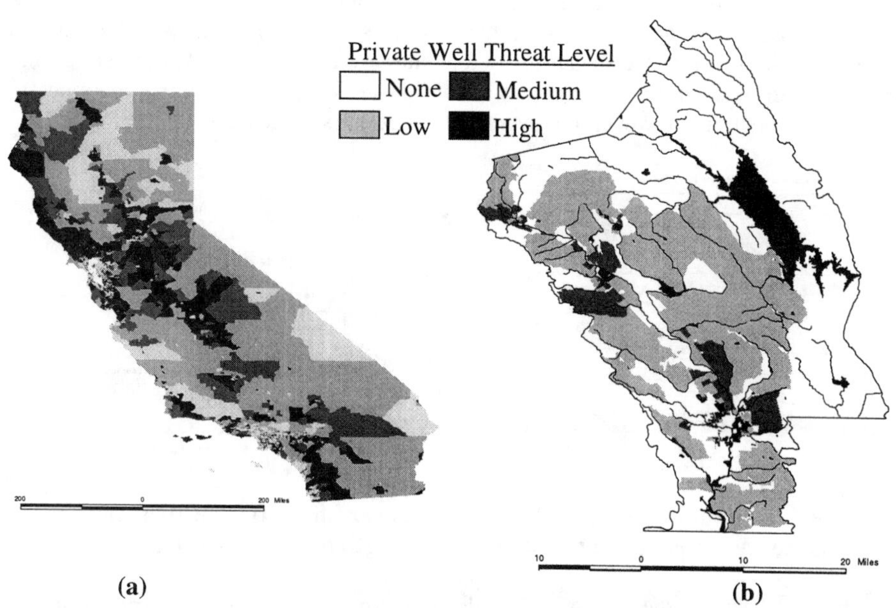

FIGURE 1. Threat levels for private well impact from LUFT sites at coarse and fine levels. (a) Coarse Level Threat at the Census Tract level for the State of California – darker areas indicate higher risk. (b) Fine Level Threat at the Census Block level for Regions of High, Medium, and Low Threat in Napa County, CA.

Methodology. In order to develop the general plume model, a Monte Carlo based approach is used in conjunction with an analytical 3-dimensional solution to the advection-dispersion equation. This general methodology (Dooher et al., 1998) can produce a plume probability domain. The inputs for this model are the appropriate ranges for the source concentration, hydraulic conductivity, gradient, and if necessary, biotic or abiotic degradation probability distribution. The resultant output gives a probability distribution of plume lengths. This is seen for benzene, compared to best professional judgement (BPJ) and non-parametric measurements, in Figure 2.

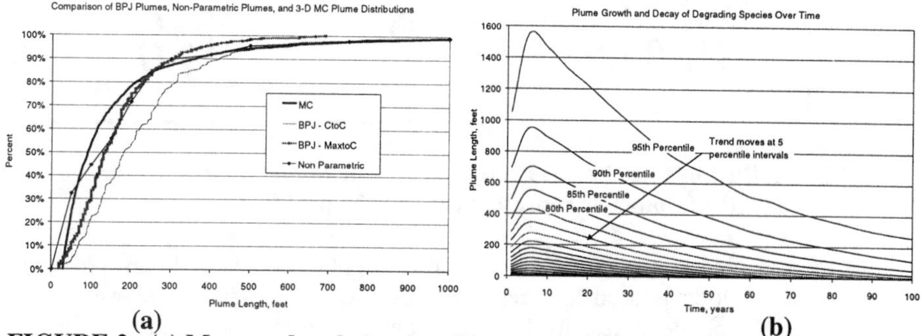

FIGURE 2. (a) Measured and simulated benzene plume length measurements. (b) Simulated benzene plume lengths over time.

Figure 3 shows the growth of a conservative species, such as MTBE, over time, using the extremely conservative approach where the source is not attenuated. This is done in order to develop a worst case scenario. Actual sites would likely not possess an infinite source. As can be seen, the 95th percentile confidence level for the plume continues to grow at 100 years time; however, the overall distribution is approaching an asymptote.

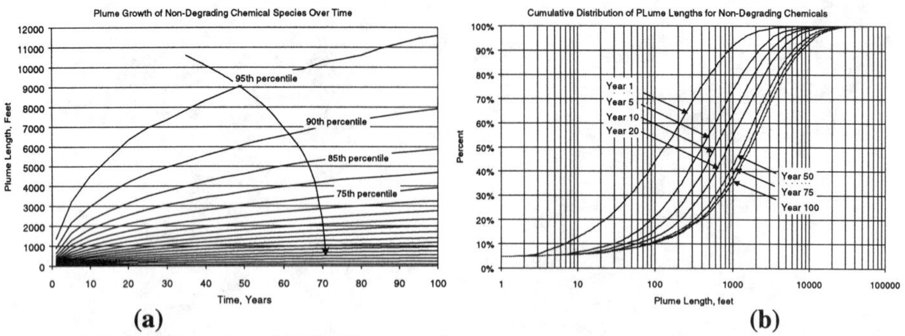

FIGURE 3. Simulated MTBE plume lengths over time. (a) Growth at various confidence levels. (b) Cumulative plume length distributions over time.

Figure 4 shows the Buffon's Needle problem, which is a geometric means of determining probability. The probability of intersecting *one* of a set of parallel lines separated by a distance D with a line (needle) of length L randomly placed between the lines is $2L/\pi D$. This method works so well that π can be calculated from it, if the needle is thrown a few thousand times.

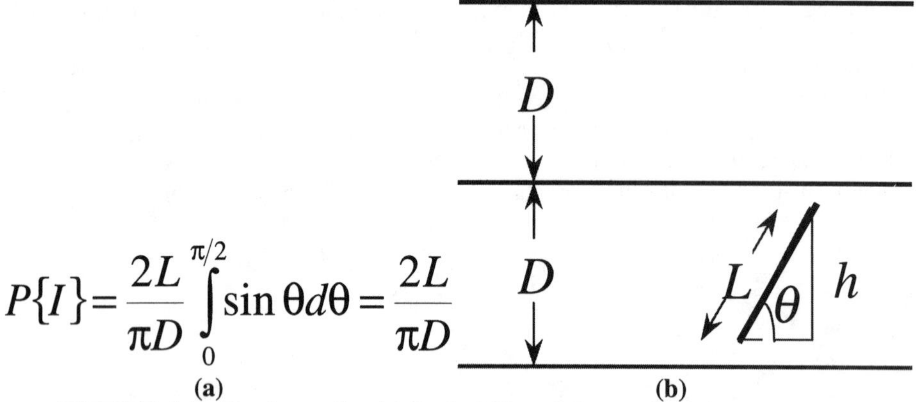

FIGURE 4. Buffon's needle. (a) Probability of the needle falling on a line. (b) Geometric representation of the problem.

A similar geometric argument may be made for a plume domain. In this problem, census block information is used to determine a density of drinking water wells in a particular area. The plume is described by the Monte Carlo method mentioned earlier. Figures 5-7 show the general principles involved in finding the relative probability. Once this probability is found, the various threat levels may be determined as seen for Napa

City, CA in Figure 7. The dots indicate LUFT sites; it can be seen that for this time frame, many of the sites pose no immediate threat. To determine the threat to the aquifer as a whole, a mass balance must be performed on the cumulative number of sites and the estimated release mass over time. This can be accomplished using such models as Visual Modflow.

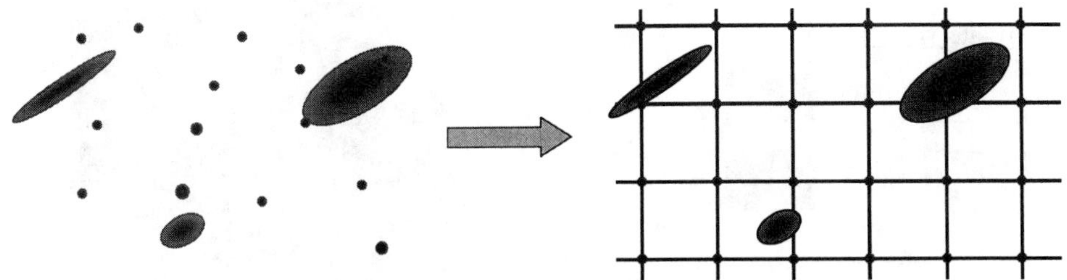

FIGURE 5. Plume probability domains (large objects) and drinking water wells. The drinking water wells are represented as randomly located information, but may be uniformly rearranged on a regular grid pattern for computational efforts.

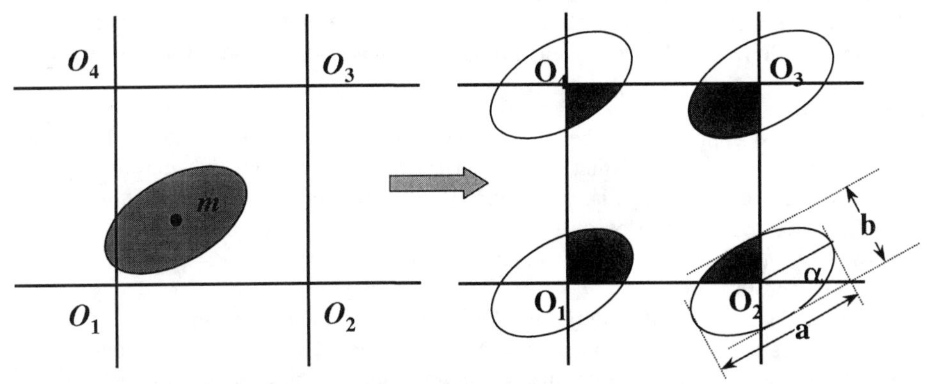

FIGURE 6. Plume probability domains rearranged in order to develop a geometric representation of the probability of impacting a drinking water well.

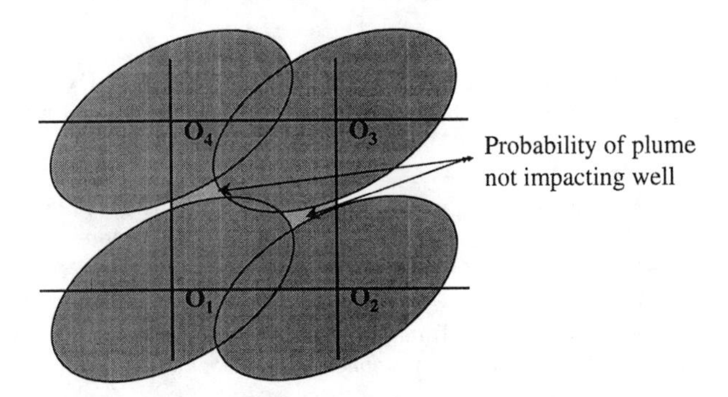

FIGURE 7. Overlapping plume probability domains where the non-impacted area becomes the probability of not impacting a drinking water well.

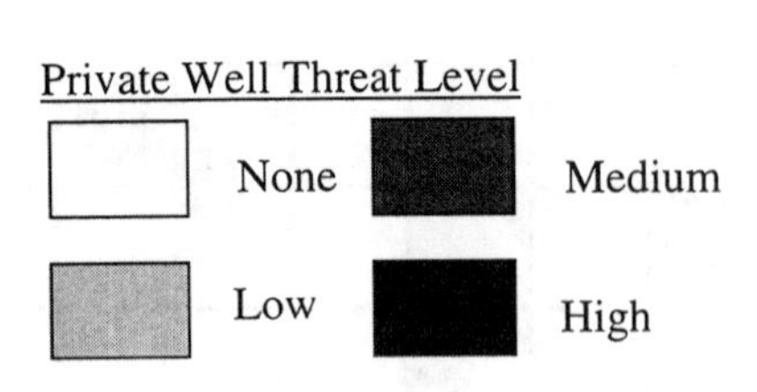

FIGURE 7. Relative areas of probability of impacting a private drinking water well around Napa City, CA. Darker areas indicate higher probabilities of actual impact.

CONCLUSION

The above approach can be used to identify areas associated with high risk to receptors. It shows that areas that have been assumed to be high risk may in fact be minimal, whereas areas where risk may initially seem low risk may in fact be high. This approach allows the focus to be on high risk areas, so that regulatory resources can be optimized. Population growth may be incorporated into the model to predict future impacts to receptors.

Further work includes defining the impact to regional aquifer and using a database of all sites to estimate mass loading in aquifer. This can only be done at the regional level. For this end we can use database management tools such as ARCView and Visual Modflow for estimation of impacts.

REFERENCES

Dooher, B.P., 1998. "Making Risk-Based Management Decisions at Fuel Hydrocarbon Impacted Sites Under Sparse Data Conditions." Ph.D. Dissertation, University of California, Los Angeles, CA.

Dooher, B.P., W.W. McNab Jr., J. Ziagos, and W.E. Kastenberg. 1998. "Evolution of Groundwater Hydrocarbon Plumes: A Comparison of Probabilistic Modeling With Statistical Analyses of Field Data." Accepted by *Journal of Contaminant Hydrology,* December, 1997.

Happel, A.M., E.H. Beckenbach, H.M. Temko, and R.R. Rempel. 1998. "MTBE Impacts to California Ground Water." Submitted to *Environmental Science and Technology.*

Mennear, J.H., 1997. "Carcinogenicity Studies on MTBE: Critical Review and Interpretation." *Risk Analysis.* 17(6): 673-681.

VOC REMEDIAL OBJECTIVES BASED ON A NEW METHOD FOR DESIGNATING GROUNDWATER BENEFICIAL USES

Gregory W. Bartow, Linda L. Spencer, David F. Leland and Richard Hiett (California Regional Water Quality Control Board, San Francisco Bay Region, Oakland, CA)
Diane K. Mims (Versar Inc., Alameda, CA)

Abstract: The beneficial use designation of groundwater ultimately affects a broad range of regulatory decisions. In California, by policy, nearly all groundwater is designated as a potential drinking water source. Experience indicates that this blanket designation limits flexibility to set clean-up levels based on realistic current and future groundwater uses. We developed a new method to better define the potential for future use of groundwater basins in the San Francisco Bay Area. The new method is consistent with regulatory decisions made at two local VOC contaminated sites. At a former Department of Defense Bay-side site, with VOC contaminated groundwater in artificial fill and fractured bedrock near the shoreline, the remedial groundwater objectives protect San Francisco Bay aquatic receptors from exposure to contaminated groundwater and human receptors from exposure to VOC vapors. In contrast, at an industrial site with high levels of VOCs located in a productive groundwater basin the remedial objectives include containment and cleanup of the contaminated groundwater. The two case examples are exceptions to the typical interpretation of existing State policy. The new method provides a viable alternative to the existing framework for establishing consistent, resource-protective beneficial uses.

INTRODUCTION

California Water Law requires responsible parties to remediate polluted groundwater to cleanup standards that protect the designated beneficial uses. The State agency that designates beneficial uses is the Regional Water Quality Control Board. There are nine Regional Water Quality Control Boards in California. This paper is based on experiences within the jurisdiction of the San Francisco Bay Regional Board (Regional Board).

The Regional Board has designated nearly all groundwater with the following beneficial uses: municipal/domestic supply, agricultural supply, and industrial process and supply. The municipal/domestic supply designation is what "drives" remediation efforts to meet drinking water standards. California groundwater policy defines all groundwaters as a "potential drinking water source", unless expressly exempted. These exemptions include a yield less than 200 gallons-per-day or a total dissolved solids concentration that exceeds 3,000 mg/L.

Toward the end of a fifteen-year period of uniformly requiring VOC remediation based on drinking water standards, the Regional Board recognized a need for more flexibility in setting and applying standards. This need was based on at least three "lessons learned". First, we grappled with the limitations of pump-and-treat technology to achieve the required drinking water standards (Kent, et.al., 1990). Our experience was echoed by the National Research Council (1994), and others who documented the limitations of existing groundwater remedial technologies. Second, we found that although groundwater had a designated beneficial use by State policy, in reality the probability of actual use of some groundwater was extremely low. For example, artificial fill deposits bordering San Francisco Bay are not viable sources for future drinking water supplies. Third, risk-based corrective action and ecological assessments emerged as powerful tools to evaluate sites and establish appropriate cleanup standards based on a more inclusive assessment of total risk. In addition, U.S. EPA Brownfield's projects were established in blighted communities; project proponents needed innovative remedial approaches to proceed with redevelopment plans.

MATERIALS AND METHODS

Based on these "lesson learned", we established a pilot project to evaluate alternatives for designating beneficial uses. Any alternative framework must continue to protect groundwater with existing or future probable beneficial use while embracing new tools such as risk based corrective action. We chose the groundwater basins in San Francisco and Northern San Mateo County peninsula for the pilot project (Figure 1). This peninsula is similar to other urbanized areas in the Region in terms of pollution types and sources. Groundwater is a primary drinking water source for some San Mateo County cities. In contrast, San Francisco currently relies on imported surface water. However, San Francisco is evaluating the future use of groundwater to augment their surface water supplies and has completed a draft groundwater master plan (S.F.Water Department, 1996; USGS, 1993).

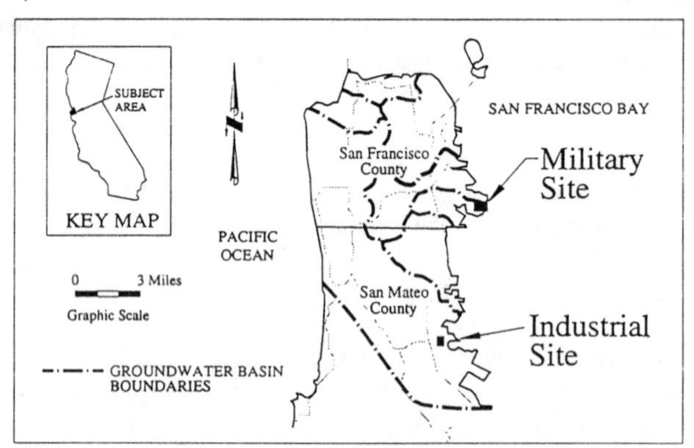

FIGURE 1. Groundwater Basins Within the San Francisco Peninsula

We developed and evaluated three alternatives to the existing framework. Our preferred alternative at this time is the Hydrogeologic Framework. It is a method of subdividing groundwater basins and their uses based on geologic materials, hydrogeologic properties, water use, and land use. The method expands on a hydrogeologic classification scheme originally outlined by Farrar and Bertoldi (1988). Beneficial use designations are assigned to the following subdivisions: bedrock, unconsolidated deposits, surface water ecological protection zones, groundwater recharge zones, bay mud, and bay-front artificial fill. Designations were made within each subdivision based on general mineral quality, recharge rates, potential seawater intrusion, potential land subsidence, extraction costs, land use, master plans for water use, and aquifer vulnerability to pollution. Our findings are summarized in a draft staff report (Regional Board, 1996) that was distributed for public review. The formal approval process for new beneficial use designations is planned for 1998. This process includes public workshops and hearings, both at a local and state level. The State must also certify an environmental impact analysis.

RESULTS AND DISCUSSION

To test the application of this approach, we evaluated two groundwater pollution sites in the pilot study area where groundwater remedial objectives have been recently established. The two sites provide examples of how an application of the Hydrogeologic Framework can result in cleanup objectives that more realistically mirror probable beneficial uses.

San Francisco International Airport. The first Site is an industrial facility located at the San Francisco International Airport (SFIA) located within the Westside Groundwater Basin. SFIA is approximately 4.5 square miles. The Westside Basin is the source of drinking water for several municipal water providers within the area including the City of San Bruno. A municipal water well cluster is located within 1/2 mile of the Site. In addition to the drinking water supply, there are ecological beneficial uses that need to be protected, since the Site is immediately adjacent to wetlands and San Francisco Bay (Figure 2).

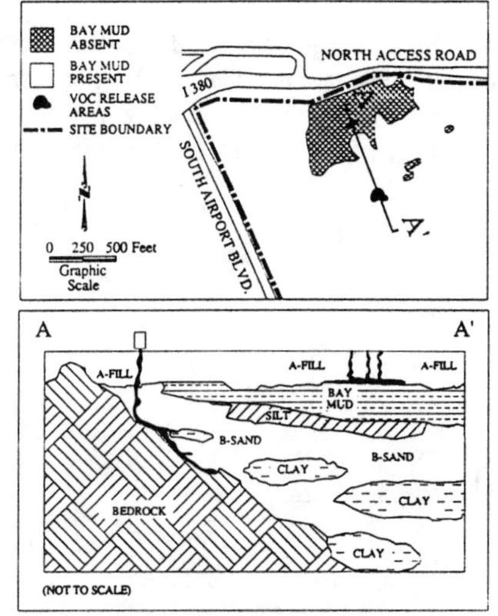

FIGURE 2. Map and Simplified Cross Section of VOC Site at San Francisco International Airport.

To establish protective cleanup objectives for the Site, a risk-based approach defined the probable risks to groundwater and surface waters. The Site is characterized by two distinct units, the upper fill (A-fill) and the lower sand zone (B-sand). The A-fill zone consists primarily of silts and clays that were used to reclaim baylands. The B-sand zone consists of multiple sand units, collectively known as the Colma and Merced Formations. The A-fill zone and the B-sand zone are separated by a low-permeability silty-clay layer referred to as the Bay Mud. An airport-wide study was conducted to verify that the Bay Mud was contiguous and to identify areas where it may absent or thin.

Based upon the hydrogeology, the Site was split into two units, the A-fill and B-sand for purposes of defining beneficial uses. The A-fill zone has low groundwater yields, elevated total dissolved solids, and if pumped could induce seawater intrusion; therefore, it was not considered a probable municipal/domestic beneficial use. The upper A-fill zone is, however, hydraulically connected to San Francisco Bay, therefore the impacts to beneficial uses of the Bay were considered in the establishment of ecological-based cleanup levels for the A-fill zone. In addition, exposure to vapors was considered in establishing human health-based criteria. Criteria were also established based on potential A-fill seepage to B-sands in areas where Bay Muds were thin or absent. The B-sand zone has a beneficial use of domestic/municipal supply as it supplies nearby wells with potable water; therefore, B-sand objectives were based on drinking water standards.

The hydrogeologic framework of the Site lead to different levels of protection based on the site-specific geology and the existing and probable use of groundwater underlying the Site. By applying different levels, it focuses the cleanup on the areas of greatest risk to water quality. It allows for the most cost-effective use of the available resources to protect beneficial uses, both in the present and the future.

Hunters Point Shipyard. Hunters Point Shipyard (HPS) is located in southeastern San Francisco, California on a peninsula bounded on the north, east, and south by San Francisco Bay within a northwest-southeast trending region of Franciscan Complex bedrock (Figure 3). HPS and the peninsula straddle the eastern boundary of the Islais and South groundwater basins (USGS, 1993).

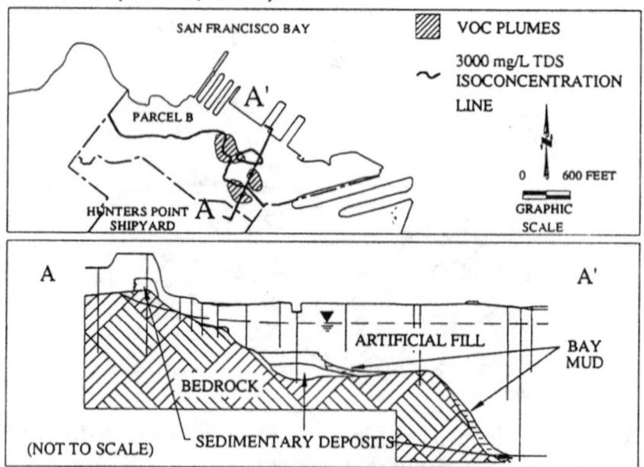

FIGURE 3. Map And Simplified Cross Section of VOC Site at Hunters Point Shipyard, San Francisco, California.

The entire HPS covers 936 acres: 493 on land and 443 under water. To facilitate the Superfund cleanup process and base transfer to civilian uses, the shipyard was divided into Parcels A through F. In Parcel B, located in the lowlands portion of HPS, ground surface elevations range from 0 to 18 feet above mean sea level. The A-aquifer in Parcel B consists primarily of bedrock-derived artificial fill, ranging from 0 to 90 feet thick, is unconfined, and discharges to San Francisco Bay. While the A-aquifer is generally underlain by Bay Mud deposits, in the southern portion of the parcel it is in hydraulic communication with Franciscan Complex bedrock or undifferentiated sedimentary deposits. There are no current or historic water supply wells in or adjacent to Parcel B. In Parcel B, chlorinated VOCs have been identified in both the artificial fill and, to a lesser extent, weathered bedrock.

A Superfund Record of Decision for Parcel B (U.S. Navy, 1997) incorporates discussion of an understanding between the U.S. Navy and the Regional Board that the A-aquifer and bedrock water-bearing zones have not been and are not likely to be used for drinking water, even though State policy defines these areas as having the potential beneficial use of municipal and domestic water supply. As a result, the Parcel B risk assessment of the A-aquifer groundwater focused on potential risks to human health from inhalation of vapors and to Bay aquatic receptors from groundwater discharges. The understanding was developed through consideration of the definition of potential drinking water source, the nature and thickness of the aquifer materials, and the potential for saltwater intrusion.

In nearshore portions of Parcel B, A-aquifer groundwater does not constitute a potential source of drinking water as defined in California groundwater policy because TDS exceeds 3,000 mg/L. Inland, TDS in some areas exceeds 3,000 mg/L, apparently as a consequence of bay water penetration along portions of the storm drain and sanitary sewer systems.

Potential yields from wells completed in the thin sections of A-aquifer fill and weathered bedrock are estimated to be limited. In areas of Parcel B where TDS is less than 3,000 mg/L, the A-aquifer thickness ranges from 0 to 60 feet, with most areas less than 30 feet thick. Saturated thickness in those areas ranges from 0 to 50 feet. Hydraulic conductivities in the A-aquifer are highly variable, with measured values ranging from 1×10^{-7} to 1×10^{-1} cm/sec. The potential drinking water source areas of the A-aquifer are limited by bedrock outcrops to the south and bay water intrusion to the north, and vary in width from 200 to 700 feet.

In addition, there is significant potential for further saltwater intrusion if long-term pumping were to occur. Tidal fluctuations in the A-aquifer have been observed up to 300 feet from the bay margin. TDS exceeding 3,000 mg/L occurs over 800 feet from the bay, indicating saltwater has already intruded into the A-aquifer at this parcel. The areas with the greatest saturated thickness and TDS less than 3,000 mg/L are located nearest the bay adjacent to waters with TDS greater than 3,000 mg/L. Thus, artificial fill with the greatest, though limited, potential for water supply development is also the most vulnerable to degradation from saltwater intrusion.

In summary, most A-aquifer material in HPS Parcel B fails to meet criteria for probable drinking water sources. In the remaining areas, the thin section and limited areal extent of water-producing materials, and existing or potential saltwater intrusion make the drinking water beneficial use of A-aquifer and weathered bedrock groundwater in Parcel B unlikely. Thus remedial objectives for A-aquifer groundwater in Parcel B were based upon inhalation exposure to human receptors and on aquatic toxicity values to protect the beneficial uses of the surface water within the San Francisco Bay.

CONCLUSION

Both of the above two examples demonstrate how cleanup objectives can be established using a new method called the Hydrogeologic Framework approach. Historically, the Regional Board has set VOC cleanup objectives at drinking water standards regardless of the whether the groundwater would ever realistically be a source of drinking water. However, in the last several years we have gained more experience in evaluating probable groundwater beneficial uses. The Regional Board has set cleanup objectives based on protection of the specific ecological and/or human health receptors on a case-by-case basis. However, the State policy framework remains unchanged. The case studies presented herein contain exceptions to the typical interpretation of existing State policy. The Hydrogeologic Framework provides a viable alternative to the existing framework for establishing consistent, resource-protective beneficial uses.

Author's Note: The views expressed are those of the authors and are not necessarily the official policy and/or positions of Cal/EPA, the San Francisco Bay Regional Water Quality Control Board, or Versar Inc.

Acknowledgments: The authors thank the San Francisco Water and Health Departments for their assistance. Steve Morse of the San Francisco Bay Regional Water Quality Control Board provided valuable editorial comments. Matthew Leedham of Versar Inc. provided valuable graphics support.

REFERENCES

California Regional Water Quality Control Board, San Francisco Bay Region, 1995, Order No. 95-136, Site Cleanup Requirements for the San Francisco International Airport.

California Regional Water Quality Control Board, San Francisco Bay Region, 1996, San Francisco and Northern San Mateo County Pilot Beneficial Use Designation Project, Draft Staff Report.

Farrar, C.D., and Bertoldi, G.L., 1988, Region 4, Central Valley and Pacific Coast Ranges, *in* Black, W., Rosenshein, J.S., and Saeber, P.R., eds., Hydrogeology: Boulder, CO, Geological Society of America, The Geology of North America, v. O-2.

Kent, M.D., Bartow, G.W., and Morse, S.I., 1990, Regulatory Outlook For Achieving Cleanup Goals For Groundwater Using Pump-and-Treat, in: Ground Water Contamination in the Santa Clara Valley, Seminar Proceedings: San Francisco Section of the American Society of Civil Engineers, San Jose, California.

National Research Council, 1994. Alternatives for Ground Water Cleanup. National Academy Press. Washington, D.C.

Phillips, S.P., Hamlin, S.N., and Yates, E.B., 1993, Geohydrology, water quality, and estimation of groundwater recharge in San Francisco, CA 1987-1992, U.S. Geological Survey, Water-Resources Report 92-4019, 69 p.

San Francisco Water Department, 1996, Draft Groundwater Master Plan.

U.S. Navy, 1997, Superfund Record of Decisions for Hunters Point Shipyard, San Francisco, CA.

COST AVOIDANCE ASSOCIATED WITH ALTERNATE CLEANUP LEVELS FOR TRICHLOROETHYLENE

Elizabeth A. Maull (Brooks AFB, TX)
Joseph A. Atchue, III (Tetra Tech EM, Inc., Vienna, VA)
Daniel Buffalo (Tetra Tech EM, Inc., Helena, MT)
Katherina J. Chaloupka (Tetra Tech EM, Inc., San Francisco, CA)

ABSTRACT: Reevaluation of the health risk assessment of TCE exposure may provide sufficient evidence for US EPA program offices (such as the Office of Drinking Water) to review and revise their current, policy based, drinking water standard for TCE to a scientifically based standard. An initial analytical basis for discussing the cost implications of using a TCE cleanup level in groundwater that is less stringent than the current level, but consistent with safe drinking water levels in the published literature, is provided. Information derived from the groundwater modeling was used to generate remedial cost estimates using the Remedial Action Cost Engineering and Requirements System. An estimated mean cost for all outcomes also was developed. A comparison of the different mean costs for three cleanup levels is made. The paper then discusses the potential effects of these changes on the overall cost to the Department of Defense (DOD) for remediation of TCE contaminated groundwater.

INTRODUCTION

EPA regulates groundwater used for drinking purposes by imposing maximum contaminant levels (MCL) for individual constituents. In 1985, EPA published its intent to include an MCL for TCE in the National Primary Drinking Water Standards and finalized in 1987. TCE was listed in the rule based on its widespread use in industry; leachability into groundwater; common occurrence as a soil, groundwater, and surface water contaminant; and carcinogenicity in animal bioassays (ATSDR 1989, EPA 1985).

The MCL published for TCE (5 ug/L, EPA, 1987) was based on both animal studies and economic and technical feasibility for detection in drinking water. Recent publications using physiologically-based pharmacokinetic models support a safe drinking water standard one to two orders of magnitude higher that the current MCL for TCE (Bogen and Gold, 1997; Clewell et al., 1995).

MATERIALS AND METHODS

Development of Parameters for the Groundwater Model. Modeling parameters for three different sizes of contaminated aquifers (5, 25, and 50 acres in area) were developed to provide a range of possible cases to examine for the cost estimation. A conceptual model was formulated to organize all assumptions so that the groundwater flow system could be analyzed more readily. The conceptual model was simplified as much as possible while retaining, sufficient complexity to simulate groundwater system behavior for the intended purposes of modeling (Anderson and Woessner 1992). Plume characteristics are provided in Table 1.

TABLE 1. TCE Plume Characteristics

Plume Size (acres)	Length of Line Source (feet)	Maximum Width (feet)	Maximum Length (feet)	Plume Length to Width Ratio (unitless)	Approximate Time of Source Activity[a] (years)
5	300	300	800	2.70	5.5
25	400	800	1,800	2.25	12.3
50	400	1,000	2,700	2.70	18.5

[a]Source activity describes the amount of time the source actively contributed contaminant mass to the aquifer. Plume concentration is assumed to be uniform and equal to 200 ug/L.

The Well Head Protection Area (WHPA) model (Blandford and Huyakorn 1991) was selected to simulate capture zones associated with pump-and-treat remedial systems. The analytical function-driven version of the Random-Walk (Prickett *et al.*, 1981) model was selected to simulate the migration of contaminants in groundwater and the progress of remediation.

Cost Estimation Model and Parameters. The Remedial Action Cost Engineering and Requirements (RACER, Version 3.2) System was used to develop the cost estimates discussed in this report. RACER is an Air Force-supported cost estimating system used by the Military Services and the Office of the Deputy Under Secretary of Defense for Environmental Security (ODUSD[ES]) to develop installation- and program-level cost estimates for completing cleanup.

Carbon adsorption (using granular activated carbon [GAC]), one of EPA's technologies of choice for remediating groundwater contaminated with halogenated hydrocarbons, was selected as the technology of choice for removal of TCE. This technology is also commonly used at military installations throughout the U.S.

The variables related to the aquifer that affect the cost of each scenario are the size of the aquifer, the specific cleanup level, and the adsorptivity of the soil matrix. These variables in turn affect the aggregate pumping rate, which is a function of the total number of wells and their individual pumping rate. These variables control the length of time that an aquifer will be pumped. All other variables were kept the same for each cost run.

Using this approach, a separate RACER model run was set up for each scenario. O&M time, the number of wells, and aggregate pumping rates for each scenario provided the major inputs used by the model to generate costs for well construction. Overall costs included O&M of the well system; O&M of the GAC system; sampling and analysis; and discharge to the publicly operated treatment works (POTW).

In order to develop an estimate of the total cost to the DOD to remediate TCE in groundwater, data from the Restoration Management Information System (RMIS) was used to determine the total number of sites where TCE was the main driver in remediation or the only contaminant (353 sites). The cost for each scenario then was multiplied by the number of sites identified. These values are used in discussions about the magnitude of total costs.

RESULTS AND DISCUSSION

While the aquifer characteristics chosen for this model may not be completely representative of the "real world", we believe that the size of the aquifers and the range of carbon adsorption (retardation) coefficients are reasonable estimations for aquifers in DOD's inventory. However, it is unlikely that either the assumed starting concentration of 200 ug/L or source removal/containment reflects the real nature of DOD TCE plumes. If the average concentration is, in fact, higher than the 200 ug/L level used for this project and the sources have not been removed, then the average length of time (and therefore the average cost) will be higher.

Analysis of Cost Components. In any remediation program, there are three major components of cost: capital costs (installation of various pieces of cleanup equipment, placement of wells, and excavation of soil); O&M costs (long-term management of a site, costs for operation of pump-and-treat systems, continued sampling of groundwater, maintenance of caps or fences); and the costs of allowances, profit, and overhead expenses. Since the same percentage was used in all cases for allowances, profit, and overheads they will not be discussed as a separate factor.

Capital costs did not play a significant role in the overall costs of remediation. Capital costs averaged 8.1 percent of the total cost, with a range of from 1.4 to 19.2 percent of total cost. Capital costs were a much larger fraction of total cost for the small (five acre) aquifer because the O&M periods for the cases of this aquifer are far shorter, on average, than for either of the other two aquifer sizes.

Operations and Maintenance costs are by far the most significant component of the total cost. O&M costs increase as the amount of time a pump-and-treat system operates increases. The average contribution of O&M cost represented 59.7 percent of the total cost with ranges from 47.3 to 68.2 percent. The difference between the sum of the capital and O&M averages set forth above represents the costs of allowances, profit, and overhead, and expenses.

Results of the RACER Cost Estimation Runs. Table 2 presents the cost estimate data generated for each case. The range of costs for the conditions modeled was $410,400 to $16,437,445, representing the smallest aquifer with the lowest soil adsorptivity, and the largest aquifer case with the highest adsorptivity, respectively. For each aquifer and cleanup level the step from the medium level of adsorptivity to the high level always resulted in the largest increase in cost. This suggests that the higher the percent organic matter in the aquifer (which, in part, controls adsorptivity) the more expensive the cleanup will be. Using mean costs (the bottom line in Table 2) the cost avoidance between the 5 ug/L and the 50 ug/L or 100 ug/L endpoints is $822,074 (19 percent) or $1,251,542 (28 percent), respectively, per site. Clearly, the differences in the cost estimates are significant in terms of dollar value. Use of 100 ug/L as the cleanup goal results in a nearly 30 percent lower mean cleanup cost. The cost estimate presented here is likely to be lower than what will actually be seen due to the assumptions used for the model.

TABLE 2. Summary Of Total Cost For Each Scenario

Size / Retardation Rate	5 ug/L Endpoint Cost	50 ug/L EndpointCost (% Change from 5 ug/L)	100 ug/L EndpointCost (% Change from 5ug/L)
5 Acre / Low	$590,550	$467,708 (21)	$410,400 (31)
5 Acre / Medium	$754,794	$630,193 (17)	$570,302 (24)
5 Acre / High	$2,879,332	$ 2,199,086 (24)	$1,885,003 (35)
25 Acre / Low	$1,181,549	$1,108,503 (6)	$1,033,742 (13)
25 Acre / Medium	$2,111,395	**$1,690,588 (20)**	$1,469,813 (30)
25 Acre / High	$8,934,700	$6,530,203 (27)	$6,183,340 (31)
50 Acre / Low	$2,664,486	$2,053,168 (23)	$1,606,994 (40)
50 Acre / Medium	$3,987,883	$3,070,909 (23)	$2,348,366 (41)
50 Acre / High	$16,437,445	$14,392,216 (12)	$12,769,401 (22)
Mean Cost	$4,393,471	$3,571,397 (19)	$3,141,929 (28)

Using the assumptions of 25 acre plumes, with medium adsorptivity (Table I, bold type), it would cost (in 1996 dollars) $1.551 billion if 353 sites were to be cleaned up to the 5 ug/L endpoint. If the 100 ug/L endpoint were used, the cost would be $1.110 billion. The potential cost savings that could be achieved by the use of the less stringent cleanup goal (100 ug/L) are on the order of $441 million.

Another way to evaluate the cost is to consider the amount of time required to remove the total mass of contaminant over time. Figure 1 is a schematic representation of the efficiency of removal of TCE for each of the aquifer sizes assuming moderate adsorptivity. The y-axis is the percentage of mass removed and the x-axis is time.

The range of time spent in reaching cleanup for each of the plume sizes was as little as 7.5 years (5-acre plume, 100 ug/L cleanup goal), to 37 years (50-

acre plume, 5 ug/L cleanup goal). For each of the plume sizes, the range of time between reaching the 5 ug/L and 100 ug/L cleanup endpoints varied by 4.5, 9, and 14.5 years for the 5-acre, 25-acre, and 50 acre plumes, respectively.

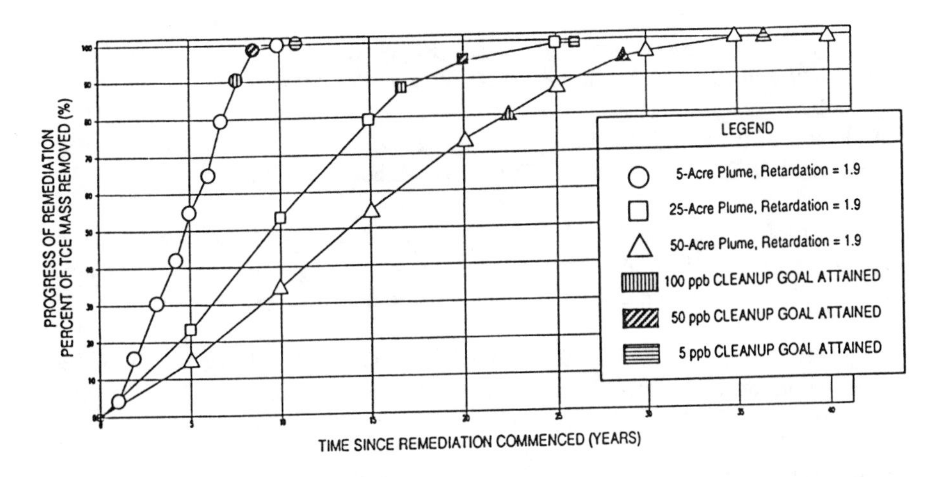

FIGURE 1. Schematic representation of TCE removal for 5-acre, 25-acre, and 50-acre groundwater plumes.

Conclusions. The savings of $441 million represents a conservative estimate of the minimum savings that could be expected with a change in the MCL from 5 ug/L to 100 ug/L, based on the assumptions used for the model. These assumptions include source removal or containment to limit continued contribution to the plume; a homogeneous plume concentration rather than a more realistic gradient of concentrations; limited plume sizes and retardation rates; constant parameters for aquifer thickness and depth; and a single, relatively low, plume concentration of 200 ug/L. While the model used here was developed to provide as much realism as possible, every assumption made for its operation has the potential to make it less descriptive than the "real world" and is likely to cause our cost estimates to be low.

Next Steps: Subsequent papers should attempt to develop model parameters that are more realistic by using a *Monte Carlo* approach to the cost estimation outputs and providing a an more realistic estimate of savings. (The *Monte Carlo* approach would allow a larger universe to be used and provide a more rigorous statistical approach.) To evaluate more thoroughly the potential benefits of the various levels of cleanup proposed for the site against the respective costs for each level, it might be helpful to conduct a cost-benefit analysis of each option.

Comparison of costs expended to achieve each level of cleanup, with the benefits (or costs avoided) of achieving that level, can produce quantified data on net costs for each cleanup option at the site and assist in determining the economic feasibility of those options.

REFERENCES

Agency for Toxic Substances and Disease Registry. 1989. *Toxicological Profile for Trichlorethylene*. ATSDR/TP-88/24. Washington D.C..

Anderson M.P., and W.W. Woessner. 1992. *Applied Groundwater Modeling, Simulation of Flow and Advective Transport.* Academic Press.

Blandford, N.T., and P.S. Huyakorn. 1991. *A Modular Semi-analytical Model for the Delineation of Wellhead Protection Area.* WHPA Version 2.0. International Ground Water Modeling Center. Indianapolis, Indiana.

Bogen, K. T. and L. S. Gold. 1997. "Trichloroethylene Cancer Risks: Simplified Calculation of PBPK-Based MCLs for Cytotoxic Endpoints." *Reg. Toxicol. Pharmacol. 25*: 26-42.

Clewell, H. J., P. R. Gentry, J. M. Gearhart, B. C. Allen, and M. E. Andersen. 1995. "Considering Pharmacokinetic and Mechanistic Information In Cancer Risk Assessments for Environmental Contaminants: Examples with Vinyl Chloride and Trichloroethylene." *Chemosphere 31*: 2561-2578.

EPA. 1985. Health Assessment Document for Trichloroethylene. Office of Health and Environmental Assessment, Washington, DC. EPA/600/8-82/006F.

EPA. 1987. National Primary Drinking Water Standards: Phase I Rule. *Fed. Reg. 52 (130):* 25090-25111.

EPA. 1996. Integrated Risk Information System (IRIS) Chemical Files. Office of Health and Environmental Assessment, Office of Research and Development. Washington, D.C.

Prickett, T.A., T.G. Naymik, and C.G. Lonnquist. 1981. "A Random-Walk Solute Transport Model for Selected Groundwater Quality Evaluations." *Illinois State Water Survey Bulletin 65*. Champaign, Illinois.

PRC. 1997. Groundwater Modeling to Support an Analysis of Remediation Costs for the Removal of Trichloroethylene From Groundwater. Unpublished Manuscript. 21 pages.

POTENTIAL RISKS ASSOCIATED WITH VAPOR MIGRATION FROM GROUNDWATER INTO BUILDINGS

Kimberley D. Cizerle (ENVIRON Corporation, Princeton, New Jersey)
Stephen Song (ENVIRON Corporation, Princeton, New Jersey)
Stephen T. Washburn (ENVIRON Corporation, Princeton, New Jersey)

ABSTRACT: Under certain circumstances, the migration of vapor phase chemicals from groundwater into buildings can affect indoor air quality. Factors which influence the extent of vapor migration from groundwater include the properties of the chemicals, the soil type, and building characteristics. In some cases, particularly when an aquifer is contaminated with volatile chemicals and is not used as a drinking water source, the potential human health risks associated with vapor migration may be important in determining risk-based groundwater cleanup criteria. This paper evaluates available options for assessing the human health risks associated with the indoor air pathway, and the advantages and disadvantages of the various approaches. A case study involving trichloroethene (TCE) is provided to highlight the various factors that can affect the significance of the pathway.

INTRODUCTION

Indoor air quality issues have gained the attention of regulatory agencies and the public in recent years for many reasons, including: heightened concern regarding potential health effects associated with exposures to relatively low concentrations of chemicals (both natural, such as radon, and primarily anthropogenic, such as formaldehyde); reports of discomfort or illness among people living and working in energy-efficient buildings constructed since the energy crisis of the 1970s; and recognition that people in the United States spend a significant portion of their time indoors (USEPA, 1991). Possible sources of indoor air contamination include chemicals within the home (e.g., building materials, wall/floor coverings, cleaning chemicals, and pesticides), as well as natural and anthropogenic chemicals from outside the home (i.e., in ambient air, soil, or groundwater). This paper addresses potential risks associated with vapor migration from groundwater into buildings. This pathway can, under certain circumstances, drive risk-based groundwater remediation decisions at a site. As an illustration, groundwater concentrations of trichloroethene (TCE) corresponding to risk-based indoor air targets are presented in a case study in this paper.

The fate and transport process for the indoor air pathway involves the following major steps:

- Chemicals in groundwater volatilize into the air-filled pore spaces of the soil above the groundwater table;
- The chemical vapors in the soil pore spaces migrate to outside the building's basement or foundation;

- Chemical vapors then migrate into the building interior, through cracks and openings in the basement or foundation floor and/or walls.

Other pathways, such as vapor migration into buildings through sewer lines, have also gained attention recently, but are difficult to address generically and will not be discussed in this paper.

For the fate and transport process described above, assessment of potential risks associated with vapor migration from groundwater into buildings typically involves estimating risks associated with inhalation of chemicals in indoor air. Estimates of chemical concentrations in indoor air are usually obtained using one of the following approaches:

- Fate and transport modeling to predict chemical concentrations in indoor air resulting from the volatilization of chemicals from groundwater;
- Collection of soil gas samples outside the building basement or foundation, followed by modeling to predict chemical concentrations in indoor air; or
- Direct measurement of indoor air concentrations inside buildings.

The advantages and disadvantages of these approaches are discussed in this paper, along with an examination of the factors which are most likely to affect the significance of the indoor air pathway.

APPROACHES FOR EVALUATING RISKS

Modeling Based on Groundwater Measurements. Several modeling approaches have been developed to theoretically assess the potential for vapor intrusion into buildings from groundwater sources. The primary advantage of modeling is that it allows for a "screening" of groundwater concentrations, assuming certain values for the many model input parameters. A few of these parameters, which will be discussed further in the next section, include: the chemical-specific Henry's Law Constant (H') and soil organic carbon/water partition coefficient (K_{oc}), soil vapor permeability (k_v), ratio of the crack area to the total cross-sectional area of the below-grade portion of the building (η), indoor/outdoor pressure differential (ΔP), building air exchange rate (ACH), and building volume (V).

Most vapor intrusion models consist of the following elements:

- Estimation of chemical concentrations in the air-filled pore space in soil directly beneath the basement or foundation ("$C_{soil\ gas}$");
- Prediction of indoor air concentrations ("C_{indoor}") corresponding to the estimated $C_{soil\ gas}$ concentration.

One of the more widely used screening models is the Johnson and Ettinger (1991) model, which has been evaluated by USEPA and several state agencies (e.g., Massachusetts, Michigan, Pennsylvania), and was cited in the nonmandatory technical appendices to the ASTM Risk-Based Correction Action (RBCA) standard for petroleum release sites (ASTM, 1995).

Disadvantages of the modeling approach include limited model validation, and the use of conservative assumptions to offset uncertainties and data gaps. Overly conservative application of the models can lead to significant overprediction of indoor air concentrations. The primary ways in which the Johnson and Ettinger model is conservative are:

- The model's $C_{soil\ gas}$ calculation does not account for the chemical degradation, depletion, and dilution that occurs between the groundwater source and the basement.
- The model assumes that all vapors released from the groundwater source directly beneath the building will migrate into the basement. As Johnson and Ettinger acknowledged in their paper, some vapors may bypass the building and migrate through the soil into ambient air.

Another common disadvantage of the modeling approach is the need to estimate many model parameters which are highly site-specific (e.g., k_v, η, and the other building-specific parameters noted above). Site-specific data for these parameters are often unavailable and difficult to obtain, and use of "generic" conservative defaults may lead to significant overpredictions. The Johnson and Ettinger vapor intrusion model is extremely sensitive to these site-specific input parameters, with results varying by an order of magnitude or more depending on the specific values selected. Therefore, careful evaluation and selection of site-specific input parameters, rather than reliance on generic defaults, can be critical.

Modeling Based on Soil Gas Measurements. One way to reduce the uncertainty associated with a modeling approach, without moving directly to the collection of indoor air samples, involves: (1) collection and analysis of soil gas samples near the basement or foundation to measure chemical concentrations in the soil pore space (i.e., $C_{soil\ gas}$), combined with (2) use of the Johnson and Ettinger model to estimate C_{indoor} concentrations based on $C_{soil\ gas}$ measurements.

Collection of soil gas samples to measure $C_{soil\ gas}$ chemical concentrations allows the risk assessor to account for the degradation, depletion, and dilution otherwise unaccounted for in a screening model. Collection of soil gas samples can also be combined with soil vapor permeability tests, so that a site-specific k_v value can be used in the C_{indoor} portion of the remaining model calculations. Thus, based on soil gas sampling, it is possible to evaluate vapor migration from groundwater without actually measuring indoor air concentrations. However, since the C_{indoor} portion of the modeling would still be required, many site-specific inputs related to the building characteristics (η, ΔP, ACH, V) must still be estimated. Thus, even with this modified modeling approach, the selection of input parameters (and the difficulty in selecting "generic" inputs) remains an important disadvantage. However, if used carefully, this method can serve as a useful site-specific screening tool.

Direct Measurement of Indoor Air Concentrations. If modeling or soil gas sampling indicates the potential for significant migration of vapors into a building, indoor air sampling can be considered. Any indoor air sampling program should be very carefully designed and implemented to ensure the collection of representative results, and to distinguish, to the extent feasible, between contributions from indoor sources, above-ground outdoor sources, and subsurface soil and groundwater. Many of the most common volatile chemicals are prevalent throughout outdoor and indoor environments and have many sources, including automobiles, cigarettes, and building materials (USEPA, 1991). Thus, it is critical to consider background levels before attempting to interpret the results of any indoor air modeling or sampling studies. Since site-specific assessment of background indoor air concentrations is difficult, USEPA references are available for determining background indoor air concentrations (USEPA, 1988, 1991).

FACTORS WHICH AFFECT THE SIGNIFICANCE OF THE PATHWAY

Regardless of the approach used for evaluating indoor air risks, three main factors affect the significance of the pathway: chemical properties, subsurface (soil and groundwater) characteristics, and building characteristics.

Chemical Properties. The indoor air pathway is potentially the most significant for chemicals with low K_{oc} and high H' values. Chemicals with these properties, such as TCE, and tetrachloroethene (PCE), are more likely to be present in water than to remain bound to soil particles (because of their low K_{oc} values) and have a high tendency to volatilize from water into air (because of their high H' values). Typical H'/K_{oc} ratios for the most volatile chemicals are 10^{-4} kg/L or higher (when using unitless H' values).

Soil Characteristics. The indoor air pathway is potentially significant at sites with relatively permeable soils (i.e., soils with k_v values higher than 10^{-8} cm^2). For soils with significant clay content and relatively low k_v values (i.e., lower than 10^{-9} or 10^{-10} cm^2), the indoor air pathway is less important. For reference, the default soil type USEPA assumed in its *Soil Screening Guidance* (SSG) (USEPA, 1996) is a loam soil, with a corresponding k_v value of about 10^{-9} cm^2 or lower.

Other soil characteristics that affect the significance of this pathway are:

- Soil water content and air-filled porosity: Soils with high water content minimize vapor migration because chemical transport by liquid-phase diffusion is much slower than vapor-phase diffusion or convection.
- Capillary fringe: The greater the height of the capillary fringe, the less important the indoor air pathway will be since chemicals must migrate by liquid-phase diffusion through this zone.
- Distance from the groundwater to the basement floor or building foundation: Theoretical modeling suggests that the depth to groundwater is not an important factor in determining indoor air concentrations. However, this is largely because the screening level models do not usually account for chemical degradation, depletion, or dilution. In reality, these phenomena can greatly reduce the extent of vapor migration into buildings. Thus, in most circumstances, the depth to groundwater is important in determining chemical vapor concentrations in buildings.
- Subsurface temperature: Since H' is a function of temperature, chemical volatility is lower in groundwater at lower temperatures. Studies of H'

as a function of temperature (RTI, 1987) suggest that H' values decrease by a factor of about 2 for 10°C to 15°C decreases in temperature.

Building Characteristics. Depending on the type of soil present between the groundwater source and the building basement or foundation, various building characteristics affect the importance of the indoor air pathway. In cases where the soil is relatively impermeable, η is the key building characteristic that determines the importance of the indoor air pathway. High values of η (e.g., 1%) represent significant crack area, and therefore significant area for vapor intrusion. For impermeable soils, the indoor air pathway becomes less important at more typical η values (e.g., 0.01%). For relatively permeable soils, ΔP, ACH, and V are the key building characteristics that affect the significance of the indoor air pathway. Site-specific values for ΔP and ACH depend greatly on how the building's heating, ventilation, and cooling system is operated.

CASE STUDY

The following case study for TCE demonstrates the Johnson and Ettinger model sensitivities to some of the key input parameters mentioned above. For this study, USEPA's SSG residential exposure factors were used, and the target risk level was set to 10^{-6} to calculate the various risk-based target TCE concentrations in groundwater for various model input values for η, capillary fringe, and k_v.

TABLE 1. Target TCE groundwater concentrations (mg/L) for the indoor air pathway based on various building and soil characteristics.

η=	1%		0.01%	
capillary fringe=	**5 cm**	**50 cm**	**5 cm**	**50 cm**
$k_v=10^{-7}$ cm^2	0.30	1.9	0.31	1.9
$k_v=10^{-8}$ cm^2	0.44	2.1	0.60	2.2
$k_v=10^{-9}$ cm^2	0.84	2.5	3.5	5.1
$k_v=10^{-10}$ cm^2	0.95	2.6	28	30

Other key assumptions for this case study include: TCE inhalation risk factor = 1.7×10^{-6} $(\mu g/m^3)^{-1}$; ΔP = 1 Pa; ACH = 1/hr; V = 5.4×10^8 cm^3 (length and width of 1056 cm; total height of 488 cm); the building has a basement (floor thickness of 15 cm), and one floor above grade; depth from soil surface to groundwater is 450 cm.

If groundwater were used as a drinking water source in this case study, the vapor intrusion pathway would not drive groundwater remediation since the MCL for TCE (0.005 mg/L, or 5 ppb) is more stringent than the concentrations presented in Table 1. However, depending on the TCE concentrations in groundwater, potential risks associated with the indoor air pathway could require evaluation if a building is present above the TCE plume. For example, under the most stringent set of input parameters evaluated in the case study, the target TCE concentration in groundwater would be about 0.3 mg/L, or 300 ppb.

CONCLUSIONS

Several options exist for evaluating potential risks associated with vapor migration from groundwater into buildings. Depending on how these risks are evaluated, groundwater risks from the indoor air pathway may drive groundwater remediation needs, particularly in the absence of the drinking water pathway. In some cases, this may be realistic (e.g., if chemical concentrations in indoor air are significantly higher than both background concentrations and acceptable risk-based air concentrations), but in other cases it may not (e.g., if generic values are used in place of appropriate site-specific values for some of the key model parameters). Thus, the risk assessor should use caution in selecting the method for evaluating the indoor air pathway, and in selecting the key model parameters.

REFERENCES

American Society for Testing and Materials. 1995. *Standard Guide for Risk-Based Corrective Action Applied at Petroleum Release Sites*. E 1739-95.

Johnson, P.C., Ettinger, R.A. 1991. "Heuristic Model for Predicting the Intrusion Rate of Contaminant Vapors into Buildings." *Environ. Sci. Technol.* 25(8): 1445-1452.

Research Triangle Institute. 1987. *Evaluation and Prediction of Henry's Law Constants and Aqueous Solubilities for Solvents and Hydrocarbon Fuel Components. Volume 1: Technical Discussion - Final Report.* USAF/ESL-TR-86-66. September.

United States Environmental Protection Agency (USEPA). 1988. *National Ambient Volatile Organic Compounds (VOCs) Data Base Update.* EPA 600/3-88/010(A). January.

USEPA. 1991. *Introduction to Indoor Air Quality (A Reference Manual; A Self-Paced Learning Module)*. July.

USEPA. 1996. *Soil Screening Guidance: Technical Background Document.* EPA/540/R-95/128. May.

PRELIMINARY RISK-BASED ASSESSMENT OF REFRIGERANTS IN GROUNDWATER

Christine Vilardi (STV Inc., New York, New York)
A. Stacey Gogos (EPM, Inc., Lake Success, New York)

ABSTRACT: Using available risk-based assessment methods, a modified approach was used for a preliminary assessment of the risks associated with several refrigerants detected in groundwater. This discussion focuses on two compounds in particular -- trichlorofluoromethane (a chlorofluorocarbon) and chloroform (or trichloromethane). Site evidence demonstrates that these chemicals were detected in the upper and lower portions of the unconfined glacial sand aquifer at levels exceeding groundwater standards. Despite high volatility, these chemicals have easily migrated downward in the aquifer due to a low soil sorption potential and high density. The dissolved plume is migrating downgradient to a urban residential and commercial area with several pumping supply wells (not potable). This preliminary assessment provides an estimate of risk-based levels associated with groundwater ingestion and serves as a useful remediation planning tool for the next phase of the groundwater remedial investigation.

INTRODUCTION

Objectives. The objectives of this preliminary risk-based assessment were to develop risk-based levels for several refrigerant compounds detected in groundwater using a combination of available risk-based assessment methods and preliminary risk and hazard indices for potential exposure pathways. The two compounds selected for this analysis were chloroform ($CHCl_3$) and trichlorofluoromethane (CCl_3F). Remediation of the dissolved refrigerant plume will be predicated upon such an assessment.

Environmental Setting. The land elevation is approximately 55 ft (17 m) above mean sea level with minor relief and a terminal moraine is situated north of the study area. The unconfined aquifer is composed of glacial outwash and till of Pleistocene age and is underlain by a clay formation at 160 ft (49 m) below grade. The aquifer has a depth to water of about 40 ft (12 m) below grade and groundwater discharges to a bay about three miles southwest of the study area. Water quality has deteriorated in this aquifer over the years due to overpumping. Although there are several public supply wells screened in the aquifer, they are pumped intermittently only for commercial or utility purposes.

Background. A hydrogeologic site characterization was conducted at the study area from 1996 - 1997. Refrigerants, a class of halogenated volatile organic compounds, were detected in upper and lower portions of the unconfined aquifer at levels which exceed state groundwater standards. The dissolved plume appears to be migrating southwest of the study area and additional investigation is proposed to delineate the plume in this area.

METHODOLOGY

Since additional investigative work will be conducted southwest of the study area, a preliminary risk-based assessment was needed to estimate potential risks and remedial needs associated with use of groundwater in this area. The most conservative approach was taken using potential routes of exposure and potential carcinogenic and noncarcinogenic effects of the chemicals even though groundwater is not used for human consumption from this aquifer.

The U.S. Environmental Protection Agency (U.S. EPA) risk assessment procedure (U.S. EPA, 1989) was used to develop estimates of risks and hazard quotients (HQs) for specific chemicals and hazard indices (HIs) for selected exposure pathways. Then, risk-based screening levels for these compounds were estimated using the American Society for Testing and Materials (ASTM) Risk-Based Corrective Action (RBCA) Standard E 1739-95 (ASTM, 1995) and then compared with the U.S. EPA Risk-Based Concentrations (RBCs) (U.S. EPA, 1996). The risk-based levels for the groundwater ingestion pathway were estimated for residential and commercial settings using default exposure values for the average adult, and a default risk probability value of 10^{-5} (unitless) or a default HI of one (unitless). Be it noted that ASTM has proposed another version of RBCA which has not been approved as of this writing and was therefore not used for this preliminary assessment (Waldorf, 1998).

Groundwater Data Evaluation. The refrigerant compounds detected in groundwater included chlorodifluoromethane, chloroform, dichlorodifluoromethane, dichlorofluoromethane and trichlorofluoromethane. Factors used to screen the compounds included detected concentrations, frequency of detection, toxicity data and groundwater standards. Background levels were based on results from an upgradient study area well where these compounds were not detected. As a result of this prescreening process, trichlorofluoromethane and chloroform were selected for further assessment, the mean and maximum detected concentrations for both of which exceeded state ground water standards (0.005 and 0.007 mg/L, respectively).

Fate and Transport. The chemical and physical properties used in predicting the environmental fate of the chemicals were solubility, relative mobility, soil sorption coefficient and density. Consequently, chloroform appears to be very mobile while trichlorofluoromethane is moderately mobile in groundwater. Both compounds easily leach from soil to groundwater and can persist in groundwater for a long time.

Site evidence also demonstrates that refrigerants can enter the subsurface environment despite a high volatility property. Trichlorofluoromethane in particular is a chlorofluorocarbon (CFC) which is an atmospheric pollutant contributing to the depletion of the ozone layer (federal mandate prohibited the manufacture and licensing to use certain CFCs on December 31, 1995).

Chloroform, on the other hand is quite ubiquitous in the environment and is derived mainly from man-made sources. Chloroform is used in the manufacture of other refrigerants. Chloroform is one of the marker trihalomethane compounds detected in chlorinated water; background levels for chloroform in public water supplies typically range from 0.0002 to 0.0044 mg/L (U.S. Department of Human Health Services, 1996). These compounds have also been used as solvents for degreasing.

The principles of groundwater flow, water level data from the nine well couplets and the retarding factor for each compound were used to predict the solute transport rate of 0.09 ft/d (0.03 m/d). The one-dimensional advective-dispersion equation was used to predict the downgradient edge and width of the plume.

Exposure Assessment. Potential exposure pathways were based on a potential future use scenario for residents (since groundwater from this aquifer is not officially used for public consumption) and a potential current use scenario for commercial users. Chemical intake rates were then estimated using the most conservative assumption for the following potential routes of exposure via groundwater: ingestion of tap water, inhalation of volatilized compounds from tap water and dermal adsorption (via bathing or showering). For example, the maximum detected concentrations of trichlorofluoromethane and chloroform (0.026 and 0.015 mg/L, respectively) in groundwater were selected and chronic exposure duration was assumed in order to estimate risk for the groundwater pathway assessment. Potential receptors in the downgradient path of the plume are in an densely populated urban area consisting primarily residences, several supply wells and a commercial area.

Toxicity. The reference dose (RfD) value for trichlorofluoromethane for the ingestion route were available from the Integrated Risk Information System (IRIS) (U.S. EPA, 1993) and for the inhalation route from the Health Effects Assessment Summary Table (HEAST) (U.S. EPA, 1992). An adjustment for absorption for the dermal absorption route was assumed since RfD values were not available. Trichlorofluoromethane is an intoxicant and causes asphyxiation in very high doses. It is on a very low order of toxicity in its stable state, and is highly volatile, but not flammable. If heated to decomposition, highly toxic vapors of chlorides and fluorides will be emitted.

For chloroform, the slope factors (SFs) for the inhalation and ingestion routes and the RfD for the ingestion route are derived from the IRIS database. An RfD for the inhalation route was not available. Again, values are not available for the dermal absorption route and an absorption adjustment factor was assumed. Based on the Agency for Toxic Substances and Disease Registry (ASTDR), a division of the U.S. Department of Health and Human Services (U.S. DHHS), chloroform may be reasonably anticipated to be a carcinogen (U.S. DHHS, 1993).

Estimated Risk and Hazard Indices. Risk values were estimated for the inhalation, ingestion and dermal absorption routes for the compounds. The estimated risk value for ingestion of chloroform was an order below the recommended risk value of 10^{-5}, while for the inhalation route it was an order above. The estimated HI value for trichlorofluoromethane for the inhalation route was slightly above unity and was insignificant for the ingestion route for both compounds. There is large uncertainty associated with the estimated values for both compounds for the dermal absorption route and so it was not considered further.

Estimated Risk-Based Levels. The estimated risk-based levels for the tap water ingestion route for the two compounds are summarized in Table 1 and were

TABLE 1. Estimated Risk-Based Levels for Tap Water Ingestion Route (All Units in Milligrams Per Liter, mg/L)

Compound/Exposure Route	Estimated Risk-Based Levels (Using Default Values)	U.S. EPA RBC for Tap Water Ingestion
Trichlorofluoromethane/Residential	1.10	1.30
Trichlorofluoromethane/Commercial	3.07	NA
Chloroform/Residential	0.14	0.00015
Chloroform/Commercial	0.477	NA

based on a combination of EPA and ASTM assumptions and procedures. The estimated levels for trichlorofluoromethane (a noncarcinogen) appear to represent a lower risk potential than the groundwater standard whereas the opposite is apparent for chloroform (a carcinogen).

The USEPA RBC levels were used for comparison even though the values are based on some additional assumptions. Children (in addition to adults) and inhalation of volatile compounds emanating from tap water (in addition to ingestion) are factored in the RBCs for the tap water (groundwater) consumption exposure scenario. Further, the RBCs apply only to residential use. The RBC values compare well for trichlorofluoromethane despite these differences but appear to be more stringent for chloroform (Table 1).

CONCLUSIONS

This preliminary risk-based assessment served as a tool to screen the potential risks associated with refrigerants dissolved in groundwater. Apparently, the potential risks of the inhalation route warrants further investigation. This assessment will be supplemented by the next phase of groundwater remedial investigation work which will delineate the horizontal and vertical extent of the downgradient plume. Data generated from the next phase will be used in a full-scale risk-based assessment. This will be compared with these initial risk-based predictions and any new risk-based methods which may become available in the near future.

REFERENCES

ASTM. 1995. *Standard Guide for Risk-Based Corrective Action Applied at Petroleum Release Sites.* ASTM Designation E 1739-95. American Society for Testing and Materials, West Conshohocken, PA.

U.S. Department of Health and Human Services. 1995. *Draft Toxicological Profile for Chloroform.* Prepared by Research Triangle Institute under Contract No. 205-93-0606 for the U..S. DHHS, Public Health Service, Agency for Toxic Substances and Disease Registry, Atlanta, Georgia.

U.S. Department of Health and Human Services. 1993. *ATSDR-TOXFAQs-Chloroform.* ATSDR Web Site Address: HTTP://ATSDR1.ATSDR.CDC.GOV:8080/.

U.S. Environmental Protection Agency. 1996. *EPA Region III Risk-Based Concentration Table.* U.S. EPA Region III Office, Philadelphia, Pennsylvania.

U.S. Environmental Protection Agency. 1993. *Integrated Risk Information System (IRIS),* Wash., D.C. , October, 1993.

U.S. Environmental Protection Agency. 1992. *Health Effects Assessment Summary Table (HEAST).* USEPA/OERR 9200.6-303(91.1), NTIS No. PB91-921199, Wash., D.C. March, 1992.

U.S. Environmental Protection Agency. 1989. *Risk Assessment Guidance for Superfund, Volume I, Human Health Evaluation Manual, Part A,* EPA/540/1-89/002.

Waldorf, H. 1998. Phone Conversation with ASTM Subcommittee E50.04 (Performance Standards Related to Environmental Regulatory Programs) Representative on January 12, 1998 Re: Status of Proposed Standard in Balloting Process (i.e., *Provisional Standard Guide for Risk Based Corrective Action*).

THE TACO APPROACH TO ESTABLISHING RISK-BASED CORRECTIVE ACTION OBJECTIVES

Monte M. Nienkerk (Clayton Group Services, Inc., Naperville, Illinois)
Jeffery L. Pope (Clayton Group Services, Inc., Naperville, Illinois)

ABSTRACT: Illinois' *Tiered Approach to Corrective Action Objectives* regulations have been used to evaluate site-specific remediation objectives for a site contaminated with methylene chloride. Remediation objectives evaluated through this approach have provided the opportunity to implement a technically feasible cleanup of the site. Site-specific soil cleanup objectives of 24 milligrams per kilogram (mg/kg) for the upper 10 feet of the unsaturated soils and 2,000 mg/kg for the underlying unsaturated soils were established. These cleanup objectives were based on 1) the use of a city ordinance prohibiting well installation and groundwater use within one-half mile of the site and 2) the use of an engineered barrier consisting of 10 feet of clean soil. Without the city ordinance and engineered barrier, the cleanup objective would have been 0.02 mg/kg; making remediation of the site technically and economically infeasible.

INTRODUCTION

Operation of a plastic film coating facility began in 1975. Facility operations ceased in 1992. Methylene chloride was the primary solvent used in a gravure coating process. Methylene chloride was stored in a 3,500-gallon above ground storage tank (AST) located at the northwest corner of the building. The fill port for the AST was located at the northeast corner of the building. A buried steel pipeline ran from the fill port along the north end of the building to the AST.

In May 1985, a corrosion leak occurred in the steel pipeline near the AST and caused a release of methylene chloride. The methylene chloride was released to an underground storage tank (UST) area and a stormwater drainage system. Emergency response activities were initiated, and the UST and stormwater drainage areas were removed and remediated.

Subsequent to the emergency response activities, a series of subsurface investigations were initiated to define the extent of methylene chloride contamination and to determine the feasibility of remedial approaches. The results of the investigations identified the extent of the methylene chloride contamination and determined the location of three source areas (the northwest corner of the building where the AST and USTs were located; the northeast corner of the building where filling of the AST and USTs occurred; and along the north side of the building beneath an exhaust fan and where the leak in the buried steel pipeline occurred.

A three-dimensional (3-D) display of the extent of methylene chloride impacted soils shown in Figure 1 was generated using a visualization software program. The figure provides a 3-D perspective of the extent and depth of impacted soils (light areas) and the three source areas (dark area) mentioned above.

The investigations determined that methylene chloride has impacted the soil not only beneath the property to a depth of 55 feet at one location but also neighboring proporties. Remediating soil to this depth is not economically practical given the silty-clay subsurface soils beneath the site and the extra-ordinary engineering requirements that would be needed to maintain the structural integrity of not only the facility building but also other nearby (within 30 feet) buildings.

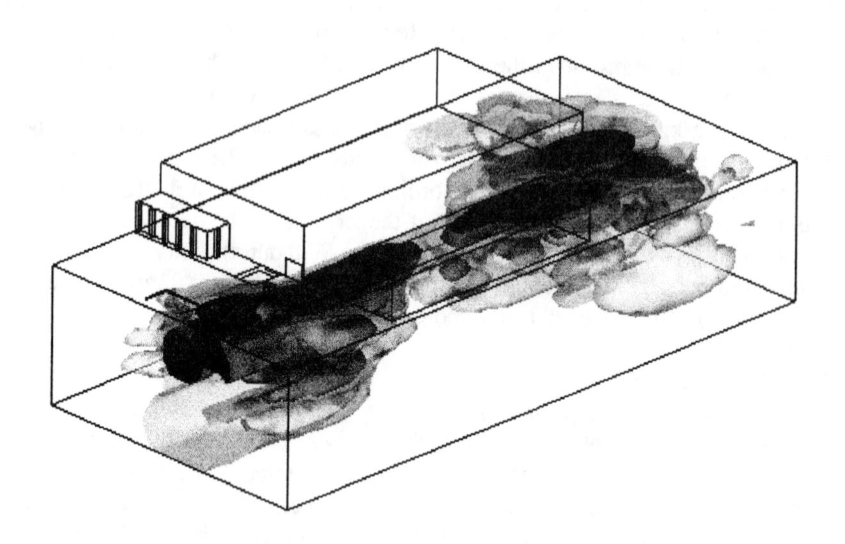

FIGURE 1. 3-D Interpretation of the Extent of Methylene Chloride in Soil

Following Illinois' newly established *Tiered Approach to Corrective Action Objectives* (TACO) regulations, site-specific remediation objectives were evaluated to determine if an economically feasible approach could be found to remediate the property. Removing all of the contaminated soil is not economically feasible. The estimated costs to do this are projected to be on the order of $5 million. The property is only valued at $1 to $2 million. The TACO approach allows for the exclusion of exposure routes provided certain conditions are met. If an exposure route can be excluded, then no remediation objectives (i.e., no cleanup) need to be developed for that exposure route. For this project, it is possible to exclude the groundwater ingestion exposure route by prohibiting the installation and use of any new water supply wells within one-half (½) mile of the site and demonstrating that groundwater contaminants will not migrate to any existing water supply wells within the ½ mile radius. This means that soil cleanup objectives only need to be developed to be protective of industrial/commercial workers and construction workers. The soil cleanup objective for methylene chloride, based on the ingestion and inhalation exposure routes, is 24 mg/kg. In contrast, a soil cleanup objective of 0.02 mg/kg is required if the groundwater ingestion exposure route can not be excluded.

SITE SETTING

The property consists of approximately 2 acres, containing a one-story, corrugated metal structure of approximately 42,000 square feet, with no basement. It is an inactive facility surrounded by light industry; located within an industrial park. There is an unincorporated residential area located approximately 400 feet north of the property. Residences in this unincorporated area rely on individual, private groundwater supply wells. The topography is generally flat, with slight manmade slopes which drain toward stormwater collection drains.

SITE HYDROGEOLOGY

Extensive data has been collected to determine the hydrogeologic characteristics of the property. Information gained from the monitoring wells at the facility indicate that there are three hydrogeologic zones. The first zone contains discontinuous saturated permeable sand lenses encountered at depths of 20 feet or less. The second zone contains saturated permeable sand seams encountered at depths of approximately 25- to 30-feet below ground surface (bgs). The third zone is encountered at a depth of approximately 70- to 80-feet bgs.

The sand lenses encountered in the first zone are thin and appear to be under confining pressure. Water levels of monitoring wells screened in this zone rise several feet above the elevation of the saturated permeable lenses encountered in the zone.

The second zone is several feet thick and appears to be more continuous than the first zone. This zone does not appear to be under as much confining pressure based on water levels being closer to the elevation of the saturated permeable seam encountered in the zone.

The third zone represents the uppermost continuously transmissive hydraulic unit underlying the property. It ranges in thickness from less than 5 feet to greater than 20 feet. All known water supply wells in the area are finished in this third zone.

Depth to bedrock ranges from 180-feet to 270-feet bgs. Bedrock in the region is Niagran series dolomite that is Silurian in age. Groundwater from the bedrock is also used as a potable water supply in the region.

EXTENT OF CONTAMINATION

In order to develop risk-based soil remediation objectives and to evaluate soil remediation options, it was necessary to determine the extent of soil contamination. Clayton completed over 100 soil borings at depths ranging from 21 feet to 55 feet bgs. Soil samples were collected and logged continuously to boring completion depth. An on-site, gas chromatograph was used to analyze soil samples collected at 5 foot intervals (644 samples). Approximately 18% of the soil samples were sent to an off-site, independent laboratory for confirmation. The analyses indicated the presence of methylene chloride at depths up to 55 feet, with concentrations ranging from non-detect to 43,000 mg/kg.

The majority of the contamination is found at depths less than 25 feet bgs. However, some areas show contamination to a depth of 31 feet and one location indicated contamination to a depth of 55 feet bgs. Concentrations generally increase with depth then drop sharply below a depth of 21 feet bgs. Highest concentrations are found in the 4 to 16 feet bgs depth range. Below 25 feet bgs, concentrations range from non-detect to less than 10 mg/kg. Using the data collected, an average methylene chloride concentration of 213 mg/kg (for all of the impacted soil) has been calculated. Based on this investigation data, the volume of impacted soil has been estimated to be 35,150 cubic yards. Using the average concentration of 213 mg/kg, this represents approximately 1,890 gallons of methylene chloride. The soil analytical data suggests that approximately 85% of the mass of soil contamination is in the depth range of 4 to 25 feet bgs. Table 1 summarizes the estimated volumes of impacted soil and volume of methylene chloride at various depth intervals.

TABLE 1. Volumes of impacted soil and methylene chloride with depth.

DEPTH RANGES	VOLUME OF SOIL	VOLUME OF METHYLENE CHLORIDE
4 to 6 feet bgs	5,551 cu. yds.	240 gallons
9 to 11 feet bgs	5,783 cu. yd.	430 gallons
14 to 16 feet bgs	5,811 cu. yds.	864 gallons
19 to 21 feet bgs	4,500 cu. yds.	43 gallons
23 to 25 feet bgs	1,500 cu. yds.	18 gallons
25 to 55 feet bgs	12,005 cu. yds.	295 gallons
TOTAL	35,150 cu. yds.	1,890 gallons

REMEDIATION OBJECTIVES EVALUATION

In-situ remediation techniques such as soil vapor extraction, bioremediation, bioventing would not be effective at this site, because of the relatively impermeable soil conditions (silty-clay). Furthermore, the presence of the facility building and neighboring building located 30 feet to the north, would make any excavation to depths ranging from 25- to 55-feet bgs technically and economically impractical. Excavations to these depths would require extra-ordinary engineering requirements to maintain the structural integrity of the buildings. This would require the installation of concrete caissons to provide the structural support for these buildings. The caissons would need to be installed to depths greater than 55 feet in many locations to support the loading from the buildings. Cost estimates for caisson installation alone exceed $3,000,000.

A risk-based corrective action approach has been taken to evaluate the remediation of the property, given the economic impracticality of excavating all of the methylene chloride impacted soil. Illinois' TACO regulation allows for the exclusion of an exposure route if certain conditions are met. The local municipality has passed an ordinance prohibiting the installation and use of any new water supply wells within ½ mile of the property. With this ordinance in place and a demonstration that no groundwater contaminants would migrate to any existing water supply well within the ½ mile radius, it is possible to eliminate the groundwater ingestion exposure route from evaluation. In order to demonstrate that no groundwater contaminants would migrate beyond or to any existing water supply well within the ½ mile radius, an equation for the groundwater ingestion route (provided in Illinois' TACO regulations) was used to show that the concentration of methylene chloride would meet the Illinois Groundwater Quality Standard for methylene chloride (0.005 mg/L) at these points.

The equation used to make this determination predicts the contaminant concentration along the centerline of a plume emanating from a vertical planar source in the aquifer. This model accounts for both three-dimensional dispersion and biodegradation. The equation is as follows:

$$C_{(x)} =$$

$$C_{source} \cdot \exp\left|\left(\frac{X}{2\alpha_x}\right) \cdot \left(1 - \sqrt{1 + \frac{4\lambda \cdot \alpha_x}{U}}\right)\right| \cdot erf\left|\frac{S_w}{4 \cdot \sqrt{\alpha_y \cdot X}}\right| \cdot erf\left|\frac{S_d}{2 \cdot \sqrt{\alpha_z \cdot X}}\right|$$

Where: X = distance from the source to the location of concern, along the centerline of the plume (7,620 cm or 250 feet - the distance from the site to the setback zone of the nearest water supply well).

$C_{(x)}$ = the concentration of the contaminant at a distance X from the source.

C_{source} = the greatest potential concentration of the contaminant of concern in the groundwater at the source of the contamination (3,030 mg/L - highest groundwater concentration level reported for methylene chloride in last 3 years).

α = dispersivity (762 cm in the x direction, 254 cm in the y direction, and 38.1 cm in the z direction).

U = specific discharge (0.17 cm/d).

λ = first order degradation constant (0.012 1/d).

S_w = width of source in the y direction (10,973 cm).

S_d = depth of source in the z direction (152 cm).

Using the site specific information indicated above, the equation predicts that the concentration of methylene chloride at a distance of 250 feet from the site (the setback zone for the nearest water supply well) will be orders of magnitude less than the groundwater quality standard for methylene chloride of 0.005 mg/L.

The remaining most stringent soil remediation objective (from other exposure pathways not eliminated) must be compared to the contaminants of concern at the site (no matter the depth of the contaminant) to determine if the site is clean. For methylene chloride, the most stringent soil remediation objective now becomes the limit for the inhalation exposure route. TACO defines the inhalation exposure route as any unsaturated soils at the site. Based on calculations for a cleanup objective for the inhalation route, these unsaturated soils will need to be remediated to 24 mg/kg. Unsaturated soils at the site range from 10 to 25 feet bgs; making remediation to this level still questionable from an economic stand point.

Therefore, limitations to the inhalation exposure route were evaluation. Illinois' TACO regulations allow for use of engineered barriers to eliminate exposure pathways. An allowable engineered barrier with respect to the inhalation exposure route can include clean soil covering the contaminated media, that is a minimum of 10 feet in depth and not within 10 feet of any manmade pathway. This is allowable provided the concentration levels of the remaining contaminated media do not exceed either the soil attenuation capacity (2,000 mg/kg) or the soil saturation limit of the contaminant of concern (2,400 mg/kg). Using Illinois' TACO regulations to develop a risk-based corrective action approach, soil remediation objectives are now being evaluated that make it feasible to remediate this site in a cost-effective and environmentally responsible manner. Soil within the top 10 feet of ground surface and within 10 feet of any manmade pathway can be remediated to less than 24 mg/kg (the Tier 1 soil remediation objective for methylene chloride for industrial/commercial properties - inhalation exposure route). Once remediated this 10-foot zone can act as an engineered barrier for the soils at depths greater than 10 feet bgs. Soils deeper than 10 feet would then only need to be remediated to less than 2,000 mg/kg (the soil attenuation capacity).

Because the contaminated soil has impacted neighboring properties, it has been necessary to involve these neighbors throughout the process. Even though, from a risk-based approach, the IEPA has found it acceptable to leave contaminated soil in place at concentrations as high as 2,000 mg/kg in some locations, the impacted neighboring properties are not so sure they can agree to those levels of contamination being left on their property.

RBCA CLOSURE AT DNAPL SITES

Joseph W. Sheahan, Groundwater Solutions, Lansing MI, USA
Roy O. Ball, ENVIRON International, Chicago IL, USA
Melinda W. Hahn, ENVIRON International, Houston TX, USA

ABSTRACT: The closure of sites with identified or suspected DNAPL under the requirements of the Resource Conservation and Recovery Act of 1976 (RCRA), or the Comprehensive Environmental Response, Compensation and Liability Act of 1980 (CERCLA) has not been well defined. With respect to RCRA, EPA has required that all contamination must be removed at closure such that no residual risk to human health or the environment remains. Therefore, even though many states administering the RCRA program have adopted or are considering Risk-Based Corrective Action (RBCA) procedures (including the statistical methods described in SW-846) for RCRA closure, the treatment of DNAPL is, at best, challenging.

The methodology for "closure" in CERCLA is described in "Risk Assessment Guidelines for Superfund (RAGS)" which requires that risks above the NCP criteria must be remediated, preferably by on-site or in-site destruction. Most risk-based Brownfield or voluntary cleanup programs do not provide any explicit allowance for DNAPL. However, while the ASTM methodology for RBCA in E1739-95 does not explicitly treat the problem of DNAPL, a basic framework for DNAPL assessment is implicitly provided.

A uniform methodology for RBCA closure of VOX DNAPL sites can be used to achieve the program objectives of RCRA, CERCLA, and Brownfield or voluntary cleanup programs. The regulatory acceptance of the application of RBCA methods to DNAPL sites will require education and discussion, but the use of a uniform methodology should facilitate that acceptance.

THE DILEMMA

Dense Non-Aqueous Phase Liquids (DNAPLs) are a class of subsurface contaminants, the existence and characteristics of which have become widely acknowledged in only the last decade or less. One of the most significant developments leading to their recognition was the observation that, at many sites of ground water contaminated with chlorinated solvents, pumping and treating the affected ground water was not achieving the predicted effect, i.e., Clean-Up Objectives (CUOs) were not being met in the predicted timeframes. The initial interpretation of this observation was that pump-and-treat as a technology does not work. More thoughtful examination of the data has led to the realization that perhaps characteristics of these particular contaminants were to blame. It is now understood that DNAPLs in the subsurface can act as essentially infinite sources of ground water contamination. This understanding, coupled with enhanced understanding of DNAPL behavior in the subsurface, has become known as the DNAPL Paradigm.

Regulatory Expectations: Most historical and many current Corrective Action Plans (CAPs) have been driven by the regulatory requirement that the quality of all ground water be returned to either pristine conditions or only slightly less stringent drinking water standards, now known as Maximum Contaminant Levels (MCLs) under the federal Safe Drinking Water Act. As has been widely documented, these criteria evolved largely from analytical laboratory detection limits which dropped significantly in the late 1970s and early 1980s. As these detection limits went down, more and more organic Constituents of Concern (COCs) were discovered in ground water. Most regulatory agencies reacted with the conservative philosophical approach that any organic chemicals in ground water are undesirable and should be removed, without any meaningful analysis of the benefit or consideration of the cost involved.

The typical historical requirement that all ground water meet these conservative CUOs has profound implications. It ignores the fact that, in many hydrogeologic regimes, much of the water present in shallow, saturated soils does not represent a useable resource and is extremely unlikely to ever affect any useable resource. This conservative, philosophical approach has, in at least one state, been expressed in the definition of an aquifer as any zone containing water which could be sampled. By this definition, water in the unsaturated or vadose zone receives the same level of protection as ground water in a true regional aquifer.

A second factor in regulatory expectations at DNAPL sites is the requirement that "product" must be removed from the subsurface. In some states, this approach persists, regardless of what the "product" comprises. This requirement logically evolved from earlier understandings of subsurface mechanisms involved in ground water contamination and remediation. These understandings - still valid in some scenarios - were that it was more cost effective to remove the highest concentrations of COCs (e.g., product in soils) before they dissolved into ground water than to pump and treat the ground water so contaminated. This model is another example of an historical approach with which the DNAPL Paradigm is usually inconsistent.

The DNAPL Paradigm: If a sufficient amount of a DNAPL COC is released to the subsurface, it migrates downward under the influence of gravity until it encounters a barrier to continued downward migration, typically a geologic boundary such as a formation contact or bedding plane in a sedimentary formation. It then migrates down the slope of that boundary until it meets another boundary or a closed depression on the boundary surface, at which time it collects in the closed depression. This scenario is somewhat independent of the presence of ground water in the pores of the soil or rock matrix, with the exception that some of the DNAPL will dissolve into the water as it passes through or displaces the ground water.

When the DNAPL has "come to rest" by this mechanism, it then exhibits the same behavior as Light Non-Aqueous Phase Liquids (LNAPLs) in "bleeding off' into passing ground water. DNAPLs differ from LNAPLs, however, in two primary ways. Because many DNAPLs are less soluble than most LNAPLs and also less amenable to degradation, a DNAPL "pool" can serve as an essentially infinite (in practical terms) source of ground water contamination. Second, because the controls on DNAPL migration are different than those which apply to ground water and LNAPLs, DNAPLs are much, much more difficult to locate and define. Largely because of this difficulty, they are virtually impossible to effectively remove from the subsurface. Even if they can be located, present technology for effectively removing them is very limited. If they are present at shallow depths, the potential may exist for excavation of the soils in which they occur, but regulatory restrictions on the handling of soils so impacted, e.g., the "Land Ban", makes such removal extremely expensive.

DNAPL recovery efforts are underway at numerous sites, generally as modifications of pump-and-treat approaches with wells and pumps designed to take advantage of the DNAPL's higher density. Current literature suggests that, even under the most favorable combination of circumstances, such efforts may approach only 50% effectiveness, with the result that, instead of pumping and treating ground water forever, one might have to do so for only half of a very long time. Even the U.S. Environmental Protection Agency has acknowledged the technical impracticality of trying to recover DNAPL.

The Dilemma: Given the scenario elements outlined above, DNAPLs pose a real dilemma to responsible parties, environmental professionals and society as a whole. DNAPLs are a class of COCs which can degrade ground water resources beyond traditionally accepted CUOs essentially forever unless the DNAPLs are taken out of the system. At the same time, we do not have effective (much less cost effective!) technologies for doing so.

The resources expended on environmental cleanups are *society's* resources - whether in the form of reduced corporate profits or competitiveness, increased costs to consumers, or higher taxes. The DNAPL paradigm has to be viewed as one of our greatest challenges. What fraction of our society's resources are we willing to continue to spend trying to clean up something which cannot be cleaned up to traditional standards?

DECISIONS, DECISIONS!

Decisions regarding the management of DNAPL sites are, almost by definition, complex. They also tend to be made by groups rather than individuals. These groups will usually comprise a combination of corporate management, legal counsel and environmental professionals. Regulators will also commonly be involved, even if primarily at an approval level. Decision theorists long ago realized that the difficulty of a decision increases exponentially with both the number of decision factors and the number of stakeholders involved. The implication of this reality to resolution of the DNAPL dilemma is that, in many cases, no rational decision will ever be made. Instead, we often make "politically correct" decisions in order to appear to be addressing a problem, knowing that our efforts will be ineffective and that our resources could be better spent elsewhere.

One - perhaps the only - way out of the DNAPL dilemma is a rational decision making process which will permit the appropriate definition and prioritization of objectives and, then, the ranking of alternatives to meet those objectives. Ideally, such a process would be designed for application by groups of stakeholders in a consensus-building mode.

HANDLING DNAPL SITES IN THE RBCA FRAMEWORK

The only area in which DNAPL sites differ from any other in the Risk-Based Corrective Action framework is in the handling of the source term. The primary issue here is the philosophical perspective that "product" must be removed from the subsurface. This can be addressed by assessing the system with DNAPL present. DNAPL can occur in the subsurface in any or all of five modes:

- adsorbed DNAPL in vadose zone soil pores
- DNAPL in vadose zone soil pores retained by capillary forces
- DNAPL migrating through the vadose zone
- DNAPL migrating through water-saturated zones
- DNAPL residing (perhaps temporarily) in a closed depression in a hydrostratigraphic boundary

The worst cases are the last two in which the DNAPL is "feeding" the ground water at the aqueous solubility limit(s) for the COC(s) present. In terms of contribution to the aqueous phase, these two are analogous to the more familiar LNAPL scenario in which floating product dissolves into the underlying ground water at its aqueous solubility limit.

In RBCA Tiers 1 and 2, risks due to ground water exposure pathways are calculated assuming a constant source concentration (for a DNAPL, this would be set at the water solubility limit) and a fully developed, steady state plume. This is entirely consistent with the behavior of a DNAPL. Care must be taken, however, not to underestimate contaminant loading to ground water in Tier 3 if a transient ground water model is applied. A schematic of the RBCA decision making process is provided as Figure 1.

SOLUTION OF THE DNAPL DILEMMA

The solution to the DNAPL dilemma is no different from that of any other problem of limited resources and multiple technological constraints. The first step is to define the status, i.e., what is known. At most DNAPL sites, the following apply:

- ground water contaminated above MCLs
- unrecoverable separate phase COCs
- timeframes for attainment of MCLs approaching infinity
- finite resources

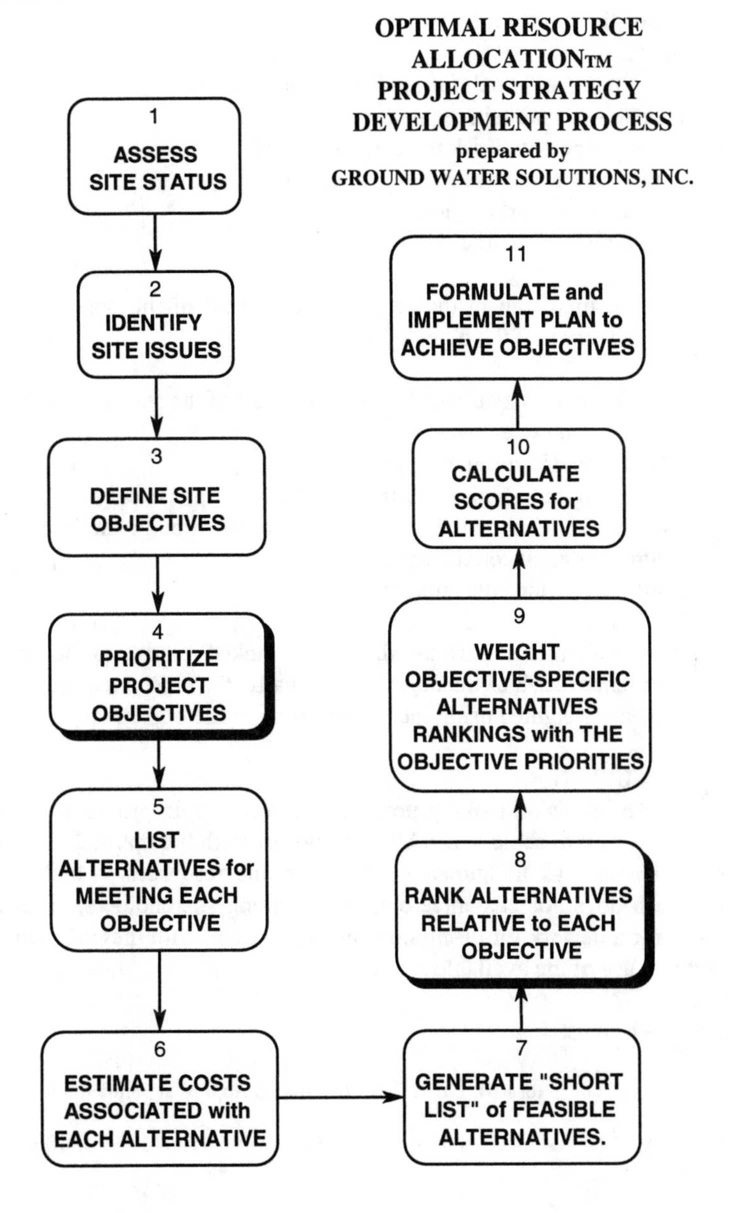

Copyright (c) 1998
Ground Water Solutions, Inc.
All rights reserved.
FIGURE 1
OPTIMAL RESOURCE ALLOCATIONTM PROJECT STRATEGY DEVELOPMENT PROCESS
prepared by
GROUND WATER SOLUTIONS, INC.
1 ASSESS SITE STATUS
2 IDENTIFY SITE ISSUES
3 DEFINE SITE OBJECTIVES
4 PRIORITIZE PROJECT OBJECTIVES
5 LIST ALTERNATIVES for MEETING EACH OBJECTIVE
6 ESTIMATE COSTS ASSOCIATED with EACH ALTERNATIVE
7 GENERATE "SHORT LIST" of FEASIBLE ALTERNATIVES.
8 RANK ALTERNATIVES RELATIVE to EACH OBJECTIVE
9 WEIGHT OBJECTIVE-SPECIFIC ALTERNATIVES RANKINGS with THE OBJECTIVE PRIORITIES
10 CALCULATE SCORES for ALTERNATIVES
11 FORMULATE and IMPLEMENT PLAN to ACHIEVE OBJECTIVES

The issues arising from this status include:

- regulatory compliance
- potentially unacceptable risk to actual or potential receptors
- public perception
- third-party actions
- use of finite resources

The objectives which follow from these issues are:

- to achieve / maintain regulatory compliance
- to reduce potential risk to acceptable levels
- to maintain a positive public perception
- to avoid third party actions
- to optimally allocate finite resources

The final step in the process is to list all of the possible alternatives to achieve any or all of the objectives.

The alternatives available to meet any or all of the stated objectives include:

- DNAPL source removal
- DNAPL source mitigation
- (very) long-term pump-and-treat for dissolved phase only
- elimination of potential exposure pathways through the use of institutional controls or engineered barriers
- institute a public education program

Once these alternatives are listed, they are ranked relative to their contribution to achievement of each of the objectives. Finally, the objectives-specific alternatives rankings are weighted using the prioritization of the objectives.

CONCLUSION

The decision making process presented herein provides interested parties with a method to achieve DNAPL site closure with the optimal reduction of actual or potential risks to human health given the resources available. While this approach does not guarantee consensus among shareholders, it does provide a rational framework for discussion and negotiation, with the goal being the highest and best use of the available resources.

REFERENCES

ASTM, 1995, Guide for Risk-Based Corrective Action Applied at Petroleum Release Sites, ASTM E 1739-95.

U.S. EPA, 1986, Test Methods for Evaluating Solid Waste, Field Methods, USEPA SW-846.

USE OF AS/SVE TO REMEDIATE CHLORINATED SOLVENTS

William Smith, P.G. (Environmental Alliance, Inc. Wilmington, DE USA)

ABSTRACT: An industrial site can be remediated economically through proper planning and implementation of a sound remediation strategy. Redeveloping the site must consider the costs to acquire the property and complete the required environmental actions, as well as the potential return on the investment. One such brownfield site was redeveloped economically by purchasing the property from the mortgage holder to settle a private CERCLA lawsuit, and by conducting remedial activities to achieve acceptable concentrations of residual site contamination, particularly volatile organic compounds (VOCs) in shallow groundwater. Air sparging in conjunction with soil vapor extraction technology was implemented to reduce VOC concentrations to acceptable conditions. Site closure goals were established by utilizing generic remediation standards for soil contamination and a site-specific risk assessment for groundwater contamination. The former industrial site has achieved closure from the State regulatory agency and is currently undergoing residential development.

INTRODUCTION

In order to settle a private CERCLA lawsuit and to redevelop the property into an economic asset, the twenty-acre site was remediated to residential standards prior to redeveloping it into a residential community. Responsible parties included the most recent tenant, as well as previous site occupants. The former tenant decided to purchase the property, remediate the areas of concern, and redevelop the property as a residential development. This decision was based on the economics of litigating the lawsuit, the cost to settle the case, the cost of conducting site remediation, and the potential resale of the "clean property". Subsequent to closure of all remediation activities and the re-zoning of the site, the property was sold to a developer in order to build a residential community.

Remediation activities for shallow groundwater contamination consisted of an air sparge/soil vapor extraction (AS/SVE) system. All environmental activities were conducted under the state's Voluntary Remediation Program (VRP). The site has received closure by the state regulatory agency. Development of the site began in the fall of 1997 and the first residential dwelling is expected to be occupied by summer of 1998.

BACKGROUND

Site assessment activities identified the areas of concern as a waste neutralization pit, PCB-contaminated soils, residual material associated with a radiological laboratory, and shallow groundwater contaminated by VOCs. The waste pit was operated in the 1960s to accept water from laboratory sinks. Its limited soil contamination by VOCs, and to a lesser extent, metals, was documented to be contained within the pit with no impacts outside the structure. Remediation of the waste pit was accomplished by excavation with off-site treatment and disposal. A discharge area from two concrete roof drainpipes was documented to have elevated levels of PCBs and to a lesser extent, SVOCs in shallow soils. Excavated PCB- and SVOC-contaminated soils were properly disposed at an off-site landfill. The residual soils indicated PCB and SVOC levels below residential standards. A former radioactive laboratory was documented to have residual levels of radiological contamination. The residual contamination in the laboratory was removed for proper off-site burial and the site approved for unrestricted use by the Nuclear Regulatory Commission (NRC).

The main contaminants of concern in the site's shallow groundwater system were identified as tetrachloroethene (PCE) and to a much lesser extent, its degradation products and petroleum hydrocarbons. An analysis of corrective action requirements and a comparison of applicable remedial technologies led to the selection of an AS/SVE system to remediate the impacted shallow groundwater, and the unsaturated soils.

The geology consists of unconsolidated sediments that appeared to be deposited as discontinuous lenses typical of fluvial deposits. A large clay lense was observed in the central portion of the contaminant plume. The shallow aquifer encountered appears to be under unconfined water-table conditions with the exception of the central portion of the site that may be under partially confined water-table conditions due to the identified clay lense. The measured groundwater table gradient indicates a groundwater flow direction from the southwest to the northeast across the site.

CONTAMINANT CHARACTERISTICS

VOCs which are lost in the subsurface travel along the path of least resistance, which is usually downward until reaching a restrictive horizon, such as a clay lense or the water table. As they move away from the leak/spill location, some of the VOCs become adsorbed onto the geologic material. When (and if) the spill reaches a restrictive horizon, the VOCs then spread laterally. If the restrictive horizon is a layer of some low permeable material, much of the VOCs may be adsorbed above the water bearing horizons of the subsurface.

VOCs typically partition into four physical phases. This partitioning typically results in a residual adsorbed phase (estimated 60-90% of the original mass loss), an aqueous dissolved phase (estimated 0-10% of the mass), a vapor phase (estimated 0-1% of the mass), and in cases of large magnitude losses, a separate phase of the VOC material. VOCs which have a specific gravity greater than water can form a separate phase which migrates vertically.

The extent of a dissolved plume is governed by the quantity of loss, its relation to biological and chemical properties of soils and groundwater, the hydraulic properties of the geologic materials, and any structural features which can act as barriers or conduits for fluids. The goals of the remedial program included remediation of the subsurface to risk-based cleanup standards and/or to practical technological limits and the completion of site closure in a safe, efficient, and timely manner.

REMEDIATION TECHNOLOGY

Pump and treat and bioremediation were not considered practical for this site due to: low aquifer yields; limited effectiveness in removing adsorbed chlorinated VOC contaminants, the large area of dilute groundwater concentrations, limiting hydrogeologic factors, and lack of co-metabolites onsite.

Rather, an AS/SVE system was selected as the most appropriate technology to address the shallow groundwater contamination. The system was designed using the data developed during the site investigation activities, as well as pilot test data. This aggressive and technologically advanced system was designed to achieve closure requirements at the site within 12 - 18 months of system startup. This comprehensive remedial system consists of two distinct elements which operate together to achieve contaminant reduction. The two main active elements of the system are the soil vapor extraction and air sparge components.

The SVE system removes adsorbed phase VOCs in the unsaturated zone and collects sparged air introduced into the saturated zone. This process removes contaminants by removing ambient air above the water table (unsaturated or vadose zone). The system remains effective as long as the induced air contacts the contaminated soils and the contaminants themselves are volatile in nature. The efficient movement of air through the subsurface is maximized by determining the adequate spacing of the soil vapor extraction points, locating discrete contaminant horizon(s), and isolating preferred flow zones within the unsaturated horizons.

The second portion of the remediation system consists of an air sparge system that volatilizes VOCs adsorbed to soil below the water table and dissolved in groundwater within the unconsolidated deposits. The air sparge points were constructed such that the screen interval (approximately two feet) was located just above the unconsolidated/consolidated zone interface, approximately 14 feet below the water table. The application of positive air pressure through the air sparge points volatilizes VOCs within the saturated zone for removal of the VOCs by the SVE system.

The removal of VOCs by use of an AS/SVE system is roughly proportional to the Henry's Law constant (H') for the VOCs of concern. A dimensionless Henry's constant greater than 0.01 generally indicates that the compound may be readily volatilized by both an AS and/or SVE remedial system.

The key to the successful operation of the combined AS/SVE system was attaining good contact between the injected air and contaminated soil/ groundwater. In the vadose zone, this is achieved by properly setting the screens for the AS/SVE points. The air sparge screen is set below the water table so that the air bubbles can freely travel vertically through the aquifer. The airflow can then strip the VOCs and be captured by the SVE system as the vent screens are usually set above the water table in the unsaturated zone.

In situations where the saturated soils consist of low permeable clays and/or low permeable clays with higher permeability silt/sand stringers such as this Site, the sparge gases will tend to flow through the permeable features created through the clay. In these situations the mass transfer of VOCs is via direct volatilization at the interface between the air and soil/water surface created by air sparging and via diffusion in the areas away from the air channels. The effects of these different permeability features can be minimized by accounting for these features during the construction of the air sparge points. Specifically, if low permeability clay layers are known to exist or are identified while drilling, these zones can be isolated such that all of the injected air is forced into this zone as opposed to following a path of lesser resistance.

REMEDIATION SYSTEM DESIGN

Air Sparge System. The air sparging system was designed to cover those areas of dissolved phase contamination located within the boundaries of the clean wells. Pilot testing was performed and used to determine that the radius of influence (ROI) of an individual air sparge point would be a minimum of 60 feet. Using this ROI, the locations of the air sparge points were arranged to provide overlapping influence of the area to be treated. A total of 42 air sparge points were installed onsite, 36 of which are co-located with soil vapor extraction points and six were air sparge-only points.

Soil Vapor Extraction System. The SVE system was designed to extract a flow rate of air well in excess of that of the flow rate of air sparge gases to ensure proper control of the air sparge system. Also, the ROI of the SVE was designed to exceed that of the air sparge process. The SVE system was installed and operated in areas beyond the sparge point installation to capture air sparge gases which may have potentially migrated horizontally away from the sparging area.

From the pilot test data, a ROI of 60 feet was found at an operating vacuum of ten inches of water. The SVE points were laid out using this ROI and in a manner such that the influence of the SVE extended beyond the air sparge system. In addition, as a precaution SVE points were installed outside of the AS system so that the AS gases which may migrate horizontally beyond the limits of the SVE system would be captured. A total of 53 SVE points were installed, 36 of which were co-located with the air sparge points and 17 of which are SVE-only locations.

As identified during the pilot test, the air sparge gases had a tendency to migrate horizontally beneath the clay layer near the water table interface across much of the site. The SVE points which were co-located directly above the air sparge points in these locations were limited in their ability to capture the air sparge gas. In order to improve air sparge gas capture, SVE-only locations were installed on the perimeter of the groundwater treatment area along the edges of the clay zone.

CLOSURE CRITERIA

According to the state's Voluntary Remediation Regulations and guidance criteria, remediation levels may be developed through a three-tiered approach.

Under Tier I, remediation levels are derived from established background concentrations. Tier II remediation levels are generic levels based on published, media-specific values (e.g., RBCs, Soil Screening Levels [SSLs], Water Quality Standards [WQSs], etc.), derived using conservative default assumptions. Tier III remediation levels are based upon a site-specific risk assessment considering site-specific assumptions about current and potential exposure scenarios for the population(s) of concern, including ecological receptors, and characteristics of the affected media. Site closure utilized Tier III remediation levels that were developed for the VOCs in the shallow groundwater and limited contaminants in site soils and surface water.

Future use scenarios must be considered when developing site-specific remediation levels. For this case, institutional controls and site deed restrictions prevented the use and installation of domestic wells on the property. The last remaining exposure pathway consisted of the potential migration of contaminated groundwater into the downgradient stream. A site-specific groundwater fate and transport modeling was performed to document that the contaminant plume will not adversely impact the downgradient stream.

CONCLUSIONS

This case demonstrates how the redevelopment of a contaminated industrial facility can be accomplished economically through proper planning and understanding of project requirements. The remedial actions must achieve acceptable risks to human health and the environment.

The site-specific risk assessment established the levels of contaminants that can remain onsite and still be protective of human health. Using the maximum exposure point concentrations, an overall site risk level of 1×10^{-4} was calculated for the dissolved shallow groundwater VOC contaminants onsite. The representative geometric mean exposure concentrations are at the 1×10^{-5} level and the predictive fate and transport model indicates the concentrations will continue to decrease over time.

Predictive simulations were performed to evaluate natural attenuation effects on remaining concentrations to determine projected rates for natural or intrinsic remediation of the residual concentrations to MCLs. The results of the fate and transport groundwater modeling were intended to support the risk-based, site-specific VOC remediation goals. The model output predicted that after 12 years, the plume size will decrease with onsite concentrations dropping to below MCLs throughout the Site and at all times the potential VOC concentrations in the stream downgradient will remain below MCLs.

Based on the achievement of the stipulated remedial goals, closure was granted for the project with a "no further action" letter. A Certificate of Satisfactory Completion provided statutory immunity for the owner of the site. Residential development of the property is now underway.

REFERENCES

American Petroleum Institute, 1991. "Technological Limits of Groundwater Remediation: A Statistical Evaluation Method". Health and Environmental Sciences. API Publication Number 4510.

Agency for Toxic Substances and Diseases Registry, Toxicological Profiles. ASTDR 1993; ASTDR 1989.

Dragun, J., 1988. "The Soil Chemistry of Hazardous Materials". Hazardous Materials Control Research Institute. pg. 77.

Keely, J.F., 1989. "Performance Evaluations of Pump-and-Treat Remediations". USEPA Ground Water Issue. ORD and OSWER. EPA/540/4-89/005.

Mackay, Donald, Wan Ying Shiu and Kuo Ching Ma, 1992. "Illustrated Handbook of Physical-Chemical Properties and Environmental Fate for Organic Chemicals" Volume I, Lewis Publishers.

McDonald and Harbaugh, 1988. "A Modular Three-Dimensional Finite-Difference Ground-Water Flow Model," USGS TWRI.

National Research Council, 1994. "Alternatives for Groundwater Cleanup". National Academy Press.

Papadopoulos and Associates, Inc., 1992. "MT3D, A Modular Three-Dimensional Transport Model, Version 1". Papadopoulos and Associates, Inc., Rockville, Maryland.

Pollock, D.W., 1989. "Documentation Of Computer Programs To Compute and Display Path Lines Using Results From the U.S. Geologic Survey Modular Three-Dimensional Finite-Difference Ground-Water Flow Model," USGS Open File Report 89-391.

Roberts, P.V., Schreiner, J.E., and Hopkins, G.D., 1982. "Field Study of Organic Water Quality Changes During Groundwater Recharge in the Palo Alto Baylands". Water Resources. 16:1025-1035.

U.S. EPA, Office of Solid Waste and Emergency Response, "Guidance for Evaluating the Technical Impracticability of Ground-Water Restoration". OSWER Directive 9234.2-25, September 1993.

U.S. EPA, 1989. "Risk Assessment Guidance for Superfund".

U.S. EPA, 1985. "Water Quality Assessment: A Screening Procedure for Toxic and Conventional Pollutants in Surface and Groundwater, Part II".

U.S. EPA, 1996. "The Risk-Based Concentration Table", Region III Technical and Programs Support Branch January - June, 1996, 4/19/96.

SELECTION OF THE LONG-TERM MONITORING REMEDY FOR AN OFF-SITE TCE PLUME

Phillip Watts, RG, CPG, Earth Tech, Inc., San Antonio, Texas
Jon Satrom, PE, US Air Force Base Conversion Agency-March AFB, California
Richard Russell, PE, US EPA, Region IX, San Francisco, California
John Broderick, California EPA Regional Water Quality Control Board, Riverside
Emad Yemut, PE, California EPA Dept of Toxic Substances Control, Long Beach
Thomas Villanueve, PE, Tetra Tech Inc., San Bernardino, California
Robert Johns, Ph.D., Tetra Tech Inc., Lafayette, California

ABSTRACT: This case study summarizes factors influencing remedy selection for an off-site TCE plume at March Air Force Base (AFB), California. March AFB was placed on the U.S. EPA's National Priorities List (NPL) in 1989 due to the presence of groundwater contaminant plumes (primarily TCE and PCE). The CERCLA Record-of-Decision (ROD) for these plumes was one of the first RODs signed by the U.S. EPA Region IX (EPA-IX), the California EPA Department of Toxic Substances Control (DTSC), and the California EPA Regional Water Quality Control Board, Region 8 (RWQCB-8) that allowed an approach other than active remediation for an off-site TCE plume exceeding the maximum contaminant level (MCL). The focus of this case study is the combination of factors existing at March AFB that made the long-term monitoring remedy appropriate for this site. These factors fall into four broad categories:

- Technical factors
- Cost factors
- Regulatory factors
- Community factors

Factors within each category are discussed, as are the site history and environmental conditions. Factors influencing regulatory agency and community approval of the remedy are also discussed.

INTRODUCTION: March AFB is located in Riverside County, California, approximately 60 miles east of Los Angeles. Military operations began at the site in 1918. As a result of these operations, soil and groundwater at the base were contaminated with chlorinated solvents and other contaminants. A Remedial Investigation/Feasibility Study (Earth Tech, 1994) and a Basewide Groundwater Monitoring Program (Tetra Tech, 1997) have documented the presence of chlorinated solvents in groundwater at the site. These contaminants were detected in on-site source areas and in a plume extending approximately 5,000 feet off-site.

The site is underlain by unconsolidated alluvial deposits consisting of alternating silts, sands, and clays. The alluvium is in turn underlain by weathered granite and granite bedrock. The bedrock surface slopes generally to the southeast, and the alluvium thickens to the southeast. Thickness of the alluvium ranges from zero in some on-site areas to a maximum of approximately 1,000 feet in the Perris Valley east of the base. Groundwater flow is generally to the southeast. Concentrations of chlorinated solvents in groundwater range from between 1,000-3,000 micrograms/liter (ug/L) in on-site source areas to a maximum of approximately 40 ug/L in off-site portions of the plume.

March AFB was placed on the NPL in 1989 due to the presence of chlorinated solvents in groundwater. In 1990, a Federal Facility Agreement (FFA) was signed by the Air Force, EPA-IX, and the State of California, outlining Federal, state, and public involvement in the March AFB environmental restoration process. In 1992, an interim remedial action was implemented, consisting of groundwater extraction wells installed at the base boundary to intercept contaminants migrating off-site. Groundwater captured by this interdiction system was treated on-site using activated carbon.

The CERCLA ROD for Operable Unit 1 at March AFB was finalized in 1995. The remedial action selected for the off-base portion of the plume was long-term monitoring. This consisted of upgrade and operation of the base boundary plume interdiction system to prevent further off-site contaminant migration, and periodic sampling of monitoring wells within the off-site portions of the plume. Most notably, the proposed remedy did not include active remediation or a demonstration of natural attenuation of the off-site plume. This was one of the first RODs signed by EPA-IX, RWQCB-8, and DTSC that allowed an approach other than active remediation for an off-site TCE plume exceeding the MCL. Since this approach may be desirable for other similar facilities, the remainder of this paper focuses on the combination of factors existing at March AFB that made the approach viable in terms of regulatory and public approval. These factors fall into four broad categories, including technical factors, cost factors, regulatory factors, and community factors.

TECHNICAL FACTORS: Technical factors include the existence of a reliable groundwater database and flow model, lack of human health risks, actions taken by the Air Force to mitigate potential health risks, and the limitations of pump-and-treat technology. These factors are discussed below.

Existence of a Reliable Groundwater Monitoring Database and Groundwater Flow Model. At the time that the March AFB ROD was finalized, the basewide groundwater monitoring program had been in place for 3 years (Tetra Tech, 1997). This program was developed in cooperation with state and Federal oversight agencies, so all parties involved had a high level of confidence in the database and flow model. Consequently, all parties involved agreed that the nature and extent of groundwater contamination had been adequately characterized, and that the groundwater flow regime and potential human health risks were well-understood. While the existence of a groundwater database and flow model is not unique to the site, lack of adequate site characterization would have made the selected remedy difficult to defend. A sound database is necessary to defend the selection of any remedy, however, it is especially critical when the remedy involves no active remediation. Therefore, an adequate database in which regulatory oversight agencies have confidence is seen as a prerequisite for consideration of long-term monitoring as a viable remedial alternative.

Lack of Human Health Risks. The RI/FS for the site (Earth Tech, 1994) documented the lack of human health risks from the off-site plume. Areas near the plume were served by a municipal system that used surface water for supply. No known drinking water supply wells were located near the site, so there were no complete pathways for human exposure to contaminants. And although TCE concentrations in the off-site plume exceeded the MCL (5 ug/L), potential risks to human health (if the contaminated groundwater were to be consumed) were determined to be in the acceptable range (10^{-5} excess cancer risk).

Alternative Approach at a Site With Potential Human Health Risks. At nearby Norton AFB, in San Bernardino, California, the Air Force was faced with a similar problem—a TCE plume in groundwater that extended off-site. However, at Norton AFB, municipal supply wells for the City of Riverside were threatened by the plume. This condition did not rule out consideration of the long-term monitoring remedy for the off-base plume, however. For many of the same reasons that applied to March AFB, active remediation of the off-site portion of the plume at Norton AFB was considered impractical. However, unlike March AFB, there were potential risks that required mitigation. The risks were mitigated by an agreement reached between the Air Force and the City of Riverside, which was strongly supported by state and Federal regulatory agencies. Under the agreement, the Air Force agreed to monitor supply wells that were potentially affected, and take action if the wells became contaminated. Actions contemplated included installation of wellhead treatment systems (granular activated carbon) and provision of additional potable water for blending.

Presence of a Hydraulic Containment System In 1992, March AFB implemented an interim removal action to prevent further spread of the TCE plume. At that time, the extent of the plume had not been defined, but off-site migration had been documented. A system of nine groundwater extraction wells was installed along the eastern (downgradient) base boundary to intercept the plume and minimize further off-site migration. Although the system was not believed to be 100% effective, it did prevent further significant spread of the plume. As part of the overall remedy for the plume, the Air Force agreed to upgrade the plume interdiction system with the goal of effective hydraulic control at the base boundary. It is anticipated that this work will be completed in 1998.

Contingency Language Incorporated in ROD. As an additional safeguard, the Air Force agreed to re-examine the effectiveness of the selected remedy periodically. CERCLA provides for such a re-examination at the five-year review. Regulatory agencies requested that the Air Force agree to re-open the issue of active groundwater remediation if the off-site plume expanded, increased in contaminant concentrations, or threatened drinking water supplies. Specific language to that effect was included in the ROD, further enhancing the overall protectiveness of human health and the environment.

Planned Source Removal Actions. As part of the Air Force's overall strategy for clean-up of the base, source areas for groundwater contaminants were targeted for aggressive clean-up. Several source areas for TCE were identified on-base, most notably IRP site 31. This site was addressed through early implementation of a pilot-scale treatment system (dual extraction of groundwater and soil vapors) concurrent with the RI/FS, followed by full-scale remedial action. Remedial actions have successfully removed most of the contaminant mass from this source area, and similar removal actions are planned or underway for other source areas. Source area removal was an important part of the overall cleanup strategy for the base, because it cut off continuing sources of contaminants for the off-site plume, which made active remediation of the off-site plume less of an issue.

Limitations of Pump-And-Treat Technology. The limitations of pump-and-treat technology in restoring contaminated aquifers to health-based cleanup standards are well-documented. Even so, many regulatory agencies have been reluctant to allow alternatives to pump-and-treat until recently. Regulatory agency personnel exercising oversight of cleanup efforts at March AFB acknowledged these limitations early-on. All parties agreed that there would be little gained by implementing pump-and-treat for the off-base plume, given the relatively low concentrations of contaminants present. This is not seen as inconsistent with the use of groundwater extraction for plume containment at the base boundary, since containment is an achievable goal of pump-and-treat.

COST FACTORS: Cost factors considered in remedy selection included not only the high cost of pump-and-treat for the off-base plume relative to other alternatives, but also the return on investment in terms of risk reduction.

Cost Comparison. Cost estimates for treating the entire off-site plume using conventional pump-and-treat technologies were developed (USAF, 1995) and compared to cost estimates for containing the plume at the base boundary and monitoring the off-site plume. Costs for treatment of the entire plume were estimated at $12M, while costs for containment at the base boundary with off-site monitoring were estimated at $2.5M.

Cost/Benefit Analysis. Since there were no completed human exposure pathways, and contaminant concentrations presented a risk within the acceptable range, meaningful risk reduction for the off-site plume would have been difficult to achieve using any technology. Given the limitations of pump-and-treat technology in restoring contaminated aquifers to health-based cleanup standards, all parties agreed that there would be little benefit in implementing pump-and-treat in an attempt to remediate the off-site plume.

REGULATORY FACTORS: Key in selection of the long-term monitoring remedy at March AFB was approval by the US EPA-IX and the California EPA (RWQCB-8 and DTSC). Project managers for each of the agencies were interviewed to provide insight into the key factors allowing selection of long-term monitoring rather than active remediation. Factors cited are discussed below.

US EPA Region IX. Key factors in US EPA approval included state agency and community acceptance of the proposed remedy, and the lack of human health risks. The RWQCB-8, which has primary state oversight of issues affecting water quality, approved of the proposed remedy and no human receptors were present to form a completed exposure pathway.

California EPA RWQCB-8. Key in State acceptance of the proposed remedy was March AFB's compliance with State Resolution 68-16, commonly referred to as the "non-degradation policy." This resolution prohibits continuing discharges to designated drinking water sources in cases where the discharges degrade water quality. Since March AFB had a hydraulic containment system in-place at the base boundary and had agreed in the ROD to upgrade the system in order to mitigate further discharges, the RWQCB-8 considered the Air Force to be in compliance with State Resolution 68-16. The RWQCB-8 also cited the lack of a completed groundwater exposure pathway as a key factor, but noted that the presence of receptors would not necessarily mandate active remediation of the off-site plume. The example of nearby Norton AFB, discussed above, was cited as an example of an alternative to pump-and-treat, as was March AFB's initial response of providing alternate water supplies to nearby domestic water users when the TCE plume was first discovered.

Air Force/Regulatory Agency Interface. In addition to the regulatory factors discussed above, regulatory agency project managers for March AFB also noted several intangibles that influenced the decision-making process. All praised March AFB for exceptionally proactive program management and open communications. As one regulatory manager put it, "the base called us in, outlined their program goals, told us where they wanted to go, and asked for our help in getting there. We were all part of the decision-making process, and felt we had some ownership of and responsibility for the success of the program. A lot of trust was built with the Air Force."

March AFB personnel in turn noted intangible attributes of regulatory personnel that contributed to the decision-making process. Again, trust was mentioned as a key factor, as were the technical expertise of regulatory personnel assigned to the project and the continuity of the project team: "This same group of people has been working together for about five years. All of the regulators assigned to March AFB are seasoned professionals, and due to their technical expertise have been able to make significant contributions to the program."

COMMUNITY FACTORS: March AFB was designated for realignment by the Base Realignment and Closure Commission in 1993. The majority of base property was transferred to the Air Force Base Conversion Agency (AFBCA) for transfer to the private sector. The community wanted the property turned over to businesses that would generate jobs and revenues to replace the government jobs lost in the realignment. It was against this backdrop that community opinions on environmental restoration activities were formed. Two key factors seen as contributing to community acceptance of the proposed long-term monitoring remedy were the Air Force's outreach/education program for the community, and the community's priorities for beneficial re-use of the available property.

Air Force Outreach/Education Program. March AFB implemented an active outreach/education program early on through the base's Restoration Advisory Board (RAB). The RAB is a group composed primarily of local citizens interested in the environmental restoration process at March AFB. Community involvement in the process was viewed as critical to success. Many of the RAB members did not have the technical background required to understand the technical issues involved, so the Air Force instituted a training program for RAB members. The program included several day-long technical workshops, as well as tours of the base's contaminated sites and soil/groundwater treatment systems. As a result of this program, RAB members were made aware of the technical and cost issues associated with pump-and-treat of the off-site plume.

Community Priorities. The community's top priority was for the Air Force to transfer properties to the private sector for beneficial re-use. There were several properties that required clean-up, and a finite amount of funding available to complete the clean-up required for property transfers. The RAB considered the cost-benefit analysis of pump-and-treat for the off-site plume vs. long-term monitoring, and realized that selection of the long-term monitoring remedy would free-up approximately $10 million for other projects. The other projects were seen as more advantageous in terms of releasing properties for re-use, and were more in line with community priorities. The long-term monitoring remedy was accepted by the RAB with little discussion.

REFERENCES

The Earth Technology Corporation. July 1994. *Installation Restoration Program Remedial Investigation/Feasibility Study Report for Operable Unit 1, March Air Force Base, California.*

Tetra Tech, Inc. January 1998. *1997 Annual Groundwater Report for March ARB, California.*

United States Air Force. December 1995. *Record of Decision, Operable Unit #1, March Air Force Base, California.*

REMEDIATION COST ANALYSIS FOR VOX AND OTHER RECALCITRANT ORGANIC RELEASES

Mary Jo Anzia (ENVIRON International Corporation, Buffalo Grove, Illinois)
David A. Schlott (The ERM Group, Vernon Hills, Illinois)
Roy O. Ball (ENVIRON International Corporation, Buffalo Grove, Illinois)

ABSTRACT: A significant issue for the land pollution control practitioner is the accurate estimation of remediation costs. This issue is most difficult for multi-plant acquisitions, where the schedule is generally short, the data and factual background are generally limited, but the parties to the transaction need to agree on an environmental cost allocation, invariably with a cost cap. The consideration of five primary factors can improve the accuracy of estimated costs to resolve identified environmental issues. These factors are: the type of issue, the types of contaminants present, the current and future uses of the property, the applicable regulatory arena, and analysis of existing data. Remediation projects at sites with adequate environmental data can be analyzed using three-dimensional kriging techniques. Three-dimensional kriging provides a quantifiable accuracy and precision, and provides the basis for accurate costing. The problem is most compelling with the presence of chlorinated compounds (VOX), and/or other recalcitrant compounds, especially polynuclear aromatic hydrocarbons (PNAs), due to their persistence and toxicity. We have analyzed the environmental cost data for three facilities. The data consist of pre-acquisition cost estimates and the actual cost of closure, including remediation. The case studies exemplify the importance of understanding the site-specific circumstances associated with these five factors.

INTRODUCTION

Due diligence associated with property transfers has its regulatory foundation in CERCLA; namely, the ability of property owners who have adequately completed the due diligence process to claim the "innocent purchaser" defense for contamination found on the site. From a monetary standpoint, due diligence allows the estimation of potential liability associated with the site in the form of dollars that may be required to resolve an identified environmental issue. These estimated dollars are frequently a significant component in the negotiation of sale price and/or cash-out dollars for environmental issues.

The data available for any given environmental issue range from none (i.e., the issue is primarily speculative and based on experience in similar situations) to substantial (i.e., identified issues have been thoroughly investigated). Such variety in data quantity and quality makes it difficult to devise a formulaic methodology for assessing the potential liability costs associated with the identified issues. However, certain key components in accurate cost estimation exist and analysis tools are available to improve the accuracy of estimates. These will be discussed herein, and three representative case studies will be presented.

COST ESTIMATION

The first item that should be considered in the development of appropriate and accurate estimated costs is the types of issues identified. The types of issues that are generally found during a due diligence process include significant compliance issues and potential environmental impact issues. The issues associated with the latter set can be divided into standard categories, such as: air emissions and permitting, potential or known contamination of soil and/or groundwater, off-site disposal liability, known or suspected contamination from sources of polychlorinated biphenyls, outside storage of smaller containers, potential or known contamination associated with underground storage tanks, and wastewater or surface water impact. Such an initial categorization assists in initiating a somewhat more systematic approach to cost estimation.

The second item that must be considered is the analytical fractions of contaminants that may be or are known to be present on the site. Compounds such as halogenated volatile organic compounds (VOX) and polynuclear aromatic hydrocarbons (PNAs) are common contaminants found at industrial sites. In our current risk-based cleanup era, such compounds can represent significant chemicals of potential concern (COPCs) that can drive the risk analysis and potential remedy. Therefore, an understanding of the COPCs and their toxicity, mobility, persistence, solubility, and vapor pressure is essential in the preparation of accurate cost estimates.

Thirdly and on a similar note, the current and future uses of the property must be considered, as the use scenario is a critical component in the risk-based approach to cleanup. Discussions with the purchaser in this regard are essential, and an understanding of any leases that may exist for the property must be obtained.

Fourthly, the estimator must possess (or obtain) a realistic comprehension of the regulatory arena in which the issues will be carried forward. The acceptance of a risk-based approach to cleanup, the level of regulatory involvement, the presence or absence of voluntary ability to address contamination, and the ability to obtain letters of no further action that document adequate resolution of the issue are all factors that can significantly affect the actual cost.

Fifthly, a concise, quick, and inexpensive method of data analysis is required for sites with some amount of previously collected data associated with the environmental issues. In most cases, insufficient dollars and, most importantly, time are available to conduct a thorough data analysis prior to the closing of the property transfer. However, data analysis tools, such as EVS©[1], can quickly and reliably provide a three-dimensional representation of the magnitude and extent of contamination, as well as a numerical and visual estimate of the uncertainty associated with the available data.

With these five factors in mind, the cost estimator can rapidly prepare estimated costs to resolve the environmental issues identified for a site.

---

[1] The Environmental Visualization System© (EVS©) software conducts three-dimensional kriging of data sets and allows the viewing of the kriging results at any iso-concentration level and oblique angle.

Development of a range of costs for each issue is appropriate, realistic, and useful from a negotiating standpoint. The higher end of the range has been termed the "A" cost, and the lower end, the "B" cost.

The "A" cost for each issue represents the maximum expected expenditure that with reasonable certainty should resolve an identified issue. While the "A" cost is not the worst case, it assumes only that the issue is not significantly exacerbated by third party intervention. The "A" cost for each issue is based on an assessment of the maximum probable extent and severity of the environmental contamination (or other exposure) associated with the issue. However, in most cases, additional sampling (or further collection of information) will be required to fully define the extent and severity of the issues involved, and the "A" cost may understate, in some cases, the expenditures required to resolve any given issue. As previously stated, it is difficult to reliably predict the expenditures necessary to correct environmental problems. However, we can assume that, in the normal course of business, the aggregate of the "A" costs does represent a reasonable expectation for the total expenditures that could resolve all the issues. An expansion of any of the major issues by third party intervention, finding additional contamination during further environmental assessments, or aggressive regulatory agency action could, of course, potentially drive total expenditures beyond the "A" cost aggregate number.

The "B" cost for each issue represents the least expenditure that can reasonably be expected to resolve an issue in the normal course of business. For some cases, the "B" cost is zero, but for most issues, the "B" cost consists of expenditures for additional sampling and testing to determine if the extent and severity of contamination is such that no further action is required. For issues not directly related to contamination, the "B" cost includes expenses necessary to further investigate the issue, if required. While the "B" cost could possibly resolve any given issue, the aggregate of the "B" costs does not typically represent a realistic expectation of the probable cost exposure of all of the issues. Environmental costs are inherently biased in their distribution, as costs are zero on the low side, but many are almost unbounded on the high side. Therefore, it is somewhat unrealistic to expect that all, or even any significant fraction of the number identified issues at the facilities will be resolved under the ideal circumstances that are inherent in the "B" costs. The "B" costs are generally meaningful only for individual issues under nearly ideal circumstances.

The following sections of this paper present three case studies that exemplify the importance of an understanding of the five items presented above.

SITE #1

Site #1 is an inter-city plant blending ink products for the packaging industry. The plant was surrounded by a wire and steel manufacturing plant who acquired the property in the early 1990s. During the initial due diligence, it became clear that one of the significant areas of concern was an underground storage tank (UST) pit containing 20 USTs that were used to dispense the liquid products for ink blending and including primarily nonchlorinated industrial solvents and petroleum distillates. During the initial due diligence, a best professional estimate was applied to this tank pit - the "A" cost for this segment of the property was set at $615,000. After acquisition, an organic vapor headspace survey was conducted on the soils within the pit. The results were used to estimate that 4,000 cubic yards of soil would require excavation and off-site disposal. Although the initial cost estimates were predicated on the application of risk-based cleanup objectives, in actuality, the stringent generic UST Standards applicable in Illinois at the time of the contract were used. A second factor not foreseen in the initial due diligence was the presence of sand seams and lenses within the otherwise impermeable clay till that served to transport chemicals from the tank pit into the surrounding soils. Finally, the contaminant levels in the excavation were determined by specific side-wall and bottom sampling directed through the use of field instrumentation rather than averaged or composited samples. As a result of these three onerous conditions, which were not predicted during due diligence, a total of 7,500 cubic yards was ultimately removed from the tank pit excavation and disposed of off site at a total cost of $1.1 million.

The soil survey samples were used to create a three-dimensional visualization (using EVS©) of the pattern of contamination, which showed that a volume of 4,000 cubic yards corresponded to an isolevel concentration of 100 vapor parts per million. However, as the actual cleanup objectives corresponded to lower vapor readings, an isolevel concentration of 10 vapor parts per million closely approximates the actual volume removed.

For Site #1, the use of environmental visualization (EVS©) would have accurately predicted the total volume of soil to be removed if the cleanup objectives were reasonably defined in terms of the corresponding vapor levels. If the parties had been aware of the volume of material with detectable contamination it is possible that a less onerous cleanup would have resulted. Fortunately, the material was able to be classified as a solid waste for disposal purposes, which limited the cost for this activity to within a factor of two of the original due diligence estimate. Had the material been disposed of as a listed hazardous waste, the cost would have been at least ten times greater than the cost of managing the material as a solid waste.

SITE #2

Site #2 consists of a former three-piece can manufacturing plant which was purchased for use as a consumer apparel distribution warehouse. A series of solvent tanks were located in an aboveground mixing room in support of cleaning and decorating the three-piece beverage cans. During the filling of one of the solvent tanks (containing chlorinated solvents), the truck operator allowed an overfill to occur which ultimately resulted in the release of hundreds of gallons of chlorinated solvent from the roof vent. This material was transported along the roof and down the roof drains where it ultimately discharged onto grassy soil adjacent to the truck dock. Vapor readings were used to determine that the material occupied an approximately 30,000 square-foot area to a depth of approximately 6 feet. The area was then subjected to an intense soil boring pattern consisting of 50 soil borings at which samples were taken at depths of 1 foot, 3 feet, and 5 feet. Based on this dense array of samples, the volume of soil to be remediated was determined (based on a cleanup standard equal to background concentrations) using a box method, in which each soil sample was considered to represent a rectangular box extending halfway to the closest neighboring samples in three dimensions. This analysis resulted in an estimate of 1,200 cubic yards of soil to be remediated at an estimated cost of $100,000, assuming that the remedy would consist of on-site land farming of unsaturated easily drained granular soils.

The excavation demonstrated that the box method underestimated the volume of the contaminated soils by a factor of almost two, as ultimately 2,400 cubic yards of soil were excavated. In addition, the character of the soil had been inadequately defined in that a significant fine-grained component was present, which significantly increased the moisture content of the soil and greatly reduced the workability and drainability of the excavated material. As a result of doubling the quantity of soil and the significantly poorer properties, the ultimate cost for land farming was $480,000 - nearly five times the original estimate.

Three-dimensional visualization of the original 148 soil samples showed that the volume was far more reliably predicted by three-dimensional geostatistical kriging (using EVS©) than by the simplistic box model. In fact, a geostatistical representation with good accuracy could be accomplished with as few as 20 soil samples, particularly if approximately 20-25% of the samples were collected based on a geostatistical uncertainty analysis. Not only would the volume of soil to be remediated have been accurately estimated, but the cost of sampling would have been approximately 20% of the actual cost. By collecting samples for geotechnical properties, such as the Atterburg indices and other measures of soil drainability, moisture content and workability prior to the design of the remediation, the expected cost of remediation would have been better predicted.

SITE #3

A printing ink facility located in the Los Angeles area was the subject of a property transfer in the early 1990s. At the time of the property transfer, a due diligence assessment was conducted, and issues regarding known soil contamination and groundwater contamination (at a depth of approximately 70 feet) were identified. A significant amount of analytical data for both the soil and groundwater was available and reviewed during the assessment. The soil and groundwater contaminants included primarily VOX and metals; and free product (a lactol spirits NAPL) was detected in one of the wells situated on the downgradient property boundary.

Both parties involved in the property transfer assumed that stringent oversight from the California Regional Water Quality Control Board and the Department of Toxic Substances Control, along with the general absence of allowed risk-based approaches to cleanup at similar sites would result in the need for extensive off-site investigation, operation of a groundwater pump-and-treat system, and NAPL recovery. Both parties independently estimated the liability associated with the soil and groundwater contamination in excess of $1.1 million (the "A" cost).

Subsequent to the completion of the property transfer, a limited amount of additional groundwater data was collected, and all data were then analyzed using the three-dimensional EVS© software. The EVS© results clearly identified source areas of the contamination, such as outside vessel washing area, UST transfer piping, and outside storage areas. The EVS© results illuminated the mingling of the contaminants from these source areas and the likelihood of off-site contamination. The visualization of risk-based levels using EVS© demonstrated that significant cost savings would be realized.

This site represents a situation where the data analysis was sufficient, although had EVS© been used, the understanding of the extent of contamination and the quantification of uncertainty would have been greatly enhanced. The most significant aspect of the closure of these environmental issues was the regulatory arena. Although the site has not yet received closure, the expected actual total cost to address the contamination is less than $500,000 or approximately half of the intial "A" cost estimate. The substantially lower cost is due to a less stringent enforcement of drinking water standards, the allowance of a risk-based approach to cleanup for the site, and the presence of naturally occurring hydrogen sulfide beneath the site, which made the shallow groundwater unusable as a water supply.

SUMMARY AND CONCLUSIONS

We have described five factors that impact the accuracy of estimated costs to resolve environmental issues identified during the due diligence process. These major factors are: the type of issue, the types of contaminants present, the current and future uses of the property, the applicable regulatory arena, and analysis of existing data. As our three case studies indicate, each factor can be dominant in determining the accuracy of estimated costs. The dominant factor cannot be presumed; rather, for each site and each type of issue, all factors must be considered or cost of remediation could be severely underestimated. In all cases, three-dimensional visualization (using EVS©) significantly enhanced the accuracy of the prediction.

OPERATIONS & MAINTENANCE...PLANNING FOR RESPONSE

Howie Aubertin, Hill Air Force Base, Utah
Wally Hise, Radian International LLC, Salt Lake City, Utah

ABSTRACT: Remedy selection and implementation are a few of the many steps toward site closure. However, the operations and maintenance (O&M) period that follows has many new challenges. These include keeping the systems operational, incorporating new technologies and technological enhancements into existing remedies, meeting remedial action objectives, addressing unanticipated problems, and funding long-term operations given the rising costs of all requirements across federal government sites. This paper presents a case study on three full-scale remedial actions in place at Hill Air Force Base (AFB), Utah where chlorinated solvents, specifically trichloroethylene, are the primary contaminant of concern. These systems are operating under a variety of geologic and hydrologic settings, technology types and contaminant concentrations. The technologies employed include pump and treat for DNAPL recovery using phase separators and steam stripping; in situ treatment using air sparging and soil vapor extraction; and pump and treat using air stripping. The systems also represent a range of site complexity that influence the expected effort, resources and costs to keep them operational.

INTRODUCTION

Hill AFB, located approximately 30 miles north of Salt Lake City, Utah, is on the National Priorities List. CERCLA restoration activities at this facility are conducted under the auspices of the US Air Force's Installation Restoration Program. A series of nine geographic operable units were established by a Federal Facilities Agreement (FFA) between the Air Force and governing regulatory agencies (USEPA Region VIII and the Utah Department of Environmental Quality). At present, most sites have been characterized and are proceeding into the remedy selection and implementation phase. To date, there are 11 actively operating treatment systems and 6 passive control systems (engineered caps).

Our experience over the past several years, like those of many others, has shown that the magnitude of technical and operational challenges, and the associated cost of operating treatment systems has been surprisingly greater than anticipated. Budgets have been stretched and frustration abounds. However, in retrospect, many of these seemingly unanticipated problems could have been foreseen and dealt with more effectively and efficiently. We are using our experiences to streamline future efforts at this facility, and reduce the life-cycle cost of remedial systems. The case studies and conclusions presented below provide insights into the importance of strategic planning, identifying potential problem areas and the allocation of personnel resources to meet critical needs.

PROBLEM STATEMENT

Many decisions for Hill AFB restoration actions are driven by the need to meet FFA milestones associated with discrete work phases (e.g., remedial investigation, feasibility study, remedial design). Given our historic eagerness to meet regulatory milestones, we spend the least amount of time adequately thinking about and planning for the longest part of the remediation process—long-term system operations.

Science and engineering disciplines are applied throughout the investigation, remedy selection and design process. Although we have adequately characterized the sites in question, and know the standard and innovative technologies that can be applied, we generally fail to accurately consider the interrelationships of all site-specific factors that may affect long-term system operations and maintenance. As a result, post-implementation problems typically result in unexpected costs that, upon analysis, we find may have been avoided or minimized with better planning during early stages of the project. There are several reasons for this; however, two major themes seem to have the greatest influence. One is that the decision-makers (project managers), considering both their engineering disciplines and experience level, are not capable of thinking in operation terms due to a lack of expertise and experience bias. That is, we don't think like operators. The other is that we have not yet embraced the concept of life cycle project management.

CONSIDERATIONS

The term "life cycle cost" of remediation systems must be broadened to account for design, construction, optimization, operation and maintenance, monitoring, and shutdown activities. Our greatest ability to impact life cycle cost is in the early stages of a project. However, we are typically 3-5 years behind in applying this approach to operations and maintenance, which results in a less than optimum start-up period. While our science and engineering skills are high, the goal of reducing life cycle costs can only be achieved through better planning and asking the question "how will our systems be operated now and in the future?" This planning should take place during remedy selection, and must be considered during the design of systems. In addition, we have found that early participation in the process by operations and maintenance personnel are a large component of the equation. Until we acknowledge this and commit the appropriate resources, we will not be able to proactively optimize remedial system performance and reduce life cycle costs.

Adequate planning and commitment of resources is central to the success of operating each system at minimum cost. Specifically, the following questions must be answered with respect to personnel resources:

- How much will you need?
- What system specific variables will most influence application of the remedy?
- At what phase of the project will you expect and need resources, and for how long?
- What type and level of expertise will be required?

CASE STUDIES

The resource requirements are unique to each site, but generally fall into categories which are a function of: 1) the complexity of the site, and 2) the type of system and technology employed. The following descriptions of treatment systems further substantiate this. A graphical representation of the resource requirements for each of the three systems in question is depicted in Figure 1. This figure shows resource demand as a function of system operating time.

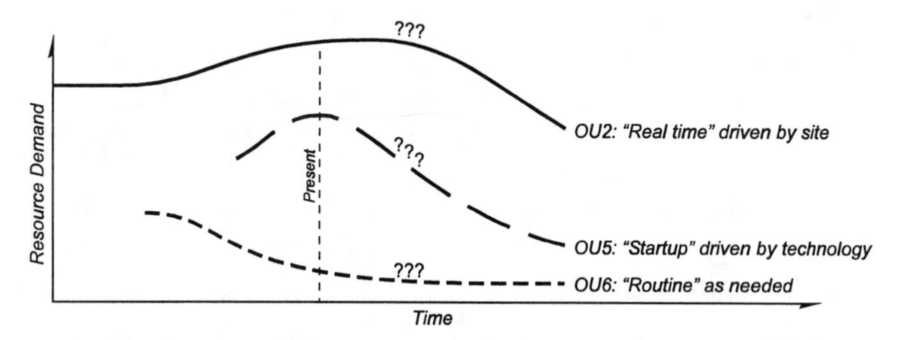

FIGURE 1. Resource demand as a function of system operating time

The Operable Unit (OU) 2 system has been in operation since 1993, and currently the technology employed at this site is well understood. Table 1 summarizes the features of this system. The significant operating challenges are tied to dynamic site conditions associated with the discovery of new DNAPL pools and innovative technology demonstrations to remediate the vadose zone. These conditions are a result of site geology and the regulatory requirements to substantiate technical impracticability. Many system enhancements have also been pursued since first operation in 1993. Originally, we did not anticipate the effects that these enhancements would have on the operation of the SRS. It shortly became evident that more attention was necessary.

We subsequently instituted intensive planning and developed a diversified team to deal with all the aspects of the project. The net effect of changing site conditions and system enhancements has required a steady high level of personnel resources that must be available for 'real time' support. As seen in Figure 1, we expect a peak resource demand to continue for the near term, then subside when enhancements are complete. The resource mix has also shifted over time as more engineering and management personnel are required in addition to the operating labor. Our conclusion is that early recognition of the problem and additional planning could have reduced conflicts, enhanced our ability to maintain schedules, and saved the cost of less than optimum system operations.

TABLE 1. Summary of the OU 2 source recovery system (SRS) characteristics and cost.

System Type:	Pump and treat
System Description:	Extraction well network, phase separators to remove free product, steam stripper for water treatment
Objectives:	Remove DNAPL and treat groundwater
Primary Contaminant and Concentration Range:	TCE, up to 1,000 ppm
Capital Cost:	$2,563,000
Annual O&M Cost:	$380,000
Treatment Cost:	$41/lb of TCE removed
Routine Maintenance and Repairs:	Acid wash heat exchanger and steam stripper, replace pump seals and boiler valve, clean tanks
Non-routine Maintenance and Repairs:	Refurbish steam stripper, automation upgrades for remote operation, replace acid tank
Significant Challenges:	Operating a system with changing input characteristics

The OU 5 system has been in operation since the summer of 1997. The subsurface conditions at this site are relatively static, but significant challenges of operating this system include system start up and verification of system treatment performance. This innovative remedial technology requires intensive levels of science and engineering resources to troubleshoot a unique system, correct design deficiencies and determine the effects of construction methods on system performance. Table 2 summarizes the features of this system. Again, as seen in Figure 1, we expect that the resource demand will decrease with time.

TABLE 2. Summary of the OU 5 aeration curtain characteristics and cost.

System Type:	In situ air sparging and soil vapor extraction
System Description:	Installed in 1600 foot trench using bioslurry (guar) during construction
Objectives:	Contain plume and treat groundwater
Primary Contaminant and Concentration Range:	TCE, 0.5-1.0 ppm
Capital Cost:	$1,500,000
Treatment Cost:	[to be determined when system stabilizes and additional data are available]
Annual O&M Cost:	$110,000 (extrapolation based on first 6 months of operation)
Routine Maintenance and Repairs:	Replace pressure and temperature gauges, lube blowers, drain SVE lines
Non-routine Maintenance and Repairs:	Replace blower air seals and oil seals, repair cap subsidence, install PVC risers to measure air flow
Significant Challenges:	Start up of a system based on innovative treatment technology

The OU 6 treatment system has been in operation since 1996. Table 3 summarizes the features of this system. Both the subsurface conditions and technology employed at this site are relatively static. As a result, the level of personnel resources needed following system start up have dramatically decreased. Regular coordination meetings with field level staff have been instituted to aid in early detection of system problems so they can be handled promptly and effectively. In addition, the resources required are typically on the level of system operator as opposed to engineering support.

TABLE 3. Summary of the OU 6 off-base treatment system characteristics and cost.

System Type:	Pump and treat
System Description:	Extraction well network, conventional pump and treat using low profile air stripper
Objectives:	Contain plume and treat groundwater
Primary Contaminant and Concentration Range:	TCE, 50-100 ppb
Capital Cost:	$522,000
Annual O&M Cost:	$45,000
Treatment Cost:	$15,100/lb of TCE removed
Routine Maintenance and Repairs:	Leak checks, oil change, inspections, pressure testing
Non-routine Maintenance and Repairs:	Replace variable frequency pump drives, re-develop extraction wells
Significant Challenges:	Dealing with unexpected operational problems

CONCLUSIONS

There are many engineering and operational challenges, both anticipated and unanticipated, associated with each of the systems described. The goal of project managers should be to reduce the number of unanticipated problems since most of them can be foreseen. We generally assume too optimistic a view of the operation period and associated challenges. We need to balance this with realism through adequate evaluation of the unique site and technology situations in which the systems operate, and then plan our personnel resources and response actions accordingly. In general, from a review of the systems described, the following steps can help to streamline system operations and reduce the life cycle cost of remedial systems:

- Set aside time to plan during all phases of work, starting with design;
- Build an integrated project team including process and field level input;
- Develop a clear understanding among all team members of the remedial system goals and objectives of system optimization;
- Maintain the continuity of project teams for technical problem solving;
- Foster proactive communication through regularly scheduled meetings;
- Ensure adequate documentation of decisions made and actions taken; and
- Integrate the operations of all treatment systems within a geographic OU.

The importance of early and continual planning and routine baselining cannot be overemphasized. Adequate planning will allow managers to meet the following objectives:

- Control and forecast future costs. Our ability to better predict and control life cycle costs starting as early as the technology selection stage will minimize or eliminate some of the anticipated and unanticipated costs. This will allow for more accurate budgeting of future expenses associated with system operations and maintenance.
- Allow "real time" management and decision making rather than making changes in a crisis mode.
- Reduce life cycle costs by reducing technical and operational inefficiencies and maintaining a concerted focus on reducing overall cost.
- Determine appropriate contracting strategy. Understanding and anticipating resource requirements can help with selection of the most qualified system operations contractor. In addition, understanding the level of required routine and non-routine repairs and maintenance will affect the type and duration of contract chosen.

STAKEHOLDER INVOLVEMENT– FREE CONSULTING THAT RESULTS IN ENDURING DECISIONS

***Gretchen E. Hund,* Battelle, Seattle, Washington, United States**

ABSTRACT: In developing and deploying new technologies, environmental technology managers have been principally concerned about technology effectiveness, with a major focus on performance. They have been less concerned about public and regulatory acceptance of technologies, and often address these issues only after the technology performance is deemed effective. In addition, technology managers often regard issues of public and regulatory acceptance as someone else's responsibility. This approach has often resulted in developing technologies that are difficult to deploy because of regulatory obstacles and/or public concern. Battelle participated in a multi-year program with the Department of Energy (DOE) to develop and demonstrate technologies for cleaning up groundwater and soil plumes. Battelle developed an approach to address the interests and concerns raised by individuals and groups who felt they had a stake in the deployment of a technology. Stakeholder issues were considered in designing demonstration tests and used in parallel with technology performance to evaluate the likelihood that a technology would be accepted and deployed. This effort, focused on fairly mature technologies, has been followed with an on-going project focused on basic science and less-developed technologies in the area of bioremediation. The goal is to identify issues and concerns early on so they may be considered and technology acceptance enhanced.

INTRODUCTION

Battelle has conducted several projects for DOE focused on involving stakeholders in evaluating innovative technologies to clean up groundwater and soil. Stakeholders were encouraged to participate in the demonstration of these innovative technologies in order to improve decisions made by DOE and its contractors about technology development, demonstration, and deployment.

One project, called the Volatile Organic Compound Arid Site Integrated Demonstration (VOC-Arid ID), was a three-year effort (1991-1993), funded by DOE's Office of Technology Development (OTD), to develop and demonstrate technologies for cleaning up VOC contamination at arid sites. A second project, called the Plumes Focus Area (1992-1995), also funded by OTD, was focused on all groundwater and soil plume contamination at all DOE sites. The contaminants of concern included VOCs, dense non-aqueous phased liquids, metals, and radionuclides.

A third, on-going project deals with stakeholders' issues and concerns regarding basic science and less-developed technologies in the area of bioremediation. DOE's Office of Energy Research is conducting the Natural and Accelerated Bioremediation Research (NABIR) program. Stakeholders may have issues and concerns about this research and with the prospects of Hanford being chosen as a Field Research Center for the program. Similarly to the two OTD-funded projects, stakeholders are being identified and engaged but, in addition, a communications protocol is being developed to help scientists communicate with stakeholders. The goal of this project is to identify issues and concerns early on in the R&D process so they may be considered before large investments are made. Stakeholder concerns can then be considered in modifying research plans to enhance technology acceptance.

METHODS

The basic approach is to identify people and organizations with a stake in the remediation process and hence in the research, development, and ultimate demonstration of innovative technologies. Stakeholders have included public interest group and environmental group representatives, regulators, technology users, Native Americans, Hispanic community members, and local elected officials. These people are invited to be involved in ways they find convenient and meaningful, and they are presented with substantive information about the research and technologies. By identifying issues and concerns, defining the kinds of information needed from the R&D and demonstrations, and assessing the acceptability of the technologies for use, these people will help DOE make appropriate investments that have the greatest promise for protecting public health and the environment.

Within the VOC-Arid ID project, the process first focused on the Hanford site in Washington State, then expanded to other western sites. Technology evaluation criteria identified by Hanford stakeholders and later endorsed by other arid site stakeholders are provided in Figure 1.

Figure 1. Technology Evaluation Criteria Developed by Stakeholders

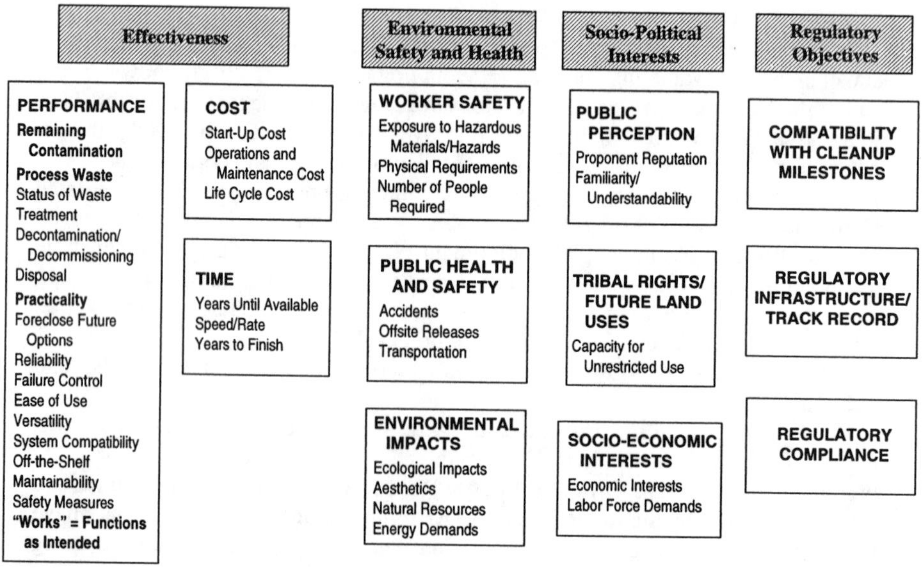

Effectiveness

PERFORMANCE
Remaining Contamination
Process Waste
Status of Waste Treatment
Decontamination/ Decommissioning
Disposal
Practicality
Foreclose Future Options
Reliability
Failure Control
Ease of Use
Versatility
System Compatibility
Off-the-Shelf
Maintainability
Safety Measures
"Works" = Functions as Intended

COST
Start-Up Cost
Operations and Maintenance Cost
Life Cycle Cost

TIME
Years Until Available
Speed/Rate
Years to Finish

Environmental Safety and Health

WORKER SAFETY
Exposure to Hazardous Materials/Hazards
Physical Requirements
Number of People Required

PUBLIC HEALTH AND SAFETY
Accidents
Offsite Releases
Transportation

ENVIRONMENTAL IMPACTS
Ecological Impacts
Aesthetics
Natural Resources
Energy Demands

Socio-Political Interests

PUBLIC PERCEPTION
Proponent Reputation
Familiarity/ Understandability

TRIBAL RIGHTS/ FUTURE LAND USES
Capacity for Unrestricted Use

SOCIO-ECONOMIC INTERESTS
Economic Interests
Labor Force Demands

Regulatory Objectives

COMPATIBILITY WITH CLEANUP MILESTONES

REGULATORY INFRASTRUCTURE/ TRACK RECORD

REGULATORY COMPLIANCE

These criteria defined the types of information stakeholders wanted about a technology before they were comfortable seeing it used. The second step was to use these criteria to provide information on specific technologies in the form of technology profiles. Three focus groups were held with stakeholders who were peers, to describe their issues and concerns about groundwater remediation technologies and suggest data requirements they felt should be considered in designing test plans for the demonstrations. Their comments were compiled in a report and distributed as background information for an integrated workshop of participants from all three focus groups. The workshop gave participants an opportunity to hear first hand what each other's issues and concerns were, and to work collaboratively with the technology developers to determine how those issues and concerns should be addressed in the demonstration. The process and results of involving Hanford stakeholders in review of the groundwater technologies have been documented in "Phase II Stakeholder Participation in Evaluating Innovative Technologies: VOC-Arid Integrated Demonstration, Groundwater Remediation System" (Peterson et. al., 1994).

By first involving Hanford stakeholders and then expanding the involvement to other arid site stakeholders, it was possible to evaluate differences and similarities in stakeholders' perspectives to broaden deployment potential of the technologies. The four additional arid sites involved were Rocky Flats, Colorado; Idaho National Engineering Laboratory, Idaho; and the Sandia and Los Alamos National Laboratories in New Mexico. A fifth site, the Lawrence Livermore National Laboratory, was asked to be involved but was unable to participate. Interviews were conducted with stakeholders at these sites to identify their issues and concerns. Discussion focused on three questions:

- What do you consider important in choosing an environmental restoration technology? What additional information would you need to evaluate these candidate technologies with confidence?
- Are there aspects of these technologies that concern you?
- What features of the technologies do you see as advantages compared to today's available cleanup technologies?

The results of this effort are documented in "Arid Sites' Stakeholder Participation in Evaluating Innovative Technologies: VOC-Arid Site Integrated Demonstration," (Peterson et. al., 1995). The feedback from stakeholders was used to enhance the design of the demonstration test plans for the relevant technologies. The results of technology demonstrations were then reported back to stakeholders to determine technology acceptance. An example of the feedback received from stakeholders following a technology's demonstration is captured in "Stakeholder Acceptance Analysis: In-Well Vapor Stripping, In-Situ Bioremediation, and Gas Membrane Separation System" (Peterson, 1996).

This approach was further broadened under the Plumes Focus Area where technology demonstration tours were conducted and stakeholders were invited from all sites that had a similar contamination problem and could potentially benefit from the technology being demonstrated. Participants were allowed to see the technology "in operation", ask questions, talk off-line to other stakeholders who shared their perspective (e.g., regulators, technology users), and share knowledge about ways to address a particular problem.

The NABIR project is just underway, and by design focuses on broadening the circle of stakeholders involved in the past to determine their interest in the overall NABIR program and to identify issues and concerns associated with Hanford being chosen as a Field Research Center. Unlike the bioremediation technology evaluated under the VOC-Arid ID (the propagation of indigenous microorganisms), this program also includes R&D of non-indigenous microorganisms. Genetically engineered microorganisms (GEM) are not within the scope of present NABIR efforts, but may be included at a later date. The basic science being conducted and R&D areas are all being described in technology area profiles that will be used to communicate with stakeholders. The focus is primarily on Hanford stakeholders, but there is also a task to engage national stakeholders. The objective of this first-year project is to alert these stakeholders about the program and determine how they would like to be involved. Furthermore, a communications guide is being developed for scientists funded under the program to use in their interactions with stakeholders. Findings from analyzing legal and social concerns about developing bioremediation technologies were used in designing this NABIR project (Bilyard et. al., 1996.)

RESULTS AND DISCUSSION

There were several results of these stakeholder involvement projects. First, stakeholders appreciated being involved in a meaningful manner. For example, in the VOC-Arid ID, stakeholders saw how their feedback was used in designing the format to describe technologies, in designing the test plans, and in designing the format used to describe demonstration results. Second, stakeholders found the technologies acceptable. There was much good will because the stakeholders felt that their input was thoughtfully considered. Third, technology developers learned quite a bit about how to communicate with stakeholders through this project. These developers were at first a bit nervous about participating in the stakeholder events. Encouragement and support from the project manager was critical, as was having the stakeholder involvement approach imbedded within the technical approach, not separate from it. The full project team embraced a parallel process, identifying and working to resolve technical issues alongside regulatory issues and other stakeholder issues. One of the technology developers began his involvement as a skeptic and by the end of the integrated workshop reported that he found the input to be free consulting. As a result of a technology user becoming knowledgeable about a particular technology, that technology was bought from the technology developer and is now being deployed worldwide.

Stakeholders have categorized several issues as "watch outs" to consider in developing technologies and selecting technologies:

- Minimize the number of steps in a treatment system to the degree possible.

- Evaluate the comprehensive treatment train (system) including all of its technology components to decrease the need to go back into a site and make more cleanup work necessary.
- Consider how co-contaminants will respond under a particular demonstration to minimize future needed cleanup.
- Destroy contaminants onsite, if possible, rather transport them.
- Minimize the transferring of process waste from one environmental medium to another.
- The more robust a technology the better; it should be able to function with minimal human intervention.
- Use the existing labor pool to operate, repair, and manage a technology; avoid being dependent on highly skilled, difficult to access personnel.
- Make sure a technology fits into the regulatory framework of the targeted sites of interest. Technology developers need to provide data applicable to all relevant regulatory drivers.

The results of the Plumes Focus Area were less positive. The stakeholder involvement program was difficult to manage, given the number of sites and a lack of direct control over how stakeholder input was solicited at sites. Demonstration tours, however, were found to be helpful and benefited the parties who participated in them. Ideas and knowledge were shared, and contacts were made among stakeholders with different perspectives from across the country.

As mentioned, the NABIR project is just underway so results are premature. The hypothesis is that stakeholders will be more interested and concerned about R&D of non-indigenous microorganisms and GEMs, if they are included in the scope, than about indigenous microorganisms. This was the finding under other cases (Bilyard, et. al., 1996). An additional hypothesis is that it may be difficult to engage stakeholders in providing feedback on basic science unless an immediate future application can be defined and described. The results of this stakeholder engagement will hopefully improve stakeholders' understanding of the NABIR program and also improve scientists' and engineers' understanding of stakeholder issues and concerns. Scientists will also be better equipped to interact with stakeholders, given the communication guidance that will be available. Finally, science and R&D under this program will hopefully be improved.

CONCLUSIONS

By using a proactive, stakeholder involvement approach in conducting science and R&D, results can emerge that have the greatest potential of being applied as a technology due to stakeholder acceptance. If conducted regionally, such acceptance can extend from one site to several sites and thereby reduce the need to re-demonstrate technologies at applicable sites.

Benefits of this approach are that it:

- Identifies major concerns so investment decisions can be made early in technology development (and potentially in earlier research),

- Enables stakeholders to become knowledgeable about science, R&D, and the technology demonstration process so they have more confidence in the results, and
- Reduces the time required to advance technology from demonstration to use by addressing technical and institutional issues in parallel.

Challenges of this approach are:

- Scientists, engineers, and social scientists must work together in a manner that is traditionally foreign to them,
- Scientists and engineers need to accept that scientific data are only part of the decision equation, and
- Social scientists need to understand the science and technology to become an active part of the technology development team.

REFERENCES

Bilyard, G. R., G. H. McCabe, K. A White, S. W. Gajewski, P. L. Hendrickson, J. A. Jaksch, H. A. Kirwan-Taylor, and M. D. McKinney, 1996. "*Legal and Social Concerns to the Development of Bioremediation Technologies*," PNNL-11301/UC-630, Prepared for the U.S. Department of Energy's Office of Technology Development under Contract DE-AC06-76RLO 1830, September 1996.

Peterson, T. S., "*Stakeholder Acceptance Analysis: In-Well Vapor Stripping, In-Situ Bioremediation, and Gas Membrane Separation System*," PNNL-10912/BSRC-800/95/021, Prepared for the U.S. Department of Energy's Office of Technology Development under Contract DE-AC06-76RLO 1830, December 1995.

Peterson, T., G. H. McCabe, B. R. Brockback, P. J. Serie, and K. A. Niesen. 1995. "*Arid Sites Stakeholder Participation in Evaluation Innovative Technologies: VOC-Arid Site Integrated Demonstration.*" PNL 10524/BSRC-800/95/003, Prepared for the U.S. Department of Energy's Office of Technology Development under Contract DE-AC06-76RLO 1830, May 1995.

Peterson, T., G. H. McCabe, P. Serie, and K. Niesen, 1994. "*Phase II Stakeholder Participation in Evaluating Innovative Technologies: VOC-Arid Integrated Demonstration, Groundwater Remediation System.*" Prepared for Thomas Brouns, Pacific Northwest Laboratory, Battelle HARC: Seattle, WA, April 1994.

A COMMUNITY-BASED, MULTI-STAKEHOLDER INITIATIVE TO REMEDIATE THE SYDNEY TAR PONDS

Maria Dober, (Environment Canada, Dartmouth, Nova Scotia, Canada)
Jim Ellsworth, (Environment Canada, Dartmouth, Nova Scotia, Canada)

ABSTRACT: A Joint Action Group (JAG) on the Environmental Clean-up of the Muggah Creek Watershed, representing community interests in partnership with three levels of government was formed to identify and evaluate remedial options for the Sydney Tar Ponds and the Coke Oven sites. JAG's mission is to educate, involve and empower the community through partnerships, to determine and implement acceptable solutions for Canada's worst hazardous waste site and to assess and address the impact on human health. JAG represents one of the most challenging multi-stakeholder initiatives that the Canadian Government has supported to date. In the 18 months since the inception of JAG the process could be described as tumultuous. It is often noisy and frustrating as we strive to develop collective wisdom and learn to exercise a collective will. However, notwithstanding the growing pains, and the technical and partnership challenges it faces, JAG has made substantial progress toward the ultimate objective of assessing and cleaning up the site(s).

BACKGROUND

Attaining public and political acceptance of complex remedial technologies is difficult. Overcoming barriers between the technical community and the lay person, is further complicated when there are controversial social and economic dimensions. In these instances, finding acceptable solutions requires the early and ongoing collaboration between lay people, the technical community and many others. Experience in such situations has shown that human needs must be met before substantive technical issues can be dealt with.

Human needs include the psychological needs of value and trust and the procedural needs of process and participation. Experience has also shown that traditional public participation processes often fail to meet these needs . Public participation methodologies are based on the underlying assumption that there is a single proponent with the authority and the resources to address the issue. These methodologies foster a "them and us" relationship between proponents and participants and segregate the lay public from the technical community.

Research has shown that the acceptance and use of information depends upon the relationship between the data producer and user, the role of the end user in producing the data and whether the user can integrate the new data with their existing knowledge and apply it readily (MacKenzie-Mohr, 1996). Experience shows that people with no stake in an issue tend to be influenced by "experts", while people with a stake in the outcome want to verify the integrity of the information.

When professionals are in situations where they are not the distant expert they sometimes feel frustrated and powerless. This can lead to what Benson as reported by Donaldson (Environment Canada, 1994) terms the “deadly sins”. They include ignorance of group processes and development, attempting to control others, a disproportionate fear of failure, comparison with dissimilar processes, attachment to old methods, detachment from the group and the process, and not accepting the knowledge of others as equal. Professionals need to be aware of the deadly sins.

In some instances, solutions require new forms of governance such as the governance model utilized by Ecosystem Initiatives. (Figure 1)

Characteristics of New Forms of Governance	
Participants:	Civil Society, Governments and the Private Sector
Interests:	Social, Economic and Environmental
Perspectives:	Local, Regional, National and Global
"Because each participant has specific interests, weaknesses and strengths, good governance must facilitate collaboration among all three." - United Nations Development Program	

FIGURE 1. Characteristics of new forms of governance.

THE JOINT ACTION GROUP ON ENVIRONMENTAL CLEAN UP

The Muggah Creek Watershed is approximately 24 square kilometers in area and is located in Sydney, Nova Scotia, Canada. A century of steel making and cokeing operations turned a pristine estuary and watershed into Canada’s worst contaminated site. Contaminants include PAHs, PCBs, heavy metals and other organics. The Tar Ponds occupy an area of approximately 35 hectares and contain 700,000 tonnes of contaminated sediments. The Coke Ovens complex is approximately 60 hectares and is also grossly contaminated. A predominant hydrogeological feature of the area is the presence of fractured bedrock which underly the sites. Both sites lie in the middle of a populated area which is often referred to as the Cancer Capital of Canada. The contaminated site and perceived health risks associated with it stifle economic growth and threaten existing livelihoods in an area with over 25% unemployment.

In 1996 the provincial government announced plans to encapsulate the Tar Ponds with slag at a cost of $25 million. This sparked a strong reaction from the community, who opposed the containment and asked that the provincial and federal governments support a community based approach to developing remedial options for both the Tar Ponds and Coke Ovens Sites. Governments listened and eventually the Joint-Action Group on Environmental Clean Up of the Muggah Creek Watershed was formed. JAG provides a forum and a framework where citizens, organizations and the three levels of governments participate as equals in a community-based process that is inclusive, open and transparent.

The participants produced a bold governance model and an innovative decision making process (Figure 2). JAG is committed to working with the larger community and the three levels of government in developing remedial solutions for the contaminated sites which are socially acceptable, economically feasible and environmentally sound. To accomplish its mission JAG puts a great deal of emphasis on community development - building the relationships, the skills, knowledge and trust necessary to make their governance model work.

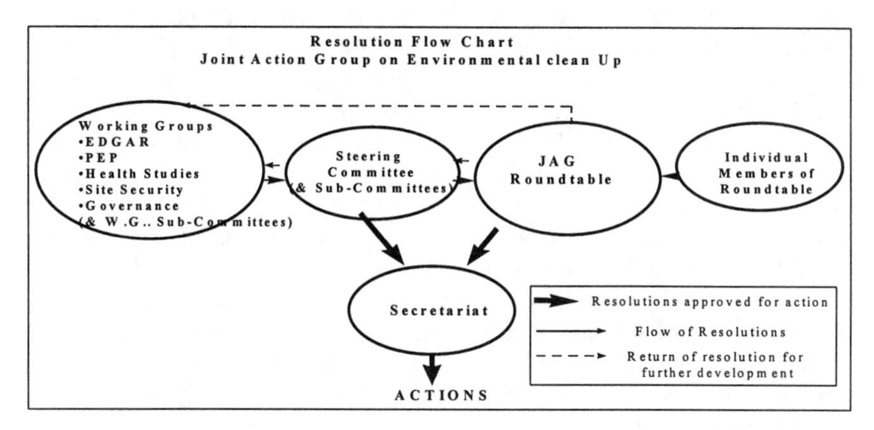

FIGURE 2: JAG resolution flow chart.

Less ambitious multi-stakeholder initiatives have been known to take up to two years to work cohesively. The complexity of the issues at this site, and the ambitious governance model adopted by JAG suggests that developing a collective identity and capacity will be difficult and time consuming. During this time there will be enormous pressures to jump to the action stage without achieving informed consent of all JAG members. The technical and partnership challenges are significant.

TECHNICAL CHALLENGES

The Sydney Tar Ponds and the Coke Oven sites represent one of Canada's largest and most complex hazardous waste sites, with issues such as Dense Non-aqueous Phase Liquids (DNAPLS) and Light Non-aqueous Phase Liquids (LNAPLS) in fractured rock environments, sediment transport, and a subsurface which contains approximately 80 miles of underground services. It is a challenge to generate and present complex technical information in a manner that is accepted and understood by the community including those with little technical background.

JAG members recognized the need to become familiar with the technical terminology, definitions and policies. To accomplish this, JAG began to build capacity both among its own members and within the community as a whole, through participation in a series of technical open houses, symposiums, workshops and conferences.

Each of the three levels of government (federal, provincial and municipal) have dedicated technical staff in support of JAG activities. These staff have served as members of the various working groups since JAG was first established. Many of the staff were raised in and still reside in the community. Local residents have earned the admiration and respect of professionals with their unwavering dedication to this initiative, and tireless efforts to move the process forward. Mutual respect and valuing all of the knowledge (scientific, local and traditional) that contribute to collective learning helps generate a common understanding and a credible, understandable data base.

The Environmental Data Gathering and Remedial Options (EDGAR) working group, one of five groups in the JAG process, is a good example of valuable new relationships between the technical community and the lay public. This group is tasked with identifying the nature and extent of contamination within the watershed, the contaminant migration pathways, and the potential risks to receptors of concern. In addition, they will evaluate and recommend clean-up criteria for potential future uses of the site. EDGAR has the lead in serving and assisting JAG and the larger community in evaluating, and recommending remedial solutions that are socially acceptable, economically feasible and environmentally sound.

These ambitious tasks have been approached with enthusiasm by the members of EDGAR, who are learning about the technical issues and the methods for assessing a contaminated sites. While the group has had difficulty, a substantial amount of progress has been made. To date, EDGAR has initiated and/or completed the following technical projects:

- a sampling program to determine residual contamination associated with derelict structures and waste material remaining on the Coke Oven site.
- the development of a methodology, for the selection and evaluation of appropriate technologies for demonstration and or treatability studies.
- the development of a comprehensive quality assurance/quality control policy
- the pre-design of a sewer collector system, designed to divert the 28 sewers which presently discharge into Muggah Creek.
- a study to determine whether leachate from the municipal landfill is migrating onto the Coke Ovens property via surface and groundwater
- a Phase I Environmental site assessment, and
- a geophysical survey of the Coke Ovens complex.

While EDGAR is following the five phases of site assessment and remediation as recommended by the Canadian Council of Ministers of Environment (CCME) (1991), attempts have been made to address immediate community concerns. As well, work is continuing on mechanisms to allow the public to participate in the review of technical contracts while still maintaining the integrity and confidentiality of the review process.

JAG is an example of action learning, still practiced by some indigenous people, where there is no distinction between teachers and students: everyone contributes and everyone learns. The approach generates a willingness to learn and apply the information by all involved. Scientific and local knowledge is combined in a single focal point and made available to everyone in an understandable format.

PARTNERSHIP CHALLENGES

Like other ecosystem initiatives deploying new forms of governance, the JAG process takes some getting used to. It engages all participants in a change process and presents all with a steep learning curve. JAG participants are challenged to think and treat one another as partners in a community of interests (Figure3) . This is a great departure from hierarchical models as the hierarchy of participants is replaced with common hierarchies of purpose, values and scale.

While participants make their way along the learning curve, create a collective identity, develop capacity, place trust in their collective wisdom and establish new hierarchies, they will continue to jockey for position on a non-existent hierarchy of participants. Governments, non-government organizations and citizens will occasionally try to dominate the process and serve as the catalyst. Various participants will occasionally form blocks or value tribes in an attempt to control outcomes, or to fill power vacuums. While internal positioning is often done with the best of intentions, it is often perceived by other participants as a demonstration of bad faith and sets off a round of positioning.

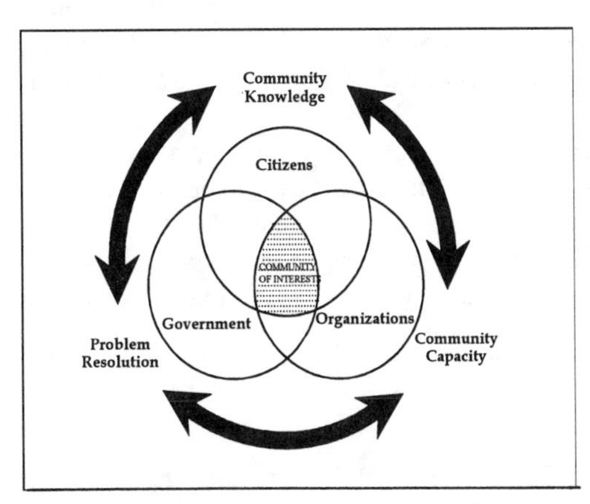

FIGURE 3: The formation of communities of interest

Participant positioning is sometimes experienced in JAG. Occasionally when blocks of participants, or value tribes, compete for the hearts and minds of other JAG participants and use the media as a tool to champion their views and values, while discrediting those of others. While a certain degree of internal conflict is unavoidable and even desirable, the JAG partnership must continue to develop its identity and capacity, and participants must agree on boundaries and decision making domains. JAG participants are still determining what decisions will be made collectively and what decisions will remain the sole prerogative of citizens, non-government organizations and government departments.

The inability to make consensus decisions is a contributing factor to participant positioning within JAG. Voting fails to reconcile the interests, concerns and/or expectations of members and often serves to create conflict when those who lose the vote seek another opportunity to have their perspective prevail. As a result, Roundtable meetings often serve as redress forums for matters the Steering Committee failed to resolve.

The core of the JAG process is the generation of informed consent through consensus decision making and JAG members have learned that consensus decision making is a developed skill that groups do not automatically possess.

CONCLUSIONS

As JAG demonstrates, the community-based approach require changes in how we do business, how we view our roles in relation to one another and how we define community. The approach requires professionals to look upon citizens as partners rather than consumers of their services and expertise. Accepting and applying a definition of community which includes government takes an effort by citizens, organizations and governments themselves. Ultimately, all participants will have to demonstrate confidence in their collective wisdom, if the community-based initiative is succeed.

The community-based approach is not necessary or desirable in all situations. In some situations, however, it is the only way to attain the critical mass of public and political support to implement remedial actions. The Muggah Creek Watershed is one such situation. It is too early in the process to say whether JAG will be successful. Considering what is at stake and the progress JAG has made in the face of enormous technical, social and economic challenges, the authors are confident that JAG will succeed and are committed to help make it happen.

Technical dimensions aside, the JAG process is largely one of developing partnerships by reconciling expectations. While many expectations have been reconciled, more group development is required to enable JAG to fulfill its mandate. JAG has made remarkable progress in managing group tasks, however, greater emphasis is required on group development, particularly consensus decision making.

REFERENCES

Canadian Council of Ministers of Environment. 1991. *National Guidelines for Decommissioning Industrial Sites*. CCME-TS/WM-TRE013E

Environment Canada. 1994. *Working in Multistakeholder Processes.*

MacKenzie-Mohr, D. 1996. *Promoting a Sustainable Future: An Introduction to Community-Based Social Marketing.* National Round Table on the Environment and the Ecomony, Ottawa, ON.

EVENTS SHAPING PUBLIC PERCEPTION OF DIOXIN AND RELATED COMPOUNDS

Betty K. Jensen, Ph.D., Staten Island, NY, USA
Richard A. Jensen, Ph.D., Hofstra University, Hempstead, NY, USA

ABSTRACT: The paper reviews the history of public communication regarding dioxin. These communications of government, national and local, as well as direct communication by members of the scientific community with the public, and media reporting, form the basis for the public's perception of dioxin. The paper reviews key events shaping this public's perception, including Love Canal in New York, Times Beach in Missouri, and Seveso, Italy. Dioxin's connection with Agent Orange has brought it further press coverage. Dioxin has been implicated as a cancer causing substance, has been tied to birth defects, and more recently has been described as an endocrine disrupter. The paper sheds light on events which shaped the public's perceptions and attitudes regarding dioxin.

INTRODUCTION

While the term dioxins refers to a family of 75 different compounds, chlorinated dibenzo-*p*-dioxins, or CDDs, the public generally uses the term dioxin to refer to the most toxic of these 75 compounds, 2,3,7,8-tetrachlorodibenzo-*p*-dioxin. Most dioxin has been a by-product in the synthesis of trichlorophenol (2,4,5-trichlorophenol), or TCP, which is an important precursor for a number of pesticides, and a number of other manufactured products. Dioxins therefore enter the environment as small impurities in other chemicals.

The draft Toxicological Profile issued for public comment by the Agency for Toxic Substances and Disease Registry (ATSDR) identifies dioxin as colorless solids or crystals. It is undetectable to the senses. Paradoxically, people exposed to dioxin at highly contaminated sites have complained of a strong overwhelming odor. The exposed people used the severity of the odor to gauge the level of contamination. The odor must have been due to the other contaminants with which dioxin occurred at these sites, rather than the dioxin itself.

The popular press often prefaces the term dioxin with the phrase "most toxic chemical known." This phrase is due to the toxic properties of dioxin when administered to laboratory animals. Different species have demonstrated different levels of sensitivity to the chemical, but its effects on humans are unclear. Numerous studies have been conducted to date. Some of the studies were inconclusive; some pointed to severe health effects, and yet others hinted at possible adverse effects. A number of government assessments were conducted on dioxin, with differing conclusions. The 1994 Environmental Protection Agency (EPA) dioxin reassessment document (EPA, 1994) indicates that "[t]hese compounds are extremely potent in producing a variety of effects in experimental animals....at levels hundreds or thousands of times lower than most chemicals of environmental interest... [H]uman studies demonstrate that exposure to dioxin and related compounds is associated with subtle biochemical and biological changes whose clinical significance is as yet unknown ..." A recent EPA press release (EPA, 1997) states that "dioxins can cause cancer, reproductive effects, immune response and skin disorders."

KEY EVENTS SHAPING PEOPLE'S PERCEPTION

All people are exposed to low levels of dioxin, mostly through their diet. A number of higher level population exposures have occurred at various times, with the more famous ones being at Love Canal in New York, Times Beach in Missouri, Seveso in Italy, and in Vietnam due to the use of Agent Orange. These four situations are described briefly below.

Love Canal. Love Canal is an abandoned canal in Niagara Falls, New York. The abandoned project was used as a disposal site between 1920 and 1953. In 1953, the canal was filled and covered with dirt, and sold to the local board of education for $1.00. "The deed contained a stipulation which said that if anyone incurred physical harm or death because of the buried waste, Hooker [the selling company] would not be responsible."(Gibbs) After home construction took place, residents began to complain of odors and various ailments as early as the late 1950's. In 1978, the state began to investigate the problems. While, the site was characterized as containing many carcinogens the public focused on dioxin.

Several months after the initial investigations, the state recommended temporary relocation of pregnant women and children under age of two. Later, 400 families were re-located, and eventually about a thousand families were permanently relocated "not because of adverse pregnancies, chromosome damage, or high chemical exposures, but because of mental anguish." (Gibbs)

Times Beach. Times Beach was a town in Missouri, covering about one square mile, on the banks of a river. Times Beach and more than thirty other sites in Missouri were contaminated with dioxin that came from the same chemical plant in Verona, Missouri. The plant produced TCP, which was used to make hexachlorophene, a bactericidal chemical. When TCP is produced, dioxin-contaminated still bottoms accumulate in the reaction vessels. Haulers were hired to dispose of these still bottoms.

One of the haulers was an independent contractor. Bliss removed 18,500 gallons of waste in 1971. Most of his business was in waste oil from service stations and others. He found out that spraying the floor of horse arenas with waste oil controlled dust, and used the same technique to keep the dust down in parking lots and unpaved roads. Applications were repeated after a few months. A total of 29 kg of dioxin were sprayed.(ATSDR)

After an indoor horse arena was sprayed in May 197 1, horses, other animals and people became ill. A Center for Disease Control (CDC) investigation led to the identification in 1974 of TCP and dioxin in the arena's soil.(Gough) Late in 1982 a reporter highlighted the problem to the community. The town was told by EPA that it would be as long as nine months before any soil testing. After public protest, the testing was accelerated, and the first round of sampling was completed early in

December. The following day the town was flooded by the river, and most of the town was evacuated. The analysis indicated high concentrations of dioxin in the soils tested. On December 23, 1982, the CDC recommended that Times Beach not be re-inhabited. In time, the federal government bought up the entire town, and the former town is now on its way to becoming a nature preserve.

Incineration was chosen as the preferred de-contamination method, and nearby residents were concerned about the safety of the proposed incineration. The incineration finally began in 1996, and was concluded in August 1997.

Seveso. An industrial facility accident in Seveso, Italy, in 1976 resulted in a release of dioxin mixed with a greater amount of TCP and caustic lime. The facility produced TCP for the purpose of making hexachlorphene. The first step in the production of TCP involves heating a mixture of chemicals in a sealed reaction vessel, an autoclave. If the heating is too rapid, the chemical reaction rate can accelerate, generating further heat and pressure, accelerating the reaction rate. When this happened at the plant, a pressure relief valve released, as per design, and the chemicals sprayed out. Estimates of the dioxin release range from half a pound, to a high of 260 pounds. (Fuller)

The release covered an area of nearly 900 acres, populated by about 40,000 people. For an entire week no public information was available regarding the specific chemicals involved in the contamination. The governor of the region called a meeting of his health officers a full eleven days after the accident, and the presence of dioxin was confirmed thirteen days after the accident. It was at then that the information was made public, and an evacuation was ordered. Five days later the evacuation zone was increased. Eventually most of the homes in the contaminated area were bought by the company and destroyed.

Seveso represents the only case of major exposure to dioxin by the general population. This population has been studied extensively, since all other major exposed groups were exposed occupationally, and tend to consist of men. Chloracne was identified among the population, with what appears to be a dose-response relations. Other adverse health effects, such as cancer, miscarriages, birth defects, and immune system effects are still being studied.

Agent Orange. Agent Orange was a 50:50 mixture of 2,4-D and 2,4,5-T in diesel oil and contained up to 20 ppm dioxin as a contaminant. Agent Orange was used by the U.S. to defoliate enemy territory and destroy enemy crops in Vietnam. Between 1962 and 1970, 11 million gallons of Agent Orange were used, and a total of 368 pounds of dioxin were estimated to have been sprayed. (Gough) A 1969 report stating that 2,4,5-T caused birth defects in laboratory mice ended the use of Agent Orange.

Only about 1300 air force personnel were assigned directly to the spraying operation. However 2.8 million Americans served in Vietnam, and a number of them were likely exposed during ground operations. Many of the veterans held Agent Orange responsible for ailments including cancer, and birth defects among their offspring. The birth defects claims were made even though animal studies indicated that birth defects occurred when pregnant animals were exposed to dioxin, not with male exposure.(Gough) The law suit against the maker of Agent Orange resulted in a settlement in 1984.

Epidemiological studies of the veterans gave mixed results, with re-analysis of seemingly the same data yielding dramatically different conclusions.

FACTORS AFFECTING RISK PERCEPTIONS

The factors affecting risk perceptions have been studied by many during the last twenty years. Some of the important attributes include familiarity, scientific knowledge, catastrophe, trust, impact on future generations, benefits, dread, reversibility, understanding, controllability, history, acute effect, salience of blame, and the issue of fairness or injustice. The level of outrage, (Sandman) which is a composite of several of the items already listed, plays a key role in determining the public's perception of a specific risk.

The least understood diseases and the ones with long latency periods are feared the most. (Jensen, 1990) Cancer and birth defects fall in these categories, as do some of the other concerns which have been associated recently with dioxin, such as endocrine disruptions. Studies have found that the public appears to have an "all or none" view that equates mere exposure to a carcinogen, regardless of level of exposure, with a high probability of being harmed.

A key factor affecting the public's perception relates to the release of risk information while the science is evolving. (Jensen, 1997) Announcing that the information is yet unknown suggests to the public that something is being hidden. Yet, withholding information until the science is more certain lends support to conspiracy theories and can potentially jeopardize public health, while releasing potentially alarming information prematurely can create inappropriate panic.

Situation-Appropriate Response. Official responses need to be appropriate to the specific situation at hand. Situations that call for an environmental response generally fall into the following categories: (i) A crisis due to either an accident such as was the case in Seveso, or to the identification of a major exposure, previously unknown, such as Love Canal and Times Beach; (ii) after the exposure has stopped, an investigation of the consequences of the exposure. The Vietnam veterans and Agent Orange present such an example; (iii) investigation of an on-going chronic low-level exposure; and (iv) anticipated exposure situation such as in the planning and permitting of a new facility.

In Seveso, the response was too slow. Nearly two weeks passed before the public was even told of the contaminant. In the second situation, an after-the- fact investigation may be allowed to consume more time, as long as the population believes that the investigation is being conducted with full faith. The Vietnam veterans returned from a war whose importance was not recognized by all in the U.S., and felt that their sacrifice was not appreciated. This added to their lack of trust of the government to investigate their concerns fully.

In the third situation, the case of low level exposure, and therefore, low level risk, again, the studies and decision-making can take time. But too often, the community does not appreciate the actual length of time required to conduct the scientific investigations. In the Love Canal situation, we suspect that while the community believed that it was in a crisis situation, local officials thought that the situation was more of the third kind, i.e., an on going chronic low-level exposure. This difference in perspective led to an inappropriate response to a specific situation. The fourth situation, before exposure, or an increase in exposure, has in fact occurred, gives the risk assessor and communicator the greatest opportunity for tailoring an appropriate communication program

CAUSES OF MISTRUST

Conflicting Messages. The public is often confused by conflicting messages it receives from the scientific community. When scientists disagree, the public tends to think that the scientist who presents a less alarming picture has ulterior motives, and the one who presents a more alarming picture is on their side. Using the logic of "better safe than sorry," the public often prefers to believe the more alarming scenario. Conflicting messages from government agencies add to the confusion. Action levels that are different among different government agencies, and in different countries provide such an example.

Official statements by a specific agency often provide "unintended" messages. A recent EPA press release (EPA, 1997) regarding its new regulations for the pulp and paper mills included a statement that its action will lead to "a 96% reduction in dioxin, resulting in undetectable levels to waterways." This statement, while correct today, suggests that the regulations will enforce "undetectable levels." Is the intent of the present regulation to automatically continue tightening emissions standards when detection limits are reduced?

Lack of Understanding of the Scientific Process. People expect science to be knowing and definitive. They often don't understand how two studies can yield conflicting results, and why the studies take so long. The Love Canal residents could not understand why, after studying dioxin for two years, and spending $63 million the scientists could not provide answers regarding the risk. In fact, the Office of Science and Technology Policy estimated that the federal government spent $1 billion on dioxin research in 1985. (Gough)

The changes in scientific understanding is confusing. The St. Louis Post Dispatch reported that "dioxin had been considered one of the most toxic substances known to man until 1987, when the EPA said it had overestimated the cancer-causing potential of the chemical." Yet, in 1994, EPA again indicated that "laboratory studies suggest the probability that exposure to dioxin-like compounds may be associated with other serious health effects including cancer." (EPA, 1994) Also, the public has difficulty understanding that in many cases, science cannot provide definitive answers, and that the scientists' comments include a scientific judgment as well.

Poor Communication. For obvious reasons, workers sent to sample and remediate potentially contaminated areas wear protective clothing. Yet the people into whose homes these workers come, find this to be most indicative of a high level of risk. The non-verbal message is strong, and seldom is the rationale for the need for worker protection explained. The story of Love Canal,(Gibbs) as told by a former resident, is full of examples of poor communications, both verbal and behavioral, and should be studied both for these examples and for insight into non-technical people's perspective.

Role of Legal System. Since many contamination situations lead to law suits, spokespersons for the involved companies are often prevented from saying or conveying messages in a manner that might be more comforting. This is unfortunate, but part of reality.

CONCLUSIONS

The case study review of events surrounding Love Canal, Times Beach, Seveso, and Agent Orange provides insight into the perspective of the potentially exposed individuals, and helps in formulating communication plans in similar instances. Tailoring the plans in advance of crisis development offers the best opportunity for a thoughtful response, and for decisive action when needed.

REFERENCES

ATSDR. 1997. *Toxicological Profile for Chlorinated Dibenzo-p-Dioxins*, Draft for Public Comment, September.

EPA. 1994. *Health Assessment Document for 2,3,7,8-Tetrachlorodiobenzo-p-dioxin (TCDD) and Related Compounds*, Vol. III, External Review Draft, August.

EPA. 1997. "EPA Eliminates Dioxin, Reduces Air and Water Pollutants from Nation's Pulp and Paper Mills," *Press Release*, November 14.

Fuller, J. G. 1977. *Poison that Fell from the Sky*, Random House, New York.

Gibbs, L. M. 1981. *Love Canal, My Story* , State University of New York Press, Albany, NY.

Gough, M.1986. *Dioxin, Agent Orange: The Facts*, Plenum Press, New York,

Jensen, B.K. and R. A. Jensen. 1990. "Responsible Environmental Communication," *1990 AWMA Annual Meeting*, 90-185.1

Jensen, B.K. and R.A. Jensen. 1997. "Risk Perception: A Comparison of the Technical Aspects of Two Health Risks and the Public's Perception of These Risks," *1997 AWMA Annual Meeting,* 97-WA90.01.

Sandman, P. 1994. Mass Media and Environmental Risk," *Risk,* 5, 251.

COOPERATIVE APPROACH IN IMPLEMENTING INNOVATIVE TECHNOLOGIES AT THE PINELLAS STAR CENTER

Mike Hightower (Sandia National Laboratories, Albuquerque, NM)
John Armstrong (Florida Dept. of Environmental Protection, Tallahassee, FL)
Paul Beam (U.S. DOE, Washington, DC)
David Ingle (U.S. DOE, Pinellas, FL)
Rich Steimle (U.S. EPA, Washington, DC)
Melinda Trizinsky (Clean Sites Inc., Alexandria, VA)

ABSTRACT: The Department of Energy's (DOE) Environmental Restoration Program Office (EM40) and the U. S. Environmental Protection Agency's (EPA) Technology Innovation Office (TIO) developed a joint program based on the Clean Sites Inc. Public-Private Partnership concept to assist in accelerating the adoption and implementation of new and innovative remediation technologies. Coordinated by Sandia National Laboratories, the Innovative Treatment Remediation Demonstration (ITRD) Program attempts to reduce many of the barriers and disincentives to the use of new remediation technologies by involving government, industry, and regulatory agencies in the assessment, selection, implementation, and validation of innovative technologies at DOE facilities. Selected technologies are used to remediate small sites and generate the full-scale and real-world cost and performance data needed to validate these technologies and accelerate their use nationwide. An ITRD project was initiated at the Pinellas Science, Technology, and Research (STAR) Center, formerly the DOE Pinellas Plant, in Largo, Florida. The project addressed chlorinated solvent contamination of soil and ground water in a shallow anaerobic aquifer. This cooperative effort provided technical and funding resources to assist the site in evaluating innovative technologies with the potential to enhance remediation efforts at the site.

INTRODUCTION

Common barriers to the use of innovative remediation technologies have been identified in many studies. These include the lack of validated full-scale cost and performance data, the lack of industry and regulatory involvement in technology evaluations, a lack of knowledge about the applicability of a specific technology at a site, the lack of funding to conduct site specific treatment studies, and a fear of penalties or fines for delays or the failure of a new technology to meet required clean-up levels. The ITRD Program was initiated in cooperation with the EPA's Technology Innovation Office in an attempt to reduce the barriers and risks associated with the application of a new remediation concepts and to accelerate the implementation of innovative remediation technologies.

To accomplish this, the ITRD program was organized based on the concept shown in Figure 1. To improve communications and teamwork, government, industry, and regulatory agencies are directly involved with a site in assessing, selecting, implementing, and evaluating innovative technologies. Sandia National Laboratories, as technical coordinator for the ITRD Program, is responsible for forming an advisory group for each project composed of technical representatives from the DOE and the national laboratories, the EPA and its laboratories, user industries with similar problems, and state and federal regulatory agencies. Clean Sites Inc. assists Sandia by contacting and soliciting industry interest in projects and tracking potential partners and their needs. The composition of the advisory group enhances communication and provides the teamwork and technical expertise necessary to effectively identify and review the variety of technologies potentially applicable at a site.

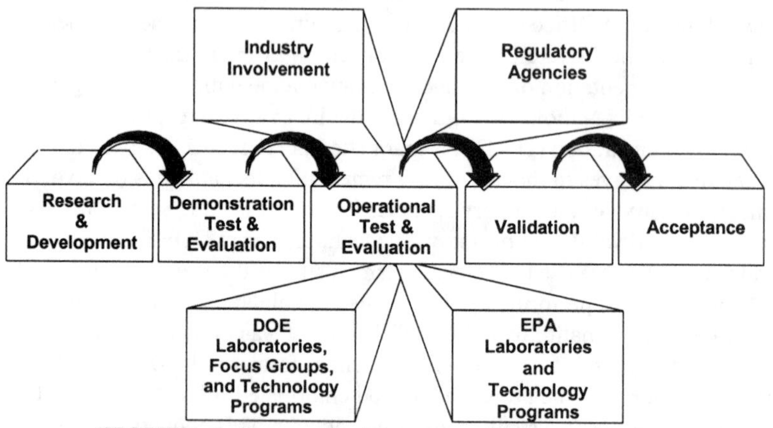

FIGURE 1. Innovative Treatment Remediation Demonstration project concept.

The technologies considered for evaluation in an ITRD project are those new or emerging technologies that often lack adequate cost and performance information, which can prevent their full consideration as remedial alternatives. Many of the technologies considered have shown promise in pilot-scale applications, but have limited full-scale cost and performance data. Technologies in this category include bioremediation, *in situ* dynamic stripping, soil washing and soil flushing, chemical treatment, *in situ* passive treatment, and advanced physical separation techniques. Following a careful review of applicable innovative technologies, the ITRD Program provides funding to conduct treatment or pilot studies needed to assess site-specific issues. These efforts help reduce the risks associated with implementing a new technology. Based on the study results, estimates of technology cost and performance are developed. From these evaluations, the advisory group recommends the best technology options that will enhance or improve remediation of the site based on overall system costs, performance, and regulatory issues.

The technologies selected are implemented to remediate small, one-two acre areas. During remediation, the operating, treatment performance, and cost data are closely monitored by the advisory group. This generates accurate operating data on the innovative technologies deployed, and provides the participating government, industry, and regulatory parties with the necessary operational data needed to validate the performance of each technology. The validated cost and performance data is used to disseminate information on the technologies that can be used to help evaluate remediation options at other sites (FRTR, 1997).

CONCEPT OPERATIONS

The first ITRD project was initiated at the Pinellas STAR Center, formerly the DOE Pinellas Plant, in Largo, Florida and provides a good example of how this kind of joint effort can be effective (Hightower, 1996). The Pinellas Northeast Site Project was organized to address chlorinated solvent contamination of ground water in a shallow, sandy, surficial aquifer at an approximately four acre site. A Corrective Measures Study had proposed a standard thirty-year pump-and-treat system as the baseline ground water remediation technology for the site. Ways to accelerate the remediation of the site and reduce overall remediation costs were important considerations.

A technical advisory group was established to identify, select, implement, and validate applicable innovative remediation technologies at the Pinellas Northeast Site. This group consisted of technical representatives from government, industry, and regulatory agencies with backgrounds in environmental remediation and technology research in soil and ground water remediation, and was intentionally very diverse. Industry participants included representatives from Phillips Petroleum, General Electric, Exxon, Clean Sites, Occidental Chemical, and Lockheed Martin. EPA participants included representatives from the Technology Innovation Office's groundwater remediation staff, the Superfund Innovative Technology Evaluation (SITE) Program, the Federal Facility Restoration and Reuse Office (FFRRO), and the R. S. Kerr Environmental Research Laboratory, EPA's groundwater research laboratory. DOE participants included representatives from EM40 and the Office of Science and Technology (EM50), the Albuquerque Operations and Pinellas Offices, and researchers from Oak Ridge, Lawrence Livermore, and Sandia National Laboratories. Regulatory participants included representatives from EPA Region IV and the Florida Department of Environmental Protection.

The advisory group identified approximately twenty innovative enabling, active, and passive technologies for investigation and review, each with the potential to enhance the proposed pump-and-treat system. Some of the emerging and newer technologies identified for review included: hydraulic barriers; *in situ* treatment walls such as iron filings, microbubbles, and zeolites; horizontal wells; hot air and steam injection for *in situ* dynamic stripping; *in situ* heating; air sparging and vapor extraction; *in situ* aerobic and anaerobic bioremediation; electrokinetics; and *in situ* soil flushing with surfactants and co-solvents.

The major contaminants of concern at the Northeast Site are tetrachloroethylene, trichloroethylene, dichloroethylene, vinyl chloride, and dichloromethane. In two areas of the site, contaminant levels exceed 1000 ppm. Controlling factors at the site are a high water table, high iron content in the ground water, a low ground water flow gradient, and anaerobic aquifer conditions. With this information in mind, each technology identified was reviewed for possible application by the technical group. Many technologies were eliminated from further consideration due to immaturity or inapplicability at the site. However, several technologies were selected for further investigation and site specific treatment studies. Treatability studies were conducted for iron dechlorination, *in situ* steam and air stripping, and *in situ* anaerobic bioremediation.

Though treatment studies can be costly and can take several months to complete, the data generated are often necessary to allow participants to determine whether a specific technology is actually a viable alternative at a site. Based on the results of the treatment studies and engineering evaluations of estimated technology cost and performance, three technologies were recommended to enhance the effectiveness of the proposed baseline pump-and-treat system at the Pinellas Northeast Site.

RESULTS AND DISCUSSION

The technologies suggested were a membrane separation system to treat pumped ground water, a mobile rotary drilling steam stripping system to treat areas of high contaminant concentration, and nutrient injection for enhanced *in situ* anaerobic bioremediation for areas of low contaminant concentration. A conventional thirty-year pump-and-treat system was expected to reduce VOC levels in the highly contaminated areas at the site to 250 to 300 ppb, which still is significantly above drinking water standards for the contaminants of concern at this site. It was estimated that implementation of the three innovative technologies suggested could reduce the ground water contaminant levels at the site to drinking water standards in three to five years at a cost savings of $5 million to $10 million, when compared to the proposed baseline design. The concept of using a series or group of technologies to first treat the source area and then treat the areas of lower contamination was developed interactively with the project participants. This combination of technologies was expected to obtain the best overall performance and be the most cost effective.

A pervaporation system was suggested as a replacement for air stripping for treating pumped ground water at the site. The pervaporation system uses a hydrophobic membrane under vacuum pressures to separate organic contaminants from the ground water. The system reduces overall air emissions and concentrates the organic contaminants in a recyclable condensate. A pilot pervaporation system underwent operational evaluation at the site and the results are presented in a Cost and Performance Report that is available from Sandia.

The mobile rotary drilling steam stripping system was used with funding support from EM40 to reduce VOC contamination levels in the areas of high contaminant concentration, greater than 1000 ppm, to levels more in line with the rest of the site, around 10 ppm to 200 ppm. In this process, hot air or steam is injected through the modified rotary drill system. While the drill agitates the soil, the injected steam strips and volatilizes the organic contaminants trapped in the soil. A hood attached to the drill rig is placed over the soil being treated to capture the volatilized contaminants and direct them to an off-gas treatment system. Implementation of the rotary drilling system was completed at Pinellas, and the results are presented in a Cost and Performance Report that is available from Sandia.

Based on the results of laboratory biological treatment studies, an *in situ* anaerobic bioremediation pilot study was funded by the ITRD Program and was completed at the site in an area of low VOC contamination, generally below 300 ppm. The area was treated using nutrient injection to stimulate existing anaerobic microbial populations to degrade the chlorinated organic contaminants of concern. The pilot bioremediation system study was designed and conducted to determine the optimum operating parameters for a full-scale bioremediation system. The results of this effort are presented in a Cost and Performance Report that is also available from Sandia.

CONCLUSIONS

As demonstrated by the Pinellas Northeast Site project, bringing together technical representatives from the EPA, DOE, industry, and regulatory agencies with the appropriate knowledge and backgrounds produces a team that can identify and define innovative and cost effective solutions for site remediation problems. This public-private partnership approach improves the communication necessary to identify appropriate technologies, generate the necessary treatment study data, and address regulatory and cost and performance concerns. This in turn helps minimize the risks and identify the advantages of applying innovative remediation technologies at a site.

REFERENCES

Hightower, M. 1996. "EPA Speeds Technology Implementation." *CHEMTECH.* 26(4): 73-75.

Federal Remediation Technologies Roundtable. 1997. *Guide to Documenting Cost and Performance for Remediation Projects.* Washington, D.C.

REMEDIATION TECHNOLOGY TRANSFER FROM FULL-SCALE DEMONSTRATION TO IMPLEMENTATION: A CASE STUDY OF TRICHLOROETHYLENE BIOREMEDIATION

Glenn C. Mandalas[1], John A. Christ[1], Gary D. Hopkins[2], Perry L. McCarty[2], and Mark N. Goltz[1]

1. Air Force Institute of Technology, Wright-Patterson AFB, OH, USA
2. Stanford University, Stanford, CA, USA

ABSTRACT: The fact that an innovative environmental remediation technology has been proven feasible in a full-scale evaluation does not assure that the technology will be commercialized and implemented. A number of barriers which may impede the transition of a technology from full-scale demonstration to implementation are explored. One of the most significant barriers that is within the purview of the technology developer to overcome is the requirement to properly gather and disseminate quality cost and performance data. Although a well conducted and documented field-scale technology evaluation is a generally accepted method of gathering data, the presentation and dissemination of that data is a crucial next step. An innovative method of presenting and disseminating cost and performance data using interactive software is discussed. A technology recently evaluated in a full-scale demonstration, *in situ* aerobic cometabolic bioremediation of trichloroethylene-contaminated groundwater through toluene injection, is presented as a case study to show how interactive software may be useful in promoting the commercialization and implementation of an innovative remediation technology.

INTRODUCTION

Cost and performance limitations of conventional groundwater remediation technologies have motivated a call for innovative technologies which can make use of the state-of-the-science to meet the regulatory goals of Superfund (U.S. EPA, 1996). Numerous reports support the need for better environmental remediation technologies and more efficient methods of commercializing those technologies (e.g. U.S. GAO, 1996). In fact follow-on legislation to the original 1980 Superfund Law, the Superfund Amendments and Reauthorization Act of 1986, specifically calls for the development and use of alternative innovative remediation technologies (U.S. EPA, 1996).

The fact that new technologies have been demonstrated to address specific environmental cleanup problems does not mean that those problems are solved. New technologies that are capable of attaining regulatory clean-up standards while satisfying time and budget constraints must be implemented in order to be beneficial. Often, new technologies are not given adequate attention by consultants, regulators, and site owners and the technologies are never used (Gierke and Powers, 1997). The unfortunate tendency among these stakeholders is to rely on well-known conventional technologies rather than implement innovative technologies, even though the innovative technology may be more efficient, less costly and, in some cases, impose less adverse impact on the environment (Cooney, 1996). Moving a technology from the laboratory to a field demonstration to full-scale implementation are challenges that must be overcome (Parkinson, 1995). Transferring an innovative technology from field-scale evaluation to commercial implementation is particularly challenging (U.S. GAO, 1994).

The purpose of this paper is to present a software product that may be useful in promoting the commercialization of one such innovative environmental remediation technology. *In situ* aerobic cometabolic bioremediation, which is currently transitioning from full-scale evaluation to commercialization, is examined as a case study. The progression from concept to commercialization of innovative environmental remediation technologies is depicted in Figure 1. Such progression is not a linear process. As information is learned during the technology demonstration phases, that information is fed back to the research and development phase to design further laboratory studies, answer additional questions, and calibrate or verify predictive models. Once a technology has been successfully demonstrated, a crucial step is its transfer to the commercial sector for implementation.

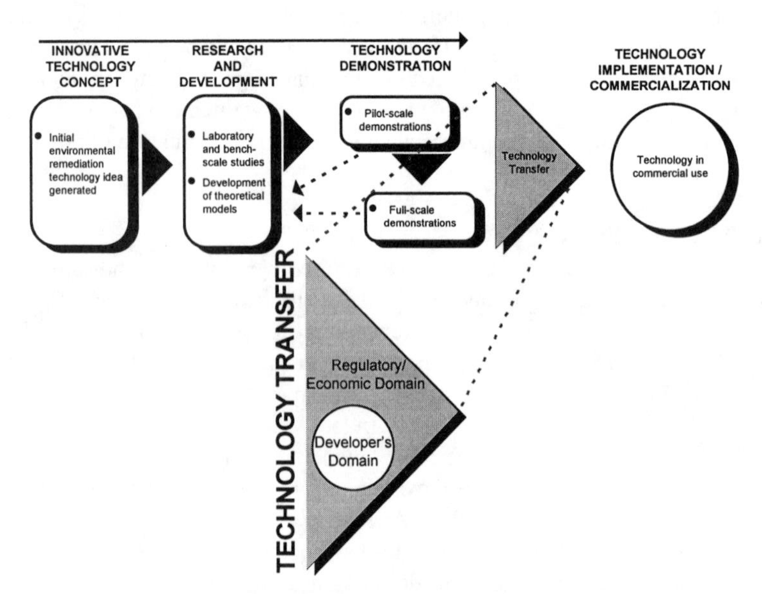

FIGURE 1. Stages of environmental remediation technology development and commercialization.

As depicted in Figure 1, technology transfer actions leading to commercialization may be grouped into two primary domains. First, there is the domain that is defined and affected by legislation, regulation, economics, and policy. This domain, which we will refer to as the regulatory/economic domain, is beyond the control of the technology developer. For example, one recommendation to stimulate commercialization put forth by the National Research Council (NRC) is that the Securities and Exchange Commission should enforce environmental remediation liability disclosure requirements (NRC, 1997). The second domain, the developer domain, is within the control of the technology developer. This paper focuses on this second domain, presenting interactive software as an example of a product that may be used to overcome a significant barrier to commercialization encountered by technology developers. Prior to presenting the software, which was developed for *in situ* aerobic cometabolic bioremediation, a brief overview of the technology and a synopsis of the barriers to environmental remediation technology transfer is presented.

IN SITU AEROBIC COMETABOLIC BIOREMEDIATION

In situ aerobic cometabolic bioremediation is a remediation technology which takes advantage of the capability of indigenous microorganisms to degrade chlorinated contaminants, such as trichloroethylene (TCE), when the microorganisms are provided an electron donor (primary substrate) and acceptor (oxygen) in the presence of the chlorinated contaminant. The presence of the primary substrate induces a non-specific enzyme which, fortuitously, cometabolically oxidizes the target contaminant while the microorganisms metabolize the primary substrate for growth and energy.

The efficacy of this technology has been demonstrated on both the pilot and field scale. Pilot-scale demonstrations conducted at Moffett Federal Airfield, CA, over several years quantified the rate and extent of contaminant biodegradation for various combinations of contaminants and primary substrates (Hopkins and McCarty, 1995). In a recent study at Site 19 at Edwards AFB, *in situ* aerobic cometabolic bioremediation was implemented at the field-scale (McCarty *et al.*, 1998). Toluene was used as the primary substrate and molecular oxygen and hydrogen peroxide were used in combination as the source of dissolved oxygen. High resolution spatial and temporal data were obtained, and it was demonstrated that TCE concentrations could be reduced up to 98% (McCarty *et al.*, 1998). These studies show that *in situ* aerobic cometabolic bioremediation is a technology with the potential to remediate chlorinated solvent contaminated sites in an efficient and cost-effective manner.

BARRIERS TO ENVIRONMENTAL REMEDIATION TECHNOLOGY TRANSFER

Basically, technology transfer of innovative environmental remediation technologies can be obstructed by either (1) a lack of, or poor presentation of, cost and performance data or (2) an unaccommodating regulatory framework (Broetzman, 1997). Barriers rooted in regulatory framework obstacles are generally overcome through actions in the regulatory/economic domain (Figure 1). On the other hand, it is within the purview of the technology developer to obtain and disseminate cost and performance data. This can be obtained through full-scale technology demonstrations (Gierke and Powers, 1997; NRC, 1997). The high resolution, peer-reviewed data gathered at the Edwards AFB demonstration to validate the efficacy of *in situ* aerobic cometabolic bioremediation are examples of the kind of information that can be obtained in a full-scale study (McCarty *et al.*, 1998). The next crucial step is making that data available in a useful form to technology users (consultants, site owners, remedial project managers, etc.).

INTERACTIVE SCREENING SOFTWARE

One novel method of technology reporting, the development of which was motivated by the full-scale demonstration of *in situ* aerobic cometabolic bioremediation at Edwards AFB, is the use of interactive screening software. The software was developed to accomplish three objectives. First, it provides a technology screening tool for site owners and consultants to use to determine whether the technology is appropriate at their site. Second, it gives site owners and consultants the ability to flexibly and simply compare cost and performance of *in situ* aerobic cometabolic bioremediation against other technologies. Finally, the software should act as a promotional tool to gain the interest of technology users and consultants. Indeed, one of the initial motivations for producing the software was testimony from a remedial project manager that similar software piqued his interest in an innovative technology that ultimately was chosen for implementation at one of the sites he was responsible for remediating.

The software requires relatively few input parameters (Table 1). Although not suitable for an actual design, it serves as a screening tool to determine whether *in situ* aerobic cometabolic bioremediation is appropriate under given site conditions. The software also provides the user with information and references regarding the technology, and how the technology complies with regulatory requirements. Copies of the software and the user's guide (Earth Tech, 1998) may be obtained by contacting the Technical Information Center of the Air Force Research Laboratory's Airbase and Environmental Technology Division, Tyndall Air Force Base, FL, email: erica.becvar@ccmail.aleq.tyndall.af.mil, phone: 850-283-6225.

TABLE 1. Input parameters.

Hydrogeological	Performance	Contaminant	Economic
• Aquifer thickness • Regional gradient • Distance to water table • Hydraulic conductivity	• Desired effluent concentration	• Contaminant type • Influent concentration • Plume width	• Project life • Interest rate

The software provides the user with cost data in multiple formats. Specifically, in accordance with recommendations made by the NRC (1997), cost data are reported as annual costs, costs per contaminant mass remediated, up-and-running costs (annual costs without consideration of capital costs), and costs per volume of contaminated matrix treated. The importance of adopting this method for reporting costs is illustrated below.

Table 2 displays costs of *in situ* aerobic cometabolic bioremediation, generated by the screening software, for two geologically different sites. Cost is reported as both an annual cost ($/yr) and a cost per mass of contaminant remediated ($/kg). Examination of the costs illustrates the importance of multiple-format cost reporting. For example, looking at Table 2, if only the annual costs are considered, technology implementation at Site B appears much more cost effective than implementation at Site A. However, upon examination of costs reported on a mass basis, it is apparent that Site A is more appropriate for technology implementation. While the annual cost is smaller for site B when compared to site A, little contaminant is being remediated at that cost. On the other hand, while the annual cost for site A appears high relative to site B, examination reveals that much more contaminant is being remediated under the site A conditions.

TABLE 2: Costs for two hypothetical sites.

Site	Depth to Water Table (m)	Aquifer Thickness (m)	Hydraulic Conductivity (cm/s)	Groundwater Average Linear Velocity (m/yr)	Annual Cost ($/yr)	Cost per Mass Remediated ($/kg)
A	5	21	5.0×10^{-2}	315	139,000	562
B	30	8	5.0×10^{-4}	3	4,970	5,280

Note: Soil porosity is assumed to be 25 percent, and hydraulic gradient is assumed to be 0.005 cm/cm. Interest rate is assumed to be 6%. Project life is assumed to be 20 years. Costs reported are based on a 150 m wide TCE plume with an influent concentration of 1000 μg/L and an effluent concentration of 5 μg/L.

CONCLUSION

Innovative environmental remediation technologies have been developed which have the potential of meeting the regulatory goals of Superfund at a reduced cost and cleanup time over conventional remediation technologies. *In situ* aerobic cometabolic bioremediation is one such technology. In order to commercialize such technologies, attention must be given to the presentation of results in order to ensure that accurate, meaningful, and clear cost and performance data are easily accessible to stakeholders. One means of providing such data is through the use of interactive screening software.

REFERENCES

Broetzman, G. G. 1997. "New Approaches Towards Promoting the Application of Innovative Bioremediation Technologies," *In Situ and On-Site Bioremediation 4*: 323-328.

Cooney, C.M. 1996. "Twenty States Join Federal Government to Facilitate Innovative Technology Use." *Environmental Science & Technology 30*(10):432.

Earth Tech, Inc., *Aerobic Cometabolic In Situ Bioremediation Technology Guidance Manual and Screening Software User's Guide*, Draft January 1998.

Gierke, J.S. and S.E. Powers. 1997. "Increasing Implementation of In Situ Treatment Technologies Through Field-Scale Performance Assessments." *Water Environment Research, 69*:196-205.

Hopkins, G.D. and P.L. McCarty. 1995. "Field Evaluation of *In Situ* Aerobic Cometabolism of Trichloroethylene and Three Dichloroethylene Isomers Using Phenol and Toluene as the Primary Substrates." *Environmental Science & Technology* 29(6):1628-1637.

McCarty, P.L., M.N. Goltz, G.D. Hopkins, M.E. Dolan, J.P. Allan, B.T. Kawakami, and T.J. Carrothers. 1998. "Full-Scale Evaluation of *In Situ* Cometabolic Degradation of Trichloroethylene in Groundwater through Toluene Injection." *Environmental Science & Technology 32*(1):88-100.

National Research Council (NRC). 1997. *Innovations in Ground Water and Soil Cleanup: From Concept to Commercialization*, 265 pp. National Academy Press, Washington, DC.

Parkinson, G. 1995. "Environmental Technology Transfer: A Two-Way Street." *Chemical Engineering 102*(6):33-39.

United States Environmental Protection Agency (U.S. EPA). 1996. *Cleaning Up the Nation's Waste Sites: Markets and Technology Trends.* Report EPA 542-R-96-005. Washington, DC.

United States General Accounting Office (U.S. GAO). 1994. *Department of Energy: Management Changes Needed to Expand Use of Innovative Cleanup Technologies*, 22 pp. Report GAO/RCED-94-205. U.S. Government Printing Office, Washington, DC.

United States General Accounting Office (U.S. GAO). 1996. *Energy Research: Opportunities Exist to Recover Federal Investment in Technology Development Projects*, 32 pp. Report GAO/RCED-96-141. U.S. Government Printing Office, Washington, DC.

ENVIRONMENTAL RESTORATION PROGRAM TECHNOLOGY INITIATIVES

Stephen W. Warren (U.S. DOE, Washington, D.C.)
Michael C. Breck (Booz•Allen & Hamilton, Germantown, MD)

ABSTRACT: Over the past two years, the Department of Energy's (DOE's) Environmental Restoration Program has conducted a number of analyses to support a program goal of deploying the right technologies at the right time at the right sites. The initiatives included: defining problem sets and environmental conditions in the Environmental Restoration Program; developing Preferred Alternatives Matrices (PAMs) and requirements definitions; and identifying performance-based contracting opportunities. In combination with other DOE and federal agency efforts, these initiatives will facilitate informed decision making.

INTRODUCTION

In 1994, the DOE Office of Environmental Management (EM) identified major problem areas on which to focus technology development activities and implemented Focus Areas to address these problems. The Environmental Restoration Program actively participates in the Focus Areas to assist in improved/innovative technology deployment and implement technologies that are commercially available to meet the program goal of deploying "the right technology at the right site at the right time."

To support this goal, a number of fundamental questions needed to be answered, including,

- what is the scope of problems requiring resolution?
- what technologies are available today?
- what are the measures for finding the "right technology"?
- where are the opportunities for deploying commercially available technologies?

TECHNOLOGY INITIATIVES

From early 1996 to the present, several initiatives were undertaken to help answer these questions. The initiatives included: defining problem sets and environmental conditions; developing PAMs and requirements definitions; and identifying performance-based contracting opportunities.

Problem Sets. The first initiative involved defining the full set of potential environmental problems facing Environmental Restoration Program sites. It was realized very early in the process that the problems would need to be grouped into manageable categories. In other words, to analyze problems in a complex as large as DOE, higher-level groupings were required that reflected common waste management problems. Hence, the concept of "problem sets" was developed. The "problem set" represents a specific combination of key data about contaminated material which includes (1) information on the type of material, or medium - such as soil, sludge, or waste water, combined with (2) data that define the contamination such as the indicated waste class (e.g., transuranic, low-level waste, organic contaminants). Problem sets were identified within the three "program-area" groupings: remediation of environmental media, waste processing of stored waste, and decommissioning of buildings and equipment. In addition, it was realized that characterization/monitoring was an area that would encompass all of the problem sets.

Preferred Alternatives Matrices. (http://www.em.doe.gov/define/) Having identified the Environmental Restoration Program's problem sets, the next step involved determining the state of commercial practice with respect to the problems, or in other words, evaluating the available technologies to resolve these problem sets. The tools that satisfied this need were the PAMs. The PAMs were developed to identify commercially available technologies and rank them on the basis of performance, risk of technology failure, and cost.

Three PAMs have been developed. The Remediation/Waste Processing PAM ranks alternatives for containing or treating contaminants in environmental media or stored waste. The Decommissioning PAM assesses alternatives for problem sets dealing with structures and equipment. The Characterization/Monitoring PAM evaluates alternatives in their ability to screen, characterize, and monitor contaminants in remediation/waste processing and decommissioning problem sets.

As expected, the problem sets provided manageable groupings for comparing environmental contamination across the sites, but these groupings lacked sufficient detail for evaluating technologies. For example, it was difficult to evaluate performance, cost, and risk of air stripping when the only "known" quantity was ground water with organic contamination. This unique challenge was resolved by defining "environmental conditions" that: 1) are present at DOE sites; and 2) narrow the focus to site-specific conditions that affect technology performance, cost, and risk. The combination of problem sets and environmental conditions allowed analysis of technologies that was broad enough to capture all Environmental Restoration Program problems without neglecting the conditions that factor into a technology selection decision.

Environmental conditions were defined based on the categorization of problem set. In general, the environmental conditions for remediation and characterization/monitoring problem sets referred to the physical setting in which the contaminant resides (e.g., consolidated vs. unconsolidated material, depth to contamination, subsurface saturation). For waste processing problem sets, the environmental conditions described the composition of the waste (e.g., presence of mercury, total organic carbon). The environmental conditions for decommissioning included the facility makeup and type of structure (e.g., floors vs. walls/ceilings, reinforced vs. unreinforced structures).

To simplify the development and improve the usability of the PAMs, the many problem sets and environmental conditions were arranged in a set of hierarchies; the hierarchies displayed the different paths that a problem holder would follow to arrive at technology solutions. As an example for remediation/waste processing, if ground water is the medium in question and containment is desired, a decision would be made as to whether the environmental conditions include consolidated or unconsolidated material, shallow or deep, and low-hydraulic conductivity or medium to high hydraulic conductivity. Following a path of unconsolidated, shallow, medium to high hydraulic conductivity would lead the user to a PAM table (Figure 1) which will contain a list of ranked technologies for those conditions.

	Unconsolidated	
	Shallow	
	Low hydraulic conductivity	**Medium-high hydraulic conductivity**
Cutoff walls		
Bentonite slurry	◑	◕
Cement-based grout	◑	◕
Chemical barriers	◔	◑
Sheet piling	◑	◕
Synthetic membranes	◔	◑
Hydraulic		
Ditches/drains	○	◕
Pumping systems	○	◕

Legend
● = Preferred
◕ = Probable
◑ = Potential
◔ = Possible
○ = Unlikely
⊗ = Not applicable

FIGURE 1. Resultant Matrix (example)

The concept of the PAMs and, subsequently, the initial technology list and rankings evolved from the Federal Remediation Technologies Roundtable's *Remediation Technologies Screening Matrix and Reference Guide* (http://www.frtr.gov/matrix2/top_page.html). The ranking of the technologies was refined through a peer review process based on available information, focused meetings, and interviews with topical experts. However, this will not be the approach to updating the PAMs. In the future, a technology's ranking or change in ranking in the PAM will be established using cost and performance data collected in a standard format. Cost and performance reports will be available for every completed environmental restoration project and will follow the format provided in guidance from the Office of Environmental Restoration in *Documenting Cost and Performance for Environmental Remediation Projects* (http://www.em.doe.gov/define/costper.html). A similar report will be available for all completed Technology Demonstrations following *Preparation Guidance to the Office of Science and Technology Innovative Technology Summary Reports.* Vendors wishing their technology's inclusion in a PAM should complete a cost and performance report and have it certified by a program manager[1] and a regulator. By collecting this data in a standard way, technologies can be credibly compared on equal footing.

The PAMs were developed to assist decision makers at DOE sites in selecting the most appropriate alternatives for individual sites and types of contaminants. They provide a tool for field personnel to focus remedy selection, expedite preferred alternatives implementation, eliminate the cost of excessive/redundant treatability studies, and allow preselection of effective, low-cost alternatives. Moreover, material in the PAMs could be used in support of the "Development and Screening of Alternatives" portion of the feasibility study. In summary, the PAMs streamline selection of a technology and shift the time and resources for remedy selection to other, more fundamental aspects of the restoration.

A key aspect of the PAMs is that they increase the flexibility in choosing alternatives which is integral to performance-based contracting. In the past, decision documents (e.g., Records of Decision) prescribed a technology which must be used to accomplish the cleanup activity. Over the past several years, the Environmental Restoration Program has attempted to change this trend and move toward performance-based contracts; these contracts outline what the contractor is to accomplish and when, while allowing the contractor to decide how the work will be done. The advantage of performance-based contracts is the recognition that the commercial sector will find the most efficient and least expensive way to perform cleanup. In order to give the contractor these opportunities, it is important to work with the regulators to develop "technology neutral" decision documents. In these decision documents, a single technology solution would not be specified, but rather a list of viable technologies would be provided from which the contractor could choose to accomplish the stated remediation goal. Additionally, a contractor may present cost and performance information for a technology not listed in the decision document, thus leading to selection of improved technology.

Developing a list of viable alternatives for technology neutral decision documents can be an arduous, research-intensive task which is impeded by significant uncertainties concerning technology cost, risk, and performance. The PAMs can streamline this process by providing a short list of good performing, low cost/risk technologies for consideration which will significantly reduce time and uncertainties.

In simplest terms, the PAMs provided an evaluation of available technologies for resolving problem sets. The next logical step was to determine which problem sets were not resolved. Thus, the PAMs form the basis for "requirements definitions" analyses which are described in the next section.

---

[1] Program managers at DOE sites are the intended users of the technologies.

Requirements Definitions. (http://www.em.doe.gov/define/) The Environmental Restoration Program developed a set of "requirements definitions" that mirror the PAM technology areas, namely, decommissioning, remediation, waste processing, and characterization/monitoring.[2] The analyses: 1) define the scope of the problem sets that comprise the Environmental Restoration Program; 2) identify which problem sets can be satisfied by commercial practice; 3) identify which problem sets cannot be addressed completely by commercial practice; and 4) identify where opportunities exist for performance-based contracting. Thus, the requirements definitions assist in developing program priorities and provide a basis for discussion between technology users and technology developers, since the analyses identify areas (or requirements, number 3 above) where technology can be improved or no commercially available technology exists.

The requirements definitions attempted to provide a better understanding of the scope of problem sets by evaluating volume, pervasiveness, cost, and risk. A series of charts and figures were developed to compare the problem sets based on these factors. In summary, 41 problem sets were defined and analyzed for Remediation, 23 for Waste Processing, and 14 for Decommissioning.[3]

With the scope of problem sets defined, the problem sets were then mapped to the PAMs to evaluate the state of commercial practice. In other words, a determination was made as to whether technologies currently existed to meet these problem sets and if the problem sets presented opportunities for performance-based contracting.

As a note, several areas lacking viable technologies were not identified through the requirements definition analysis but instead were identified as the result of EM Focus Area Meetings and other technical forums. These areas included: remote handling/robotics in high radiation areas and remediation of dense non-aqueous phase liquids (DNAPLs) in ground water.

Performance-Based Contracting Analysis. In response to questions from Congress on the utilization of technologies, the Environmental Restoration Program surveyed the DOE sites about their future projects and planned technology use. This survey was unique in that future projects were identified and defined at a lower level of detail than was previously evaluated in the requirements definitions (i.e., included the environmental condition hierarchy, not just problem sets). With projects and problem sets better defined, the Environmental Restoration Program was able to take a closer, more precise look at performance-based contracting opportunities.

Using the requirements definition results, 678 opportunities were identified and evaluated. Of these 678 opportunities, it was determined that,

---

[2] Note: The Characterization/Monitoring Requirements Definition is in draft form and is expected to be complete in early 1998.

[3] Characterization/Monitoring problem sets, which total 78, are the aggregate of the other three program areas.

- 28% have good opportunities for performance-based contracting;
- 54% have possible opportunities for performance-based contracting; and
- 18% have limited or no opportunities for performance-based contracting.

In addition, the analysis included a break-out of the good, possible, and limited/no opportunities for each site. The break-outs included project information such as the problem set, environmental conditions, volume, and technologies applicable for each opportunity. Hence, site program managers and bidding contractors can use this information as a guide for identifying potential performance-based contracting opportunities.

Not only did this survey enhance our understanding of the scope and extent of the Environmental Restoration Program, it gave us the first detailed analysis of the utilization of technology. The results were as follows,

- Past projects: for 4% of release sites/facilities (RS/Fs)[4], the technology selected was developed by EM, and for 18% of RS/Fs, innovative technology was selected.
- Pending projects: for 4% of RS/Fs, the technology selected was developed by EM; for 32% of RS/Fs, innovative technology was selected; and for 10% of RS/Fs, innovative technology is still under consideration.
- Future projects:[5] for 21% of RS/Fs, the technology selected was developed by EM, and for 30% of RS/Fs, there is potential for innovative technology selection.

CONCLUSION

Numerous tools and analyses have been developed by the Environmental Restoration Program in an effort to assist sites to make technology decisions, gain a better understanding of requirements that cannot be met, and evaluate performance-based contracting opportunities. In combination with other EM and federal agency efforts (listed below), the initiatives discussed here facilitate informed decision making in an effort to get the right technologies to the right sites at the right time and perform cleanup as efficiently and safely as possible.

Additional Programs/Projects. Office of Science and Technology (http://em-50.em.doe.gov/); Federal Remediation Technologies Roundtable (http://www.frtr.gov); Innovative Treatment Remediation Demonstration (http://www.em.doe.gov/itrd/index.html); TechCon Program (http://www.ead.anl.gov/~techcon/); Site Technology Coordination Groups (http://em-52.em.doe.gov/ifd/stcg/stcg.htm); Technology Information Exchange (http://www.em.doe.gov/tie/index.html)

---

[4] Release sites are defined as unique locations where hazardous, radioactive, or mixed waste release has occurred or is suspected to have occurred.

[5] The EM Office of Science and Technology has linked innovative technologies to Environmental Restoration Program problem sets and environmental conditions.

CONSENSUS-BASED CLEANUP POLICY: THE WASHINGTON STATE MODEL TOXICS CONTROL ACT EXPERIENCE

Patricia J. Serie (EnviroIssues, Seattle, Washington, USA)
J. Daniel Ballbach (Landau Associates, Edmonds, Washington, USA)
Amy J. Grotefendt (EnviroIssues, Seattle, Washington, USA)

ABSTRACT: Washington's current contaminated sites cleanup law and regulations resulted from a citizens' initiative in 1988. In 1995, the legislature created a statewide policy advisory committee charged with review and, if appropriate, overhaul of the Model Toxics Control Act or MTCA. The scope of the resulting revisions included increasing the use of risk-based decisionmaking, clarifying the remedy selection process, and adding administrative and policy enhancements to encourage cleanups. Consensus recommendations supported successful legislative action. Translation of the new law and the consensus recommendations into regulation and policy, however, illustrates the challenges of participatory rulemaking. Factors that are critical in achieving consensus-based results in a regulatory environment are suggested.

INTRODUCTION

The controversial citizens' initiative known as Model Toxics Control Act (MTCA) taxed industry to pay for cleanups. There was an early consensus process to turn legislated objectives into a comprehensive regulatory framework for cleanups. Key elements of MTCA, which distinguish it from the federal cleanup program, have included encouragement for cleaning up sites independent of regulatory oversight and inclusion of petroleum cleanups within MTCA's purview.

Defining the Problem. Extensive flexibility existed under MTCA in how risk could be used in setting cleanup levels and how remedies could be selected to incorporate factors such as cost. Nevertheless, many parties instead chose the straightforward cleanup levels specified in the rule to achieve "walk-away" cleanups. Washington State Department of Ecology (Ecology) site managers were not always flexible in their application of MTCA requirements, and liable parties were not satisfied with how the more complex and sophisticated elements of MTCA were being used, or often not used, to expedite cleanups. While MTCA requires Ecology to protect "human health *and the environment*," application of ecological risk assessment and resulting protective measures were not clearly defined or consistently carried out. The regulated community complained that established methods were too conservative, and going beyond those well-understood methods resulted in unwilling regulators, increased uncertainty, and unnecessary costs. Only highly experienced consultants and attorneys tended to use the existing flexibility to a significant degree, and they often encountered site managers with less understanding of the process who could block workable cleanup solutions.

Catalyst for Change. The catalyst for change was an industry-based push to have the Legislature change MTCA to allow the use of site-specific risk assessment. Its roots came from a specific site with extensive soil and groundwater cleanup required, and from the liable party's desire to force more flexibility in how cleanup levels were set. The result was extensive argument in the legislative session, and a mandate to Ecology to establish a study committee with broad authority to examine MTCA in its entirety, and to report back with specific recommendations for change within 18 months. By the time the process was over, far more than site-specific risk assessment was addressed and changed.

METHODS AND APPROACH

Committee Formation and Composition. As in the formation of any broad-based advisory committee, there are challenges of identifying all relevant interests and finding the right groups and individuals to sit at the table. The MTCA Policy Advisory Committee combined representatives of large and small business, legislators and local government representatives, agriculture and the financial community, the petroleum industry, and a range of environmental interest groups. Ecology's Science Advisory Board was represented. Ecology held one seat at the table, and participated on equal footing with the other interests. Many participants had worked closely with Ecology on site cleanups; some participants were involved in the original drafting of MTCA and its implementation.

Committee Structure and Administration. A presiding officer for the committee was named, supported by a professional facilitator. The presiding officer's knowledge of MTCA from its origins to its current implementation was key in structuring committee activities, and his credibility and willingness to play a leadership role helped establish a trusting atmosphere. The facilitator and support team were also knowledgeable about MTCA and experienced in group interaction processes. This allowed an in-depth analysis of the issues and resulted in detailed and specific recommendations.

Scoping of Committee Issues. Opening up the entire law and regulation to scrutiny and potential change provided a flurry of issues at the policy and technical levels. The group focused its issues into major categories, which served as the basis for a subcommittee structure – risk assessment, remedy selection, independent cleanups, and a set of institutional and implementation issues. There were several highly contentious issues such as reconsideration of joint and several liability that the group did not take on, presumably because the chance of consensus was slight. What resulted was a daunting set of issues that needed to be addressed individually, yet also considered in the overall context of implementing site cleanup.

Case Study Method. To create that context, the group considered several case studies – hypothetical sites requiring cleanup that incorporated all of the issues on the table. It became clear that there were widely divergent levels of knowledge and experience with MTCA among the participants. The case study approach allowed all members of the group to come to an even playing field of knowledge about how MTCA actually works in a way that made the issues real for them.

This relatively time-consuming approach resulted in some discomfort among the members and outside observers, but proved important to building trust as a group and to illuminating the issues. As it became clear that more information sharing was needed, study sessions on specific topics were held (e.g., ecological risk assessment, petroleum cleanups) to broaden understanding. Meetings were also held in different locations throughout the state, which allowed community-specific examination of issues. For example, a meeting in Wenatchee, Washington, in the eastern part of the state, considered historic contamination of orchard soils from lead arsenate and the resulting issues of changing land uses and allowable exposures.

Subcommittee Structure. The hard work was only beginning. Subcommittees considered dozens of issues related to the major topics of risk, remedy, independent cleanups, and institutional arrangements. Monthly and bimonthly subcommittee meetings augmented monthly all-day meetings of the entire committee. Ad hoc work groups added to the level of study and the level of administrative complexity, meeting frequently on tough issues such as areawide contamination, ecological risk assessment, petroleum cleanup standards, and public participation. As issue papers were written and considered, proposed recommendations were prepared and rewritten, and the pieces of a consensus package of recommendations began to come together.

A key negotiating technique was considering the committee's product as that full set of recommendations. No one decision was reached in isolation; decisions were logged and set aside for final consideration as part of the overall package. Tradeoffs were acknowledged and tracked, and the group moved slowly toward agreement on many fronts at once. Consensus was the goal, and was reached on many issues. Broad agreement was enough, however, and the few dissenting opinions were ultimately documented as minority reports. The result – a detailed report analyzing all of the issues addressed, and providing a set of recommendations for legislative, rulemaking, and policy implementation – went to the Legislature in December of 1996. Legislative recommendations were acted on in the 1997 legislative session, with remarkable bi-partisan and joint House and Senate sponsorship. The new law formed the basis for a participatory rulemaking process that is now nearing completion.

KEY ISSUES

New ground was forged in several areas of MTCA. Site-specific risk assessment, the original catalyst for action on MTCA, was designed to allow more flexibility for establishing site cleanup levels and application in the remedy selection process. Site-specific exposure assumptions could be proposed for measurable factors such as how the human body absorbs contaminants, and for physical aspects of the site such as soil type, hydrological conditions, or measurable characteristics of the contaminants. These elements, if quality of information requirements are met, can be considered in establishment of site-specific cleanup levels. Other factors that may vary based on human behavior may also be changed and used in establishing remediation levels.

Remedy selection is based on results of the remedial investigation and risk assessment, and analysis of potential cleanup remedies. The process of remedy screening, evaluation, and selection is now much more clear. The practice of setting cleanup action levels that guide remedy selection had not been clearly explained within the original MTCA, and is now clarified and renamed as "remediation levels." The statutory preference for remedy permanence to the maximum extent practicable remains. The rewritten regulation will include an understandable process for balancing factors of cost, practicality, timeframe, and public concerns.

The need for ecological risk assessments can now be more easily evaluated using a consistent tiered roadmap approach. Only if certain conditions exist and the site does not quality for the off-ramps, will an ecological risk assessment be required. Many sites are expected to be screened out of the need for such an assessment using this tiered approach.

If a site has been contaminated as part of an areawide problem, or is a "brownfield," alternatives that will speed cleanup and expedite redevelopment are now offered. Ecology, in conjunction with local government or potentially liable parties, may develop areawide solutions, including investigation work plans, model remedies, or area-wide determinations of groundwater potability. Legal protections for landowners who have met their cleanup responsibilities, with corresponding ongoing obligations, apply to successor landowners. Landowners overlying contaminated groundwater plumes will be exempted from liability if their property is not the source of the contamination, they did not cause or make the problem worse, and they allow access for cleanup. Enhanced public participation goals that include sustainable economic development and environmental justice will be developed for certain "brownfields" projects.

Independent cleanups, with no Ecology oversight, are allowed today but there will be greater certainty because Ecology can provide written, though nonbinding, feedback on plans and results as the site proceeds through cleanup. This provides added confidence in the assessment of risks and identification of protective remedies and adds to the ability of third parties (e.g., lenders) to understand true site liabilities and rely on independent cleanup results to a greater extent than possible today.

The way in which Ecology, liable parties, and the public interact during site investigations and cleanups will focus on clarifying expectations, providing full information to all parties, supporting the site manager with peer review and feedback, and providing needed guidance documents, training, information access, and outreach on MTCA. An informal dispute resolution process will be defined clearly for times when the communication measures discussed above do not solve a problem. After two years, the approach will be reviewed to determine if a more formal dispute resolution process is needed.

Citizens will be assured of early notice and more effective participation in site cleanup decisions affecting their quality of life. Citizens living in the vicinity of the site will be able to apply for public participation grant funding to support local public involvement, review, and information needs through a more simplified process. For a complex and regionally significant site, increased grant funds may be made available. Citizens may call on the capabilities of a third-party "ombudsperson" who can help interpret technical issues and materials.

If the site contains petroleum contamination, a new interim approach is now available to more accurately reflect the characteristics and risks of the contamination within the existing MTCA framework. This approach is expected to provide more options to persons conducting petroleum cleanups and make some cleanups less expensive, while remaining protective of human health and the environment. Over the longer term, there are ongoing efforts to fully develop a petroleum cleanup approach for the State of Washington and work with Ecology to change the rules.

Under current law, if the selected remedy calls for leaving contaminants onsite, institutional controls are called for (e.g., fencing or capping; restricting future land uses through deed records or requiring public notification). MTCA now includes an emphasis on evaluating the cost and reliability of those controls to see that they continue to provide a protective remedy.

TRANSLATION TO REGULATORY RULEMAKING

The committee thought that coming up with comprehensive consensus recommendations was hard work, and it was. In many cases, however, even more detailed analyses and decisions were needed to translate the policy decisions into actual rule language. Ecology committed to continue the collaborative process through rulemaking, and convened essentially the same group to act as an external advisory group to rulemaking. The presiding officer and facilitator continued in similar roles, though less intensively. And this is where the even harder work began.

Ecology rule-drafting staff continued to consult actively with committee members and other outside advisors to resolve remaining issues and draft language. Some of the changes that worked well at the policy level proved more difficult to integrate as rule. Adding flexibility, as some of the recommendations entailed, created discomfort unless that flexibility was completely explained and caveated. Site managers in the field, not involved on a day-to-day basis with the committee process, questioned some of the recommendations in terms of their real-world regulatory responsibilities. External reviewers wanted flexibility for themselves, but constraining language for the site managers. Everyone argued about what the recommendations had really meant, and memories sometimes proved short or selective.

The trusting relationships built up over the 18-month committee process proved invaluable, and continues to contribute to progress. Where there are disagreements, established channels of communication can address the conflicts. Participants with special expertise or interest continue to focus the input in those areas, and are trusted by other members to represent their interests. The case study mindset continues to be important, testing out draft concepts and language as to how it would really work at a site. And the history of consensus is empowering, though it needs a great deal of maintenance and nurturing to remain an effective tool toward agreement.

SUGGESTIONS TO OVERCOME BARRIERS

Many lessons were learned throughout this process and can be applied to similar challenges. First, a process of this depth and complexity taxes the patience of the participants and the adequacy of the support structure. The task was to formulate meaningful top-level policy decisions. The group could have skated across the surface of the law and the rule; but to make truly substantive changes, it was necessary to examine MTCA in depth and understand how it is applied.

Second, the group learning process contributed both knowledge and trust to the participants. Some members had extensive experience implementing MTCA; others barely understood it or had focused on only one issue. The use of study sessions, cases studies, issue papers, and subcommittee and work group activities let participants share their knowledge and learn to work together constructively. Time spent on learning and building trust, while a source of frustration for some, proved well invested when negotiation of consensus recommendations came along.

A combination of strong leadership and skilled facilitation enabled the committee to stay on top of the many players and issues. A two-person team mediating the process helped to share responsibility for tracking and resolving issues, for spotting areas of conflict and defusing them, and for effectively negotiating consensus with different participants. Logistical and administrative support were critical – accurate documentation of discussions and decisions, coordination of virtually hundreds of meeting locations and agendas, and responding to the information needs or complaints of participants.

Lastly, the transition from active advisory committee role to more arms-length rulemaking advice is perilous. That long-developed trust and involvement, and ownership for the work products of the group (issue papers, recommendations, analyses), are key to maintaining momentum and support. The time needed to translate the recommendations into detailed rule language, and for the agency to complete its internal communication and analysis, results in a gap in people's investment, and opens the door for doubts, fault-finding, and regression. Ambitious negotiation is needed to keep the consensus on track and move it to implementation. Committee members and agency staff need to be brought back to the earlier agreements, to the established communication channels, and to the trust that forged a highly visible and publicized consensus.

PRESUMPTIVE REMEDIES FOR GROUND WATER CLEANUP: A LEGAL PERSPECTIVE

John H. Johnson, Jr.
Douglas A. Henderson
Troutman Sanders LLP
600 Peachtree Street, Suite 5200
Atlanta, Georgia 30308-2216
United States of America
404-885-3000 (telephone)
404-885-3990 (fax)

ABSTRACT: In recent years, Congress has been unsuccessful in enacting substantial reforms to the Comprehensive Environmental Response, Compensation, and Liability Act (CERCLA or Superfund). While the legislative debate has centered primarily on proposals which would make drastic changes to the CERCLA liability framework, efforts have also been directed at simplifying and expediting the CERCLA site assessment and remediation process. In part to counter these legislative efforts, the U.S. Environmental Protection Agency (EPA) has launched a series of regulatory initiatives aimed at accelerating cleanups and making the CERCLA program more efficient. As part of this reform, EPA has established guidance for remedial decision-making at sites containing contaminated ground water.

This presentation analyzes EPA's guidance for remediation of contaminated ground water at CERCLA sites, known as the "Presumptive Response Strategy and Ex-Situ Treatment Technologies for Contaminated Ground Water at CERCLA Sites." To place the guidance in context, the presentation initially provides a brief overview of EPA administrative reforms to the CERCLA program. The presentation then examines major features of a primary component of the guidance for contaminated ground water sites - EPA's Presumptive Response Strategy. The presentation concludes with a discussion of the potential benefits and shortcomings of the Strategy. As discussed, although EPA describes the Presumptive Response Strategy as an effort to make the remedy selection process more expeditious and cost-effective, the Strategy does not effect fundamental modifications to EPA's cumbersome remedial investigation/feasibility study (RI/FS) or remedial design/remedial action (RD/RA) process and may, in fact, serve to increase both the cost and duration of remedial action at many contaminated ground water sites.

EPA's SUPERFUND REFORM INITIATIVES

Since the early 1990's, in reaction to growing criticism of the Superfund program from interested constituents, there have been numerous efforts in Congress to make sweeping changes to CERCLA. In an effort to head off broad statutory changes, EPA has initiated a number of programmatic reforms which, according to the Agency, are intended to improve the efficiency, pace, cost, and fairness of the Superfund program. The Superfund Reforms are comprised of various initiatives and pilot programs that are being implemented primarily through guidance documents, rather than through changes to the Superfund statute or regulations. Since 1993, EPA has initiated three rounds of reforms. The third and final round, commenced in October 1995, consists of twenty reforms, several of which are designed to achieve cost-effective cleanups. Of these, EPA, in its Superfund Annual Report for Fiscal Year 1997, has described the following reforms as being designed to fundamentally change Superfund:

- establishment of the National Remedy Review Board
- updating of remedy decisions at select sites
- clarification of PRPs' role in risk assessments
- consideration of response actions prior to an NPL listing
- deletion of clean parcels from the NPL
- promotion of risk-based prioritization of NPL sites
- issuance of presumptive remedy guidance

As part of the presumptive remedy guidance initiative, in October 1996, EPA issued a guidance document entitled "Presumptive Response Strategy and Ex Situ Treatment Technologies for Contaminated Ground Water at CERCLA Sites." Unlike previously-issued presumptive remedy guidance documents, which apply to specific types of operations (e.g., wood treatment sites, municipal solid waste landfills), this guidance is applicable to all sites with contaminated ground water. Also, instead of establishing one or more presumptive remedies, the guidance sets forth a "Presumptive Response Strategy." As described in the guidance, this Strategy "integrates site characterization, early actions, remedy selection, performance monitoring, remedial design and remedial implementation activities into a comprehensive, overall response strategy for sites with contaminated ground water." The guidance notes that "[a]lthough this response strategy will not necessarily streamline the remedial investigation/feasibility study (RI/FS) phase, EPA expects that use of the presumptive strategy will result in significant time and cost savings for the overall response to contaminated ground water."

MAJOR FEATURES OF THE PRESUMPTIVE RESPONSE STRATEGY

In discussing the Presumptive Response Strategy, the guidance specifies the following objectives for response actions at sites with contaminated ground water, listed in the sequence in which they should generally be addressed:

- prevent exposure to contaminated ground water above acceptable risk levels
- prevent or minimize further migration of the contaminant plume (plume containment)

- prevent or minimize further migration of contaminants from source materials to ground water (source control)
- return ground water to beneficial uses wherever practicable (aquifer restoration)

The guidance proceeds to note that in certain cases, most notably where nonaqueous phase liquids (NAPLs) are present in ground water, aquifer restoration to drinking water or similar standards may not be feasible using current technologies and, in such instances, response action objectives may be different for different portions of the ground water contaminant plume at the site, e.g., containment of NAPLs and aquifer restoration for remaining portions of the plume.

In order to both ascertain whether aquifer restoration is a feasible objective for the entire ground water plume and expedite plume containment, the Presumptive Response Strategy encourages implementation of "early response actions" during the RI phase, in addition to typical site assessment activities. These early actions may encompass such activities as ground water extraction and treatment, excavation or in situ treatment of soils, and removal of free product. The guidance specifies that these early actions are interim in nature and, after sufficient site information has been developed, these actions should be followed by more comprehensive action (e.g., an expanded extraction and treatment system) consistent with the long-term remedy for the site.

Mirroring this approach, the guidance encourages the use of phased ground water response actions, either through implementation of two separate actions, where site characterization data are insufficient to determine the likelihood of achieving aquifer restoration, or through implementation of a single action in more than one phase, where site characterization data are sufficient for making a determination on the achievability of long-term objectives. Consistent with the emphasis on early actions, the guidance notes that the latter approach, in which phased response actions occur only after issuance of the Record of Decision (ROD), should be used only in those limited instances in which attainment of long-term objectives (e.g., aquifer restoration) is likely. The guidance states that in the standard case, in which RI phase early actions will be followed by a post-ROD long-term remedy, "the early or interim action may need to operate for several years."

The Presumptive Response Strategy contains specific recommendations for sites at which ground water is contaminated with dense, nonaqueous phase liquids (DNAPLs), noting that these sites "pose special cleanup difficulties." Noting that subsurface DNAPLs may be present at up to 60 percent of the Superfund National Priorities List sites, the guidance specifies that the RI for such sites should be designed to delineate the potential extent of DNAPL zones (i.e., where free-phase DNAPLs are present either above or below the water table), as well as the extent of the aqueous contaminant plume. Significantly, the guidance recognizes that where such zones exist, removal of DNAPLs is often impracticable, and no treatment technologies are currently available which would allow attainment of risk-based cleanup levels. The guidance concludes, accordingly, that "it is expected that [applicable or relevant and appropriate (ARAR)] waivers due to technical impracticability will be appropriate for many DNAPL sites, over portions of sites where non-recoverable DNAPLs are present." The guidance goes further in noting that for some sites, "ARAR waivers may also be appropriate for all or portions of the aqueous plume when supported by adequate justification."

EVALUATION OF THE PRESUMPTIVE RESPONSE STRATEGY

The Presumptive Response Strategy is significant, and potentially promising, in that it reflects an acknowledgment by EPA that many contaminated ground water sites, and most sites containing DNAPLs, cannot feasibly obtain ARARs and risk-based cleanup standards and, where these conditions are demonstrated, waivers due to technical impracticability may be appropriate with respect to all or a portion of the ground water contaminant plume. Unfortunately, the strategy establishes a cumbersome, time-consuming, and potentially expensive protocol for documenting that these conditions exist. For example, by promoting "early actions," in conjunction with traditional RI sampling and analysis, the time required for completion of the RI phase can be substantially lengthened and, factoring in capital and operational and maintenance costs for the early activities, the costs of the RI phase can be substantially increased. If the guidance more strongly suggested that use of these early actions might result in an expedited cleanup, this time and expense might well be justified. However, the guidance emphasizes that the early actions typically will need to be expanded upon and refined as the long-term remedy is developed. Also, in stating that different response objectives may be applicable to different portions of the ground water plume, the guidance suggests that, as to those portions of the site for which aquifer restoration is deemed feasible, early actions such as extraction and treatment could be required indefinitely in an effort to achieve stringent risk-based cleanup standards.

The discussion of DNAPLs in the guidance indicates that, before technology waivers may be considered, it must be demonstrated on a site-specific basis that feasible efforts to remove DNAPLs have been undertaken (e.g., through removal of free-phase DNAPLs) and migration of subsurface DNAPLs has been controlled. These requirements suggest that findings of technical impracticability may not be made until time-consuming and costly efforts have been undertaken to remove and/or control portions of the plume for which aquifer restoration is not feasible. Perhaps most significantly, where technical impracticability is demonstrated, the guidance expresses a preference for bifurcation of remedial objectives for different components of the ground water plume, rather than for granting a technical impracticability waiver for the entire plume. In view of the fact that, in many cases, remediation of the aqueous plume to risk-based levels cannot be accomplished as long as source materials, such as DNAPLs, remain in the subsurface, this cleanup objective will, as a practical matter, be impossible to achieve.

Given the predicates to a determination of technical impracticability and the fact that such determinations may apply to only a portion of the contaminant plume, the guidance, while attempting to address the difficulties inherent in many ground water cleanups, may have the effect of substantially increasing the costs of ground water cleanup. EPA's Presumptive Response Strategy was issued in recognition of the fact that the "pump and treat" remedy which has been implemented traditionally at contaminated ground water sites has, in many cases, been ineffective in achieving stringent risk-based cleanup standards. However, this guidance, by bifurcating cleanup objectives, could in many instances continue to require indefinite use of extraction and treatment technologies to achieve aquifer restoration, but, in addition, require earlier implementation of such remedies, as well as extensive assessment and remediation activities directed at containing and, to the extent feasible, removing the NAPL component of the ground water plume.

DEVELOPMENT OF U.S. EPA REGION 4 TECHNICAL IMPRACTICABILITY GUIDANCE

William N. O'Steen (U.S. EPA, Atlanta, Georgia)

ABSTRACT: The U.S. Environmental Protection Agency (U.S. EPA) Region 4 is developing a regional guidance document on the technical impracticability of ground-water restoration. Core principles have been developed by the U.S. EPA Region 4 which will frame the regional technical impracticability guidance. The regional guidance will address issues of both data generation and data interpretation in technical impracticability evaluations. The guidance will also define how the U.S. EPA Region 4 will address an area to which technical impracticability applies, based on cost and resource valuation considerations. The regional guidance will require that regardless of the timing of a technical impracticability request, the same questions applicable to the technical impracticability evaluation must be answered in a technical impracticability demonstration submitted to U.S. EPA Region 4.

INTRODUCTION

The U.S. Environmental Protection Agency (U.S. EPA) Region 4 is developing a regional guidance document to supplement the *Guidance for Evaluating the Technical Impracticability of Ground-Water Restoration* (U.S. EPA, 1993). The regional technical impracticability (TI) guidance is planned to define a defensible decision process for identifying a site-specific ground-water remedial time frame beyond which it is technically impracticable to restore ground water to drinking-water standard quality. It will define the requirements for a TI demonstration and request for consideration submitted to the U.S. EPA Region 4. The regional guidance will define the interrelationship between the ground-water remedial time frame, the cost of a remedial action to attain regulatory objectives, and the valuation of the ground-water resource. The regional guidance will also better define the monitoring data and data analysis techniques necessary to adequately demonstrate the conditions meriting a technical impracticability waiver. This aspect of the regional TI guidance will include requirements or recommendations for the duration and location of monitoring activities, and the relevant methods of data interpretation necessary to complete a TI evaluation in the U.S. EPA Region 4.

U.S. EPA REGION 4 EXPECTATIONS AND TI PHILOSOPHY

The U.S. EPA Region 4 believes that a comprehensive definition of the Region's perspective on the technical aspects of the TI decision process are necessary in order for regulated parties to prepare an appropriate petition for a TI waiver. The U.S. EPA Region 4 has defined several core principles which will frame the regional TI guidance:

The regional TI guidance must be consistent with the national TI guidance document, and all applicable Federal regulations.

The hydrogeologic setting and the valuation of the ground-water resource (potentially including a TI-oriented risk analysis), as the principal factors, will be considered by the U.S. EPA Region 4 in order to address issues regarding the long-term management of the TI zone (that portion of an aquifer, or aquifers, where the drinking-water applicable or relevant and appropriate requirements (ARARs) are waived on the basis of technical impracticability). Depending upon these factors, the TI zone may either be addressed by long-term containment (to the maximum degree technologically achievable), partial containment, or by some degree of active remedial action. Where indicated, active remedial action in the TI zone will be designed to either (1) attain less stringent ARARs (such as Federal rather than more stringent state ARARs) inside the TI zone, (2) attain certain ARARs and reduce risks inside the TI zone, or (3) attain ARARs and reduce risks outside of the TI zone.

Given sufficient time, all ground-water contamination can be remediated to meet ARARs, provided that the continued contaminant mass flux to the saturated zone is eliminated. Thus, demonstration of technical impracticability must include quantification, with uncertainty considered, of the time required to remediate ground water following elimination of contaminant mass flux to the ground water.

Ground-water remedial technologies have made impressive gains over the several years since the U.S. EPA's national TI guidance was developed. The U.S. EPA Region 4 anticipates that because of the continuing improvements in remedial technologies, an ever-decreasing number of sites will meet the requirements for demonstration of the technical impracticability of ground-water restoration. Within a few years, most, if not all, TI demonstrations are anticipated to be made on the basis of hydrogeologic factors, rather than on the presence of ground-water contaminant factors, such as dense nonaqueous phase liquid (DNAPL) located below the water table. Thus, the regional guidance must incorporate flexibility into the requirements for assessing technical impracticability on the basis of contaminant-specific factors.

The U.S. EPA Region 4 recognizes that a working definition of "engineering feasibility and reliability", the primary criteria for determining the technical impracticability of achieving ARARs (U.S. EPA, 1993), must incorporate either a remedial time frame, or a cost factor. Both time and cost are related; however, that relationship will vary based on the nature of contamination and the hydrogeologic setting.

The U.S. EPA Region 4 intends to approve requests for TI waivers to ground-water ARARs by considering the relationship between remedial costs and the value of the ground-water resource, and may consider a TI-oriented risk evaluation, where appropriate. Typically, there will be a remedial time frame minimum threshold below which health risk will be a minor factor in the determination of technical impracticability. Under those circumstances, the cost of the remedial action and value of the ground water will be factors used to determine if ground-water remediation to ARARs is technically impracticable for a contaminant plume. The U.S. EPA Region 4 may establish either a specific time or a range of minimum threshold times below which technical impracticability waivers will not be considered. Figure 1 shows a conceptualization of the approach the U.S. EPA

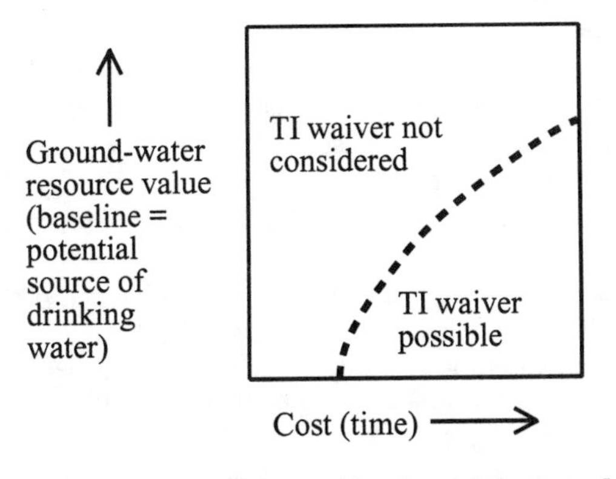

FIGURE 1. Conceptualization of Approach for Determining if a TI Waiver is Possible

Region 4 is considering for determining where a TI waiver is possible.

The U.S. EPA Region 4 intends that the TI zone established for a site be limited to the smallest area possible. The U.S. EPA Region 4 recognizes the considerable difficulties in establishing the boundary of a TI zone. Therefore, the same type of evaluation used to define the TI zone size and the remedial response appropriate for the TI zone should also be considered in defining the scope of investigations needed to establish the TI zone boundary. The requirements for definition of the boundaries of the TI zone will primarily be based on consideration of contaminant-specific conditions such as DNAPL, and fixed hydrogeologic factors present at a site. However, the environmental risk, remedial costs and time frame, and ground-water resource valuation may also be factors which define the scope of investigations to reduce uncertainties or define factors relating to the TI zone determination.

The U.S. EPA Region 4 expects that where a TI demonstration is made on the basis of failure of an ongoing ground-water remedial action to meet remedial goals, the cause of such failure will be demonstrated.

"FRONT-END" TI DEMONSTRATIONS

A "front-end" TI decision is made prior to remedy implementation (U.S. EPA, 1993). The "front-end" decision must be made without the benefit of long-term ground-water monitoring data which demonstrate the inefficiency of an ongoing remedial action, suggesting the technical impracticability of ground-water remediation. However, the U.S. EPA Region 4 will require that the "front end" TI request and demonstration answer the same questions as would a TI request made on the basis of inefficiency of an ongoing remedial action. The "front end" TI request will be included in the Feasibility Study for a National Priorities List (NPL) site, and for a site under Resource Conservation and Recovery Act (RCRA) regulatory authority, the "front end" TI request will be included in the Corrective Measures Study.

The "front end" TI request must quantitatively define the reason why attainment of a ground-water ARAR in a reasonable time frame is not possible. If, for example, a low mobility organic contaminant is present in the saturated zone such that it provides a reservoir of contamination which cannot be physically removed, both the rate(s) of desorptive mass transfer of the contaminant to the dissolved phase and the mass of sorbed contaminant will need to be quantified, in order to demonstrate technical impracticability.

The time necessary to attain ARARs within the proposed TI zone must be estimated, and the uncertainty of that estimate must be discussed. Additionally, the time frame to attain ARARs outside the TI zone must be estimated, for conditions where the TI zone is fully contained, is partially contained or treated (if full containment is not possible), or where the TI zone is addressed as is the rest of the contaminant plume.

The "front end" TI request will include a plan for the containment, partial containment, or treatment of the TI zone. The plan must be such that for NPL sites, program goals, expectations, and threshold remedy selection criteria are met, consistent with Section 300.430 of the National Oil and Hazardous Substances Pollution Contingency Plan (U.S. EPA, 1990).

The "front end" TI request will define the part of the aquifer where the request applies.

The "front end" TI request will fully consider any remedial option that could be used to remediate ground water in the proposed TI zone to meet ARARs, regardless of cost. For example, if, at extraordinary cost, the contaminated aquifer in the TI zone could be dewatered and the contamination removed from below the ambient water table, the cost, effectiveness, and implementability of such action would have to be fully presented.

The "front end" TI Request will include an assessment of the valuation of the ground-water resource in the area proposed for a TI ARAR waiver. At present, there are many uncertainties related to ground water valuation (National Research Council, 1997). The regional supplemental TI guidance will acknowledge this problem, and will thus likely specify a framework for primarily qualitatively evaluating the value of the ground-water resource in terms of its current use, influence on the quality of receiving bodies of surface water, and other related factors.

MONITORING OF ONGOING GROUND-WATER REMEDIAL ACTIONS TO DEMONSTRATE TI

Ongoing ground-water remedial actions involving perceived failure of "pump and treat" systems to attain ARARs typify the type of problem which may be addressed by a TI waiver. The U.S. EPA Region 4 supplemental TI guidance intends to require that for this sort of circumstance, the reason the remedial action is believed to be incapable of attaining the specified remedial objective must be defined before the Region approves a request for a TI waiver. This requirement will necessarily mean that a projected time frame for the remedial action, as operated, to attain the remedial objective must be determined. Additionally, a comprehensive review of the operations of the extraction system or other remediation system must be performed, to determine if any changes in well placement, pumping rates, extraction well design, systems operation, remedial technology, or other factors would substantially improve the ability of the remedial technology to attain remedial objectives. Alternative remedial strategies that appear more effective for attaining remedial objectives must be assessed as to their implementability, reliability, cost, and other factors considered by the U.S. EPA when selecting a ground-water remedial technology.

One goal of the regional TI guidance is to better define the appropriate placement of monitoring wells, amount or duration of ground-water monitoring data, types of monitoring data or testing procedures, and statistical data analysis approaches which must be used to (1) best define the projected remedial time frame for a ground-water remedial action to attain ARAR concentrations, (2) best define the remedial action's progress toward meeting secondary remedial action objectives such as contaminant mass reduction, and (3) best define the DNAPL zone, or other portion of a plume, for which a TI waiver would be approved by the U.S. EPA Region 4. It has been the Region's experience that where initial estimates of ground-water remedial time frames have been overly optimistic, failure of the remedial action (typically "pump and treat" remedial actions) to achieve such unrealistic goals has been equated with the technical impracticability of ground-water restoration. The U.S. EPA Region 4 TI guidance is intended to conceptually and, to the extent possible, quantitatively discuss the causes of such perceived failures, and define those apparent "failure conditions" which may be used to demonstrate TI.

REFERENCES

National Research Council, 1997, *Valuing Ground Water: Economic Concepts and Approaches*, National Academy Press, Washington, DC.

U.S. EPA, 1990, National Oil and Hazardous Substances Pollution Contingency Plan, Code of Federal Regulations, Chapter 40, Part 300.

U.S. EPA, 1993, *Guidance for Evaluating the Technical Impracticability of Ground-Water Restoration*, Office of Solid Waste and Emergency Response, Directive 9234.2-25, Washington, DC.

EPA POLICY ON THE TECHNICAL IMPRACTICABILITY OF GROUND-WATER CLEANUP

Kenneth A. Lovelace, Jr. (U.S. EPA, Washington, DC)

ABSTRACT: In responding to sites contaminated with toxic chemicals, cleanup of contaminated ground water has proven to be one of the most difficult problems facing the U.S. Environmental Protection Agency (EPA). Contaminated ground water has been found at approximately 85% of hazardous waste sites being remediated under EPA's Superfund program, and at a large proportion of facilities undergoing Corrective Action under the Resource Conservation and Recovery Act (RCRA). One of the most important lessons learned during the implementation of these programs is that complex site conditions are more common than previously anticipated.

A major goal of EPA's environmental cleanup programs is to return contaminated ground waters to a level of quality appropriate for the current or future uses of these resources. Drinking water standards generally are used as cleanup requirements for ground waters that are current or potential sources of drinking water. Although both programs have been successful at reducing the immediate health and environmental threats posed by the contaminated ground water, experience has shown that achieving these cleanup levels over the entire plume may not be possible at many sites with currently available remediation technologies. Where it can be demonstrated to EPA that these requirements cannot be attained due to specific site conditions, both the Superfund and RCRA Corrective Action programs allow cleanup requirements to be waived due to technical impracticability.

EPA developed the *Guidance for Evaluating the Technical Impracticability of Ground-Water Restoration* in September 1993. This guidance established evaluation factors for determining whether ground-water cleanup requirements are technically achievable at a particular site, and how to establish an alternative cleanup strategy that fully protects human health and the environment.

Site-specific data should be used to describe site conditions that make attainment of drinking water standards technically impracticable (TI). A TI waiver does not necessarily apply over the entire contaminant plume, and should only apply to that portion of the site where drinking water standards cannot be attained. Most importantly, a TI waiver is not a "walk away" solution. An alternative cleanup strategy must be selected that adequately protects human health and the environment, and may involve measures that have significant cost. An alternative cleanup strategy generally will be expected to: 1) prevent exposures to contaminated ground water; 2) control sources of contamination; 3) restore as much of the contaminant plume as possible; and 4) prevent continued plume migration.

INTRODUCTION

Contaminated ground water has been found at approximately 85% of hazardous waste sites being remediated under EPA's Superfund program, and at a large proportion of facilities undergoing Corrective Action under the Resource Conservation and Recovery Act (RCRA).

Cleanup Goals. A major goal of EPA's environmental cleanup programs is to return contaminated ground waters to a level of quality appropriate for the current or expected future uses of these resources. For the Superfund program, this goal is stated in the National Oil and Hazardous Substances Pollution Contingency Plan (Federal Register, 1990), as follows:

EPA expects to return usable ground waters to their beneficial uses wherever practicable, within a timeframe that is reasonable given the particular circumstances of the site. When restoration of ground water to beneficial uses is not practicable, EPA expects to prevent further migration of the plume, prevent exposure to the contaminated ground water, and evaluate further risk reduction.

Federal and/or State drinking water standards generally are used as cleanup requirements for contaminated ground waters that are a current source or future source of drinking water. Water quality standards for surface water may also used as cleanup levels for contaminated ground waters that discharge to surface waters. In general, these cleanup levels are to be attained within the aquifer, throughout the zone of contamination but not including areas of the plume beneath waste management areas.

Lessons Learned From Site Remediation. The most important lesson learned during implementation of Superfund and other remediation programs is that complex site conditions are more common than previously anticipated. As a result of these site complexities, restoring all or portions of the contaminant plume to drinking water or similar standards may not be possible at many sites using currently available remediation technologies (U.S. EPA, 1992; National Research Council, 1994).

RESTORATION POTENTIAL

The restoration potential is defined as the likelihood that appropriate cleanup levels can be achieved at a given site. Examples of site conditions that affect the restoration potential fall under two major categories: 1) contaminant-related characteristics, and 2) hydrogeologic characteristics.

Conceptual Site Model. Site conditions affecting the restoration potential should be considered in the **conceptual site model**, which is an interpretive summary of the site information obtained to date. A conceptual site model is a three-dimensional representation that conveys what is known or suspected about sources of contaminants, release mechanisms, and the transport and fate of those contaminants. A "conceptual site model" is **not** a "computer model;" however, a computer model must be based on a sound conceptual model of site conditions to provide meaningful results. Thus, the conceptual site model provides the basis for assessing the restoration potential of the site.

Multiple Sources of Information. For every site, existing data should be reviewed or new data should be collected to identify factors that could increase (or decrease) the difficulty of restoring ground water to the appropriate cleanup levels. For many sites, data collected from site characterization activities alone may be insufficient to estimate the restoration potential with any degree of certainty. This uncertainty can be reduced by using remedy performance data in combination with site characterization information to assess the restoration potential.

Assessing Restoration Potential. By implementing a ground-water remedy in more than one step or phase, performance data from an initial phase can be used to assess the restoration potential and may indicate that additional site characterization is needed. In addition to providing valuable data, the initial remedy phase can be used to attain short-term response objectives, such as preventing further plume migration. Phased implementation of response actions also allows realistic long-term remedial objectives to be determined prior to installation of the comprehensive or "final" remedy. For this reason, the restoration potential should be assessed prior to establishing final remedial objectives for contaminated ground water at every site (U.S. EPA, 1996).

INFORMATION SUPPORTING A TI DECISION

EPA developed the *Guidance for Evaluating the Technical Impracticability of Ground-Water Restoration* in September 1993 (U.S. EPA, 1993). This guidance established evaluation factors for determining whether ground-water cleanup requirements are technically achievable at a particular site, and how to establish an alternative cleanup strategy that fully protects human health and the environment. In general, determinations of technical impracticability (TI) will be made by EPA based on site specific data and analysis of the following factors:

1. Specific cleanup levels or other standards to be waived;
2. Spacial area of the site over which the TI decision will apply;
3. Conceptual site model that describes site hydrogeology, contamination sources, potential environmental receptors, and the transport and fate of contaminants;
4. Evaluation of the restoration potential of the site, including:
    a) Demonstration that sources have been or will be controlled to the extent practicable,
    b) Appraisal of the performance of any previous or ongoing remedial actions,
    c) Prediction of the time frame necessary to attain the required cleanup levels, and

d) Demonstration that no other remedial technologies could reliably or feasibly attain the cleanup requirements; and

5. Estimates of the cost of the alternative considered for attaining requirements and the cost of alternative remedial strategies, in the event cleanup standards are waived due to TI.

It should be also be noted that the presence of a particular site condition (such as dense nonaqueous phase liquids (DNAPLs), light nonaqueous phase liquids (LNAPLs) or fractured bedrock) generally will not provide sufficient grounds on which to justify a TI determination. Adequate site characterization data must be presented to document that a particular site condition exists **and** to explain why the condition prevents restoration of ground water using available technologies.

Timing of TI Decisions. For EPA programs, TI decisions may be made either at the time a site decision document is completed (e.g., RCRA Statement of Basis or Superfund Record of Decision) or after the remedy has been implemented and monitored for some period of time. Since it is often difficult to predict restoration potential based on site characterization data alone, EPA believes that TI decisions will often need to be supported by performance data from a full scale remediation system, either a pilot-scale treatability test or the initial phase of a ground-water remedy. For some sites, the information available at the time of the decision document may be sufficient to support a TI determination. Data and analyses presented for "front-end" TI decisions should thoroughly document site conditions which limit ground-water restoration, since the decision will be based on predictive analysis rather than actual remedy performance.

ALTERNATIVE REMEDIAL STRATEGIES

If cleanup standards are waived due to TI, EPA requires that an alternative remedial strategy be developed that protects human health and the environment. An alternative remedial strategy for a given site generally will be expected to: 1) prevent exposures to contaminated ground water; 2) control sources of contamination; 3) restore as much of the contaminant plume as possible to the appropriate cleanup levels; and 4) prevent continued migration of the plume. These objectives are discussed in more detail below. Thus, obtaining a TI waiver is not a "walk away," since an alternative remedial strategy will be required and may include measures that have significant cost.

Exposure Control. Protection of public health demands that, at a minimum, measures be implemented to prevent ingestion and other exposures to contaminated ground water. Preventing exposure to contaminants may be accomplished by either removing the contaminants from ground water prior to use, developing an new water supply, or preventing use of the contaminated ground water.

Source Control. Preventing the continued release of contaminants from sources to ground water could include removal, treatment or containment of source materials.

In general, EPA prefers that sources be removed or treated where practicable. Where sources cannot be completely treated or removed, containment of source materials will be necessary if ground water is to be restored to the appropriate cleanup levels.

Containment of sources has several benefits. First, it will limit the migration of source materials that are potentially mobile, such as DNAPLs or LNAPLs. Second, effective source containment may allow restoration of the portion of the plume outside the containment area. Third, containment of the source area may be necessary prior to implementation of measures designed to remove or treat source materials, because some in-situ treatment technologies tend to increase the mobility source materials.

Plume Restoration. In general, EPA recommends that as much of the plume as possible be restored to appropriate cleanup levels. Where sources cannot be effectively removed or treated but can be effectively contained, the portion of the plume outside of the containment area generally should be restored to the required cleanup levels. For some sites, restoration of the entire plume may be technically impracticable, because site conditions do not allow sources to be effectively controlled, or other physical limitations prevent such restoration.

Plume Containment. In general, preventing plume migration is necessary to prevent uncontaminated ground waters or surface waters from becoming contaminated; protect drinking water supplies, protect habitat for aquatic or terrestrial life, or prevent adverse impacts to other natural resources. For this reason, EPA generally requires that continued migration of the contaminant plume be prevented, to the extent practicable.

EPA EXPERIENCE WITH TI DECISIONS

Relatively few TI determinations have been made to date, in either the Superfund or RCRA Corrective Action programs. For Superfund a total of 26 TI decisions were made through 1996. A recent poll of EPA Regional staff indicated that only two TI decisions have been made to date in the RCRA Corrective Action program. Although there appears to be a lot of interest in TI for ground-water cleanup, it is not completely clear why so few TI waivers have been requested.

Decision Process. In general, a TI Evaluation report is prepared by the party or parties responsible for cleaning up the site (e.g., site owners, State Agency, EPA, other Federal Agency) and submitted to the EPA site manager for review. The TI Evaluation report should explain the site conditions that prevent restoration of ground water to appropriate cleanup levels, based on site specific data and analysis of the TI evaluation factors defined in U.S. EPA, 1993 (summarized above). EPA has not developed a standard format for this report.

In most cases, the TI Evaluation report is reviewed by a panel of EPA staff, including the EPA site manager, site managers from other State or Federal Agencies responsible for cleanup, and technical experts from the relevant EPA Regional Office and from EPA Headquarters. In some cases, experts from EPA's research laboratories are also requested to provide technical review. If the review panel determines that the TI Evaluation report provides adequate justification for waiving cleanup requirements, then a decision document can be completed which includes the TI decision (TI waiver) and selects the revised remedy (alternative remedial strategy), as discussed above.

Recommendations for TI Evaluations. In the TI Evaluation reports submitted to date, it is often unclear what remedial alternatives were considered in an attempt to restore ground water to cleanup levels appropriate for the site. EPA reviewers generally want to know what technologies or combinations of technologies were considered for controlling sources and restoring ground water, and what specific site conditions prevent restoration of ground water at this particular site. The TI evaluation factors (see above) should not be discussed in the abstract, but should relate to the methods considered for fully restoring site ground water.

For reasons discussed in the above paragraph, a TI Evaluation is similar to a Superfund Feasibility Study (FS) or RCRA Corrective Measures Study (CMS). Also, a TI Evaluation can be easily incorporated into the FS or CMS. However, it is often unclear to EPA reviewers whether a remedial alternative discussed in the FS (or CMS) is intended to fully restore ground water or not. It is highly recommended that the FS (or CMS) include separate sections for discussion of: 1) remedial alternatives intended to fully restore ground water; 2) why these alternatives cannot fully restore ground water (TI evaluation); and 3) remedial alternatives to be considered in the event a TI waiver is issued (alternative remedial strategy).

REFERENCES

Federal Register. 1990. "National Oil and Hazardous Substances Pollution Contingency Plan; Final Rule" (NCP). Volume 55, No. 46, March 8, 1990; 40 CFR Part 300.

National Research Council. 1994. *Alternatives for Ground Water Cleanup*. National Academy Press, Washington, DC.

U.S. EPA. 1992. *Evaluation of Ground-Water Extraction Remedies: Phase II, Volume 1 Summary Report*. Office of Solid Waste and Emergency Response (OSWER) Publication 9355.4-05 (NTIS publication PB92-963346), February 1992.

U.S. EPA. 1993. *Guidance for Evaluating Technical Impracticability of Ground-Water Restoration*. OSWER Directive 9234.2-25, EPA/540-R-93-080 (NTIS publication PB93-963507), September 1993.

U.S. EPA. 1996. *Presumptive Response Strategy and Ex-Situ Treatment Technologies for Contaminated Ground Water at CERCLA Sites, Final Guidance*. OSWER Directive 9283.1-12 (NTIS publication PB96-963508), October 1996.

ESTABLISHING PRACTICAL CLEANUP GOALS WHILE TRANSFERRING FROM SITE INVESTIGATION TO REMEDIATION

Randolph C. Brandt (Dames & Moore, San Jose, California)
S. Tariq Hussain (Dames & Moore, Santa Ana, California)

ABSTRACT: This paper reviews the practical aspects of establishing realistic cleanup goals for a site impacted by chlorinated and other recalcitrant compounds. The data presented in this study are based on real site conditions and present the difficulty of establishing goals for remediation, purely on the basis of soil and/or groundwater sampling.

INTRODUCTION

The subject site had been impacted by several chlorinated compounds; however, the distribution of these chemicals in the subsurface was complex and difficult to characterize. The complexity arose because of the nature and timing of the release, the various locations the releases had occurred, and the geologic and hydrogeologic conditions prevailing in the subsurface. The site investigation data in conjunction with computer modeling data underestimated the volume and mass of the chemicals in the soil by several orders of magnitude. Fate and transport modeling predicted very little impact to groundwater because of the concentrations of the chemicals in the soil. However, empirical data suggested that the groundwater had been significantly impacted. Results of the feasibility study indicated that the most appropriate method of soil remediation was to employ the technology of soil vapor extraction (SVE).

As the project moved into the remediation phase, the requirements for establishing remedial action objectives and cleanup goals became a priority. Initially, the regulating agency mandated that a clean-up level of non-detectable concentrations be established. The agency believed that significant attenuation of the concentration of chemicals found in the subsurface with time was not considered a factor in the attainment of remedial action objectives; rather an overall reduction in chemical mass was deemed more appropriate. However, given the site conditions, a cleanup goal of non-detectable concentration levels was not a practicable solution. Reduction of the mass of the chemicals in the subsurface could only be achieved by employing an engineered solution. It is our opinion that, if nonattainable goals had been accepted as a basis for cleanup it would have rendered the whole remediation process into a meaningless and futile exercise as the technology of SVE would never be able to achieve this objective. On the basis of this background we proposed a different approach to establishing cleanup goals. We believed that goals, which were based on the limitations of the accepted technology employed rather than numerical targets that are not practicable and impossible to achieve. This paper presents the proposal in hopes of providing an additional tool for negotiating and establishing remedial goals based on practical experience gathered on this project.

BACKGROUND

With the publication of the Lawrence Livermore Report (LLNL, November 16, 1995), the process of natural attenuation of petroleum hydrocarbons has been validated and widely accepted. Now industry can, and readily does, invoke the scientific principals presented in that report to justify closure of a site without the application of an engineered remedial action. Unfortunately the same principals cannot apply in the case of sites contaminated with halogenated compounds. Based on our current understanding of the basic principles of the degradation, transformation and or decomposition of these compounds within the subsurface environment, these processes are considered extremely slow and often result in more toxic daughter products. For this reason the regulatory community considers a natural attenuation solution without an engineered component for these chemicals unacceptable.

The most common halogenated compounds that were widely used by industry and thus ended up in the subsurface media include carbon tetrachloride, chloroform, dichloroform, trichloroethylene, trichloroethane, tetrachloroethane and vinyl chloride. These chemicals were and are produced in large quantities and were commonly used as solvents for cleaning and chemical synthesis (Cookson, 1995).

In the subsurface environment these chemicals are difficult to detoxify or degrade effectively or efficiently. These compounds are known to undergo abiotic transformation in the subsurface, these transformation processes include chemical substitution, dehydrohalogenation, and reduction. Dehydrohalogenation results in elimination of the halogen, forming an alkene (non-halogenated hydrocarbon). The rate at which dehydrohalogenation occurs increases as more chlorine atoms are attached to the carbon. However, polychlorinated species hydrolyze less readily. In the absence of biological activity these reactions proceed extremely slowly. A variety of transition metals, including nickel, iron, chromium, and cobalt, can reduce the halogenated compounds.

The in-situ application of engineered solutions applying one or more of these processes to degrade and/or detoxify halogenated organic compounds has not developed to a stage in which we can say that it is a technology that the regulatory agencies readily accept as a long-term solution. In instances where both petroleum hydrocarbons and halogenated hydrocarbons are present, the focus is mostly on the halogenated compounds because of the fact that these chemicals are considered more tenacious and longer lasting. In most cases, when this occurs, the regulatory driver for establishing clean-up goals is usually the halogenated compounds.

SITE DATA

In the subject case study, the nature and extent of chemical constituents in the soil were evaluated through a two-phased Remedial Investigation. The analytical results from the Phase I work indicated that soils in three discrete areas of the site were impacted mainly by chlorinated volatile organic compounds (VOCs), specifically Trichloroethene, 1,1,1-Trichloroethane, Tetrachloroethene, their derivatives, and some jet fuel. Soil borings drilled to a depth of 80 feet below ground surface show that elevated concentrations of VOCs under the two former dry wells extend to 75 feet below the surface. In the third area, intermittent lenses of VOCs were detected to a depth of 125 feet below ground surface in borings drilled to a depth of 139 feet. The soil borings were terminated when field tests, did not detect VOCs. None of the borings reached the water table, which is at a depth of approximately 250 feet below ground surface. Investigation of groundwater beneath the Site during the Phase II investigations indicated that groundwater has been impacted by VOCs from past onsite activities. Some of the chemical constituents currently exceed state and/or federal standards in groundwater beneath the site. Sources upgradient and possibly crossgradient of the site may also be causing concentrations of chemicals in groundwater to exceed some groundwater standards.

A Baseline Health Risk Assessment was conducted to: 1) identify potential chemicals of concern at the Site, 2) identify exposure pathways from chemicals detected at the Site to human or environmental receptors; and 3) assist in developing Remedial Action Objectives for soil and groundwater. Exposure and risk calculations associated with chlorinated and non-chlorinated VOCs as well as one SVOC (the identified chemicals of concern) were carried out for three potentially exposed groups: 1) current and future facility workers; 2) current offsite residents; and 3) potential future offsite residents. The potential for migration of chemicals from subsurface soils, to both groundwater and the air, and subsequently to the three potentially exposed groups was considered. For current and future facility workers, inhalation of vapors, dermal contact, and soil ingestion were considered. Groundwater ingestion, dermal contact, inhalation of vapors while showering, plant ingestion, and inhalation of soil vapors were considered for current and future offsite residents.

To evaluate possible future impacts to groundwater quality from VOCs in site soils, a modeling analysis was conducted using information generated in the RI and from scientific literature. Since the primary potential risk posed by VOCs in Site soils is to groundwater, the modeling effort focused on predicting whether VOCs in soil would impact groundwater quality above State or Federal maximum contaminant levels (MCLs). After screening five methods, two analytical methods (the Summers Method and the Organic Leachate Model or OLM) and one numerical model (VLEACH) were selected to perform the evaluation of potential impact of Site soils to groundwater quality.

Results from the most reliable method (the VLEACH model) indicate that existing residual VOCs in soils would not cause concentrations in groundwater to exceed MCLs in the next century for any parameter detected at the site. These findings are corroborated by the results of the OLM method. By contrast, the Summers analytical method predicts that the existing residual maximum observed VOCs in soil could cause TCE, PCE, and 1,1-DCE to exceed MCLs in groundwater. The Summers method over-predicts leachability of residual VOCs in soils with very low organic carbon content such as those found at the Site. Its predictions are not corroborated by the other more reliable methods (VLEACH and OLM), that find that current VOC concentrations in soil will not cause groundwater levels to exceed MCLs.

Based upon the evaluation of predictive tools used to determine if residual soil concentrations could result in an impact to groundwater, it was concluded that the VLEACH model is the most reliable and appropriate model for the Site. The results from this model demonstrate that the observed maximum concentrations of VOCs beneath the Site do not pose a threat to groundwater quality (i.e., cause an exceedance of MCLs). The OLM method results corroborate those of VLEACH. Based upon these models, remediation of soil in order to protect groundwater quality would be unnecessary at this Site. However, due to the uncertainty cast on the analysis by the Summers Method, the uncertainty in soil characterization data, and the lack of a confirmed connection between soil impacts and groundwater impacts, the regulatory agency directed the owner to remediate VOCs in the soil.

Once a decision that soil remediation was required, a Feasibility Study was conducted to identify the most appropriate technology for remediation. The results of the feasibility study identified SVE as the most appropriate technology to remove VOC mass from the soil. The agency and the owner both agreed that SVE would be implemented at the site.

ESTABLISHING CLEAN-UP GOALS

During the Feasibility Study, a significant focus turned to establishing realistic clean-up objectives and goals and identifying the process by which achievement of clean-up goals had been met. The agency took the position that clean up should proceed until non-detectable concentrations, as verified through a comprehensive subsurface soil-sampling program, were achieved. We believed that this approach was overly conservative and impossible to achieve for the following reasons:

- Fate and transport modeling suggested that future impacts to groundwater would likely not occur thus suggesting that remediation, to any level, was not warranted from a human health or groundwater protection perspective;
- The site was so heterogeneous and the VOCs so randomly distributed that any amount of soil sampling would likely turn up some concentration of residual VOCs;
- The proposed treatment technology was designed to remove VOCs entrained in the soil gas, not remove VOCs adsorbed onto the surface of soil grains.

As an alternative approach, we suggested establishing a goal and verification method that was consistent with the treatment technology chosen. Because the SVE system was designed to extract VOC mass by collecting volatilized constituents entrained in the soil gas, we believed a reasonable goal would be to operate the system until VOC concentrations in the extracted soil gas reached a low, asymptotic level. Likewise, because the SVE operated on the principal of extracting soil gas, we believed that a reasonable confirmation sampling approach was to sample soil gas to verify that the asymptotic level was reached. We believed this approach was both scientifically appropriate as well as protective of human health and the groundwater for the following reasons:

- VOCs that could become mobile and available to migrate to and possibly impact groundwater are those that could become easily volatilized. If easily volatilized, they would be captured by an appropriately designed SVE system;
- Residual VOCs adsorbed onto the soil grains likely would not be significantly available to migrate to groundwater, especially if surface water infiltration was eliminated;
- Operational monitoring of the SVE system focused entirely on its ability to extract soil gas. Likewise, confirmation sampling of the soil gas should be used to verify that cleanup objectives were achieved.

We did acknowledge that this approach had limitations as follows:

- SVE could be ineffective at affecting the entire VOC impacted area due to channeling and preferential pathways for air movement;
- Because VOCs volatilize at different rates, making the assumption that clean-up goals are achieved once an asymptotic level had been reached for total VOC concentrations might be pre-mature, as the opportunity for additional mass removal of certain chemicals could be possible.

However, if the system was properly designed and operated, we believed that these limitations could be overcome. For example, by designing the extraction wells with multiple extraction zones, the possibility for channeling or preferential pathways would be minimized. The issue of differing volatilization rates could be addressed by first operating the system continuously until an asymptotic concentration level had been reached. After that point, the system would be operated in a pulsed mode until the mass reduction of VOCs reached a consistently low, asymptotic level.

CONCLUSIONS AND RECOMMENDATIONS

It is the opinion of the authors that clean-up objectives are often selected to be over-conservative and do not consider the technology limitations of the remediation technology selected. We challenge the regulatory and scientific community to consider establishing clean-up objectives and confirmatory sampling plans in the context of the environmental benefit that is being achieved as well as the limitations of the remedial technologies employed. Approaches to balancing these issues could include:

- Thoroughly understanding the limitations and degree of certainty in the site characterization data;
- Thoroughly understanding the elements of human and environmental risk posed by the site;

- Thoroughly researching, carefully selecting, and appropriately designing remedial alternatives for site restoration;
- Establishing clean-up objectives that are consistent with the technology selected; and,
- Designing a confirmatory sampling program that is focused on the appropriate media and constituents that the remediation technology is designed to address.

We point out that establishing clean-up objectives and the corresponding confirmatory sampling methodology must consider the technology employed and the degree of human health and environmental that may exist from residual constituents.

REFERENCES

Cookson, John, D., Jr. 1995. *Bioremediation Engineering*. McGraw Hill, Publishers, New York. 1995

Lawrence Livermore National Laboratory. 1995. California Leaking Underground Fuel Tank (LUFT) Historical Case Analyses. Submitted to the California State Water Resources Control Board. November 16, 1995

REMEDIATION OF CHLORINATED-COMPOUND-CONTAMINATED GROUNDWATER IN OREGON

Michael McCann, Bill Mason, and Matt Clouse
Oregon Department of Environmental Quality, Eugene, Oregon, USA

ABSTRACT: The 1995 Oregon Legislature re-wrote Oregon's environmental cleanup law to require the lowest cost protective remedial action. Consequently, it is expected that future cleanups will be more likely to consider in situ alternatives, natural attenuation, and long-term monitoring. An evaluation of actions conducted at chlorinated-compound-contaminated groundwater sites in Oregon found that the majority of remedies implemented to date have involved groundwater extraction with above-ground treatment. However, in the past two years, an increasing number of less costly, less intrusive, and yet fully protective remedial actions have been implemented. These have included funnel-and-gate, in situ bioremediation, phytoremediation, and natural attenuation monitoring.

INTRODUCTION

During its 1995 session, the Oregon Legislature re-wrote Oregon's environmental cleanup law (ORS 465.200-.900), elevating the importance of cost as a factor in the selection of remedies at State-led cleanup sites (non-NPL sites). The revised cleanup law requires cleanup actions to consider current and reasonably likely future land and water use, and requires treatment, if feasible, of "hot spots" of contamination. The law also establishes a preference for selection of the lowest cost protective remedial action for contamination which does not constitute a "hot spot." Administrative rules implementing the revised cleanup law were subsequently developed by Oregon's Department of Environmental Quality (DEQ) and a public advisory committee (OAR 340-122-010 - 140).

Concurrent with revising Oregon's cleanup rules, the environmental industry has been developing alternative approaches to extraction and treatment needed for the revised cleanup law to be successful at sites with contaminated groundwater. Sparging, biodegradation, and reactive treatment walls have all been established as effective in-situ alternatives to groundwater extraction. In addition, there have been a number of efforts to establish a protocol for applying natural attenuation to sites contaminated with chlorinated solvents, including one developed for the U. S. Air Force Center for Environmental Excellence (Wiedemeier, T. H., Wilson, J. T., et al. 1996).

An evaluation of remedial actions conducted at chlorinated-compound-contaminated groundwater sites in Oregon found that the majority of remedies implemented to date, both as interim actions and under Records of Decision (RODs), have involved groundwater extraction with above-ground treatment. However, in the past two years, an increasing number of less costly, less intrusive, and yet fully protective remedial actions have been implemented in response to both the mandate of the new cleanup law and the availability of innovative technological approaches.

BACKGROUND

DEQ currently coordinates investigation and cleanup activities at approximately 315 sites around the state. It is estimated that roughly half of these sites are contaminated with chlorinated compounds. The sites are tracked on a site information database which currently contains information on approximately 1,500 sites, of which up to a third may be impacted by chlorinated compounds. The database contains records of remedial activities and contaminant concentrations observed.

Non-NPL environmental cleanup projects in Oregon are managed through the Voluntary Cleanup Program, Site Response Program (State Superfund), and Orphan Site Program. All sites are managed in accordance with Oregon's Environmental Cleanup Rules which specify the requirements for all aspects of environmental cleanup work from site discovery through remedial action. The statutory basis for the rules is contained in the Oregon Revised Statues.

During its 1995 session, the Oregon Legislature enacted the Recycled Lands Act (Oregon Legislative Assembly House Bill 3352) which extensively re-wrote the statutory framework for the environmental cleanup rules. The act was the result of an industry-led effort to streamline the cleanup process by enabling cleanup projects to be completed more quickly and at less expense. Following passage of the revised cleanup law, DEQ convened a twelve member advisory committee of interested parties to direct the development of companion rules for the revised statute. Advisory committee members included environmental professionals, attorneys, industry representatives, environmental advocates and natural resource managers. With the advisory committee's backing, DEQ issued proposed rules for public comment in late 1996. Following the incorporation of public comment, the revised rules were adopted for implementation by DEQ's rulemaking body, the Environmental Quality Commission, in January 1997.

OREGON'S REVISED CLEANUP RULES

The Recycled Lands Act was an attempt by the Oregon Legislature to deal with the frustration felt by responsible parties, industry, and the public, that environmental cleanup work is too slow, too costly, and too uncertain. The particular components of the cleanup process targeted by the act included land and water use evaluation, risk assessment, and remedy selection. These were also the areas that received the most attention during the rulemaking process.

Under the previous rules, remediation was evaluated within the context of the highest beneficial use (typically resulting in efforts to protect all water resources as potential future drinking water source), whereas the revised rules require an evaluation of current and reasonably likely future beneficial water uses. In addition, risk assessments are now required to take into consideration the exposures associated with beneficial water use determinations.

Similarly, the previous rules required cleanup to background or lowest feasible concentrations with a stated preference for treatment, while the revised rules provide bright-line standards for acceptable risk (for example, one in one million increased cancer risk for individual compounds) and limit the preference for treatment to "hot spots." For groundwater, hot spots are defined as an exceedance of a relevant standard for the given beneficial use. For example, an exceedance of an MCL would constitute a "hot spot" for a current or reasonably likely future drinking water source. The revised rules also state that protection may be achieved through removal, treatment, engineering controls or institutional controls.

In many cases, the primary impact of these revisions to the cleanup rules is to change the remedy selection process from one of source (or risk) reduction to one of risk management. Containment and use restrictions as remedial measures have been made equal, and even preferable because of cost, to more conventional extraction and treatment schemes for contamination which does not constitute a "hot spot." In addition, the renewed emphasis on cost as a factor in the selection of a remedy often bodes well for less traditional treatment alternatives, even for contamination constituting a "hot spot."

GROUNDWATER REMEDIAL ACTIONS

As discussed in the introduction, we expected the focus of groundwater remedial measures since the passage of the Recycled Lands Act to shift to less costly and less invasive methods. In order to evaluate this theory, a survey of the chlorinated-compound-impacted sites contained within DEQ's site database was conducted. The survey focused on sites impacted by chlorinated solvents and pentachlorophenol. Both interim remedial action measures (IRAMs) and RODs were evaluated. The results were separated into pre- and post-Recycled Lands Act actions.

Pre-Recycled Lands Act. Through 1996, DEQ had issued seven RODs for sites with groundwater contaminated by chlorinated solvents. Four of those RODs called for active extraction and treatment of groundwater. The other three were for sites where no current nor future exposure was expected. Those RODs required long-term monitoring and deed restrictions against future groundwater use.

DEQ also has records on 23 IRAMs at chlorinated solvent-contaminated groundwater sites currently in the Remedial Investigation/ Feasibility Study (RI/FS) process. Thirteen of the IRAMs included extraction and treatment. The others included source removal (6) and providing an alternative water supply (3).

The results are similar for sites impacted by pentachlorophenol, where the three pre-Recycled Lands Act RODs included non-aqueous phase liquid (NAPL) and/or groundwater extraction. Five of seven IRAMs completed at pentachlorophenol sites employed extraction and treatment with the other two relying on source removal.

Post-Recycled Lands Act. In the past two years, DEQ has signed two RODs allowing for natural attenuation of chlorinated solvents in groundwater, although one of these also calls for the continued operation of an extraction system installed in the source area as an IRAM. Two RODs have also been signed allowing for natural attenuation monitoring of pentachlorophenol (PCP) plumes in conjunction with source removal or control.

Also during this time period, IRAMs employing sparging (2), phytoremediation, and a reactive treatment wall have been implemented to address chlorinated solvents in groundwater. Groundwater extraction and treatment has been implemented twice in the past two years as IRAMs for chlorinated solvents and pentachlorophenol.

ANALYSIS OF RESULTS

Although it is too early to say with any certainty that there has been an overall shift away from extraction remedies towards in situ and natural attenuation remedies since the implementation of Oregon's new cleanup law, the early data appear to indicate a trend in that direction. However, this move appears to be limited to cases involving impacts to groundwater where ecological impacts are not significant and where: a) current and reasonably likely future beneficial uses of water do not include drinking water; or b) sites where less invasive measures are capable of restoring or protecting water uses which are determined to be reasonably likely to occur in the future. In contrast, sites with impacts to existing drinking water use of groundwater are likely to require extraction-based remedies or other alternative means in order to more quickly achieve protectiveness.

Of the four RODs that include natural attenuation as a component, three have a potential beneficial use as a drinking water source and one has an ecological use (discharge to surface water). None of the sites have currently impacted drinking water supplies. The two sites where phytoremediation and funnel and gate technologies have been employed have impacted drinking water supplies. Well head treatment is being used at both sites to protect the water users, and the IRAMs are being used to address the contaminated groundwater near the sources. Bottled water is being provided to impacted residents at one of the sparging sites, and the other sparging site has an ecological endpoint with no drinking water use.

Another question to be evaluated is whether or not the Recycled Lands Act has made a difference in remedy selection. The development of natural attenuation protocols, and funnel-and-gate and phytoremediation technologies would have added these options to the mix of potential remedial alternatives with or without the Recycled Lands Act. The results indicate that these options are generally being used in conjunction with either source removal or well head treatment to achieve the protectiveness standard required in remedy selection.

Source removal (soil, NAPL, or groundwater) as a means of protecting groundwater and downgradient beneficial users has been and will continue to be employed where possible because, from a cleanup perspective, it provides the greatest benefit per unit cost.

CONCLUSIONS

We expect that as Oregon implements the revised cleanup law, the trend toward less invasive and less costly cleanup alternatives for chlorinated-compound-contaminated groundwater will continue.

The revised cleanup rules call for an evaluation of current and reasonably likely future beneficial uses of groundwater. As a result, future cleanup actions in Oregon are less likely to rely on conventional pump and treat technologies for addressing groundwater contamination in areas where there is no current beneficial use. This will enable DEQ to give stronger consideration to remedies like in situ bioremediation, phytoremediation, treatment walls and natural attenuation monitoring when such measures are capable of restoring or protecting future beneficial uses of water. These technologies often will be favored where there is no current beneficial use impacts.

Where a current beneficial use has been significantly impacted (e.g., for drinking water at levels exceeding MCLs), we expect to see a continued reliance on extraction, particularly since it is perceived by the general public as a more active response. The same holds true for IRAMs in areas where a current groundwater use is threatened or impacted; extraction continues to be preferred as a reasonable and proactive response to the identified contamination.

REFERENCES

Hazardous Waste and Hazardous Materials Statutes Pertaining to Removal or Remedial Action. Oregon Revised Statutes 465.200-.900.

Hazardous Substance Remedial Action Rules. Oregon Administrative Rules 340-122-010 through 140.

Oregon Legislative Assembly. 1995. Enrolled House Bill 3352- The Recycled Lands Act.

Wiedemeier, T. H., M. A. Swanson, D. E. Moutoux, J. T. Wilson, D. H. Kampbell, J. E. Hansen, and P. Haas. 1996. "Overview of the Technical Protocol for Natural Attenuation of Chlorinated Aliphatic Hydrocarbons in Ground Water Under Development for the U.S. Air Force Center for Environmental Excellence." In Proceedings of the *Symposium on Natural Attenuation of Chlorinated Organics in Ground Water (Dallas, TX; September 1996)*. EPA/540/R-96/509, pp. 35-59.

CLOSURE CRITERIA MODEL FOR SOILVAPOR EXTRACTION SYSTEMS NORTHERN CALIFORNIA SITES

Richard S. Makdisi, R.G. (Stellar Environmental Solutions)
Ron Gervason, R.G., and Stephen Hill
(California Regional Water Quality Control Board)

ABSTRACT: Operation and post operation data from 13 Soil Vapor Extraction (SVE) systems in Northern California Bay Area with oversight by the San Francisco Bay Regional Water Quality Control Board (RWQCB), which were granted closure between 1993 and 1996, are evaluated to identify common closure criteria for use by future sites seeking SVE closure in order to develop a closure model. The case studies include SVE operation from six months to over five years. Common physio-chemical properties from the 13 SVE site closures were examined and evaluation criteria developed for use in assessing future applications for closure. A subset of 3 of the 13 case studies were selected for further analyses bases on their representativeness of the Bay Area SVE systems and the completeness of the data collected for them. Characteristics of the contaminants, regulatory cleanup standards, soil type and depth to groundwater, SVE operation and performance, mass recovery and groundwater impact criteria are examined to determine the granting of site regulatory closure status. SVE closure has often been based solely on post remediation soil sampling. Variability in the vadose zone makes this an uncertain predictor of adequate soil remediation. Performance factors, such as attainment of asymptotic level, are relatively simple to assess and a good predictor of adequate soil remediation and the effective limits of the SVE system.

INTRODUCTION

Soil Vapor Extraction (SVE) systems have become a remedial technology of choice in California for cleanups of sites with soil contamination by volatile organic compounds (VOCs). SVE technology is widely applied, particularly in the Northern California Bay Area environment, because of the positive climatic and geologic environment, relative simplicity of design and installation, soil contaminant mass recovery rates achievable, and comparative speed of the remediation effort relative to other alternatives.

VOC are released into the environment from and acted on by numerous forces that influence the degree and rate at which they can be recovered. The extent to which the released VOCs partition into the vapor phase for recovery versus being bound in solid or soluble states is dependent on the nature of the contaminant, the subsurface environment, and the elapsed time since the release occurred. The manner by which the released product behaves in the subsurface unsaturated zone forms the technical underpinnings of development of closure criteria for SVE systems. The factors or mechanisms important to determine the effectiveness of SVE applications are: contaminant characteristics like the vapor pressure, water solubility, Henry's Law, and soil vapor coefficient; the soil environment such as permeability, porosity, soil structure, residual saturation, air-water permeability as a function of water content, ganglia formation, subsurface conduits, water content, and preferred soil pathways; and the characteristics of vapor flow in subsurface such as vapor transport, pressure gradients, mass-flux to groundwater, diffusion constant, flow under vacuum conditions and radius of capture dynamics (1, 2, 3).

Site closure is the primary goal of any remedial treatment system. Closures can be achieved only when residual levels of chemicals or an acceptable "minimus" in the contaminated media are acceptable to regulators. With respect to soil pollution, regulatory agencies typically require remediation until VOC concentrations no longer present a threat to shallow groundwater or the health of site users. The Bay Area Regional Water Quality Control Board (RWQCB) is the regulatory entity that has provided the oversight of SVE remediation efforts on many Bay Area sites which were granted closure between 1994 and 1996. The RWQCB commonly sets a soil remediation target of 1 mg/kg total VOCs to prevent leaching. This is referred to as their to be considered ARAR.

REVIEW OF BAY AREA SVE SITES GRANTED CLOSURE

The RWQCB staff approved closure of about 16 SVE systems during the period 1993 through 1996. Thirteen of these were examined to identify common features and factors cited in the closure decisions (Tables 1 and 2). These 13 represent a range of soil and site conditions found in the Bay Area with chlorinated VOCs being the principal contaminant type. Only one of the sites represents a case where hydrocarbons are remediated through SVE.

The traditional regulatory method of assessing progress towards SVE closure has involved the collection of soil samples within the area of known soil contamination for comparison with stipulated compound-specific, concentration-based standards. Remediation is considered complete when these standards have been attained within some statistical framework. A key flaw in the traditional regulatory closure criteria is determining what, on a given site, will constitute confirmation sampling, particularly with the inherent biases of standard soil sampling methods employed in evaluating residual contaminant mass in the vadose zone.

The soil types present at the 13 sites are highly variable, ranging from clays to sandy gravels. All are alluvial soils typical of Bay Area contaminated sites. Most of the sites are within the Santa Clara Valley area known as Silicon Valley and the sites have a mix of soil types typical of an alluvial vadose zone with associated differing permeability with respect to SVE. Depth to groundwater is also highly variable, ranging from 10 feet below ground surface (bgs) to almost 100 feet bgs, although depths of less than 30 feet bgs are most prevalent. Pollutants include chlorinated solvents, non-fuel hydrocarbons, and tetrahydrofuran. TCE is the most common pollutant.

SVE systems at the 13 sites operated for periods ranging from two to seven years. Mass removals range from 4 to more than 3,000 pounds. VOC mass removals do not correlate closely with time of operation, but tend to be much greater for sites with some coarse-grained soils. Pilot testing was used at more than half the sites. In most cases, the SVE system's zone of influence fully encompassed the area of known soil pollution. Most SVE systems were operated on a continuous basis; only two featured cyclic operation. Table 1 presents the pertinent data on the 13 sites.

Volatile organic compounds, typically industrial cleaning solvents, are the dominant class of contaminants remediated by SVE in the Bay Area. This is mainly due to the predominance of computer-related industry in the area. Only one the sites is a hydrocarbon-related cleanup. Comparison of the maximum soil and groundwater concentrations detected generally follow the expected pattern of a SVE candidate site, with the higher concentrations being in the soil. However, the groundwater concentrations were significantly higher at two of the sites reviewed (sites 9 and 12 in Table 1). In cases like these, particularly when the vadose zone is thin, the SVE also captures contaminant mass from the groundwater.

TABLE 1. Data From SVE Sites Granted Closure (All Concentration Data in mg/kg for Soil and mg/L for Groundwater)

Site	Contaminant	Lithology	Depth to GW	Max. Soil Conc.	Max. GW Conc.	SVE Operation	Soil Cleanup Goal (mg/kg)
1	VOCs	Silt/Clay	35	2000	7.0	1989-95	1
2	TCE	Silt	95	4.3	1.5	1991-92	1
3	TCE	Silt	20	1	13.0	1988-94	1
4	Freon	Silty Sand	36	4.2	55.0	1990-92	1
5	NFH	Clay-Silt	30	14	45.0	1993-94	1
6	TCE	Grav-Sand	70	10	33.0	1988-93	1
7	PCE	Silt	52	NK	0.9	1992-95	1
8	Freon/TCE	Silt	11	110	1.2	1991-95	1
9	THF	Silt/Clay	25	58000	200000	1991-95	150
10	TCE	Silt/Clay	11	19	1.6	1989-94	2.5/0.5
11	TCE	Clayey Gra	23	5	3.8	1991-95	1
12	Xylene/TCE	Silt/Clay	12	2	43.0	1993-94	1
13	TPH	Silt/Clay	20	1200	29.0	1992-94	100

Notes: NK= Not Known

Table 2 shows a compilation of the SVE operational and closure characteristics for the 13 sites. While some of the data points are missing because the data was not presented, the trends that emerge show that only half of the sites had pilot tests performed and only three performed cyclical pumping tests where the SVE is shut down for one to three months to evaluate potential VOC rebound. All of the sites, to a greater or lesser degree, showed a mass removal equilibration or asymptote reached.

SELECTED SVE CLOSURE CASE STUDY/REVIEW

Three of the thirteen sites closed were selected to analyze in greater detail. These three sites were selected based on the extent of data available and their representativeness. Sites 4 and 6 are more typical of the SVE closures, being associated with the remediation of TCE in Santa Clara Valley; Site 13 was selected as the one representative SVE at a TPH contaminated site and is located in the Brisbane Hills south of San Francisco. Table 3 summarizes the data of interest for the three sites.

San Jose VOC Site. The soil vapor extraction system at Site 4 is assumed to address contamination from a single source. The soil contamination is confined to the upper 25 feet of soil above the water table. Deeper contamination is addressed as groundwater contamination, not soil contamination. The dominant soil type is clayey silt although clayey sands and clays are also present. Estimates of mass of Freon 113 in soil based on

TABLE 2. SVE Operation and Closure Data

	SVE closure features:					SVE operation features:			
Site	Period of Operation (years)	VOC mass removal (lb)	Pilot SVE?	Cyclic operation?	Adequate coverage?	Reached asymptote?	Minimal rebound?	Met soil target? (1)	Closure Approval Date
1	6	4	Yes	Yes	NK	Yes	Yes	Yes	6/18/95
2	1	170	No	No	NK	Yes	Yes	No	8/19/93
3	7	46	No	No	Yes	Yes	Yes	Yes	12/1/94
4	2	1,500	Yes	No	Yes	Yes	Yes	Yes	3/24/94
5	1	1,020	Yes	No	NK	Yes	Yes	No	3/9/95
6	5	3,025	Yes	No	Yes	Yes	Yes	Yes	5/14/93
7	4	NK	Yes	Yes	Yes	Yes	Yes	NK	6/14/95
8	4	110	Yes	No	Yes	Yes	Yes	NK	7/17/95
9	4	2,970	No	No	Yes	Yes	Yes	Yes	8/13/95
10	6	82	No	Yes	NK	Yes	Yes	Yes	2/7/95
11	5	1,360	No	Yes	Yes	Yes	Yes	Yes	6/9/95
12	2	2,800	No	No	Yes	Yes	Yes	NK	3/6/95
13	2	1,700	No	Yes	Yes	No	Yes	No	3/6/97

Notes: (1) = RWQCB to be considered ARAR ; NK=Not known

TABLE 3. Three Case Studies for SVE Closure Criteria

Site I.D. Site Criteria	Site 1: San Jose VOC Site	Site 2: Cupertino VOC Site	Site 3: Brisbane TPH Site
Soil ARAR's Met?	YES	YES	NO
Initial Concentration/soil vol	4.2 ppm Freon	10 ppm TCE	1400 ppm/2000 CY
Closure concentration/soil vol	< 2 lbs/day	< 1 ppm TCE	190 ppm/120 CY
Mass Removal %	99%	99%	95%
Mass Removal lbs	1500	3025	700
Cyclic Pumping ?	YES	NO	YES
Rebound Observed ?	YES	YES	NO
Asymptote Observed	YES	NO	NO
GW Impact	<0.7 mg/L Freon	< 0.3 mg/L TCE	<0.1 mg/L TPH
Engineer Controls	Surface Cover	Surface Cover	Surface Cover/DN

Notes: DN = Deed Notification; lbs = Pounds; soil vol = volume of soil in cubic yards (CY)

analysis of pre-remediation soil samples and basic assumptions about the soil properties indicate a mass of approximately 2000 pounds of Freon in soil. Post remediation soil mass based on the four confirmation samples that were collected would indicate that all mass had been removed since no volatile organic compounds were detected in the confirmation samples.

A pilot system was installed that included a total of six extraction points. Three of these points did not produce either adequate volume or concentration and were replaced with three new extraction points closer to the assumed source area. The system was operated for 132 days and then operated in cyclic mode for the next year, with a short down cycle of two weeks followed by collection of vapor samples from the six extraction points for analyses. Since minimal rebound in vapor concentration was empirically observed this was followed by a longer system shutdown cycle of approximately one year. Since minimal rebound in vapor concentration was again empirically observed after this down cycle the site was approved for closure of the soil vacuum extraction system. During this period of operation the soil vapor extraction system removed approximately 1,500 pounds of Freon. This is similar to the estimate of total mass present in the sub-surface. Additional mass was removed during the excavation phase.

Initial closure criteria for this soil vapor extraction system was based on mass removal rates. The criterion for removal of Freon 113, which proved to be the only significant contaminant, was a mass removal of less than two pounds per day. This criterion was first achieved in September 1991. The system continued operation for another three months, then went into cyclic operation to confirm that the closure criteria had been achieved. Confirmation samples were collected in the areas with soil samples that had the highest initial concentration of Freon 113. All four samples did not detect any volatile organic compounds. After cyclic operation, the removal rate had stabilized at less than one pound per day and the cumulative mass removal curve had reached asymptotic conditions. While the initial closure criteria were met based solely on daily mass removal Site 4 also demonstrates the importance of consideration of other system performance criteria.

Cupertino VOC Site. Review of the data for site 6 indicated that the subsurface soil had been contaminated to depths of up to 60 feet below ground surface. The original sources of contamination are assumed to have been from two to four separate source areas. The area of investigation was divided into two primary and two sub areas for investigation. An intensive soil sampling program was instituted to characterize the subsurface contamination with over 50 soil borings installed in an area of less than 240,000 square feet with over half of the area covered by structures.

The maximum soil contamination detected in a soil sample was 10 mg/kg of trichloroethylene. The mass of contaminant present in the subsurface was estimated based on soil volumes and soil types assuming normal soil saturation and porosity. The estimates for this site are highly variable due to the possibility of multiple source areas and high variability in depth of contamination. The estimated mass of contamination present in the subsurface based on analysis of soil samples collected prior to remediation would range between 10 and 30 pounds.

The soil vapor extraction system at this site included at total of seven vapor extraction wells. The system operated with some combination of these extraction points for approximately 1,740 days. The system removed 3,025 pounds of TCE during this period of operation. The mass remaining in place after the system was in operation for approximately 1,650 days was less than one pound, based on confirmation soil samples with an average of 0.005 mg/kg and the same assumptions used to estimate the initial mass.

Examination of the data showed the cumulative mass removal curve flattens appreciably after approximately 1,000 days of operation (February 1991). The cumulative mass removal curve however, appears to not be truly asymptotic. The mass removal rate had declined to less than a pound per day by February 1991. The initial removal and cumulative removal rates are dominated by relatively high influent concentrations (greater than 500 part per million by volume) in two of the vapor extraction points. However, by the end of the operation of the soil vapor extraction system all vapor extraction points had similar vapor concentrations, with all extraction points have vapor concentrations near ten part per million by volume.

Brisbane TPH Site. Site 13 illustrates the application of SVE at a hydrocarbon contaminated site. The main hydrocarbons contamination was associated with a 10,000 gallon mineral spirits leak. An estimated 1,860 pounds of contaminant was removed during the excavation stage of the project. Although mineral spirits is less volatile and less soluble than gasoline the groundwater was located at a relatively shallow depth of 20 feet below ground surface in mainly silty clay alluvial deposits although sand intervals of one to two foot were located at a 10-foot depth and in the area of the capillary fringe at about 18 feet bgs. Groundwater contamination was reported at 29.0 mg/L TPH as mineral spirits. Although there was only one source, the spread of the contamination over an estimated 2,000 cubic yards made an extensive excavation project unattractive. Average concentrations of TPHms in the soil were estimated at 360 mg/kg with the maximum being at 1,400 mg/kg TPHms near the former excavation.

A SVE systems consisting of five extraction wells was installed with a design capacity of 140 CFM although during the initial operation the actual flow rates from each well were lower than predicted and the combined flow was measured at 75 CFM. The lower flow rates were likely due to the presence of the fine-grained sediments and higher than expected soil moisture in the available pore spaces. The system began operation on in December 1992 and operated until July 1994 when it was shut down after an evaluation of its effectiveness which included a 3-month trial shutdown to evaluate rebound. Groundwater mounding occurred during the operation of the system and remediation of the groundwater was a byproduct of the operation.

Initial mass removals were estimated at between 4 and 5 pound a day. After 515 days of operations the mass removal rate was calculated at 0.02 pounds per day. There was no rebound observed in the vapor concentrations after the 3-month shutdown of the system. The groundwater concentrations in the five monitoring wells were all below 1 mg/L, down from a high of 29.0 mg/L in the last six months of operation. Confirmation samples had an average of 74 mg/kg TPHms but one of the six samples collected showed 190 mg/kg. Although the cleanup goal was 100 mg/kg TPH closure was allowed. About 120 CY of residual contaminated soil is estimated to remain on the site and a deed notification was required to be placed. As part of the closure evaluation, oxygen and carbon dioxide content of the influent air stream were collected before the final shutdown of the SVE. Chemical analysis results reported the presence of oxygen at 1 percent and carbon dioxide at 3.6 percent. The average atmospheric content of oxygen is 20 percent and carbon dioxide is 3 percent so the difference suggests that biorespiration/transformation is occurring in situ, causing a decrease in oxygen and rise in carbon dioxide.

SUMMARY

The determination of the acceptability by the regulatory community has increasingly moved away from finite cleanup standards and is more site-specific, relying on a more complex assessment of baseline conditions, technical achievability, indications of diminishing returns of contaminant mass capture and evaluations of residual risk, as well as regulatory cleanup goals. Use of these criteria, developed out of the review of the case studies presented, in future evaluations of SVE system closures should simplify the closure petition process for the applicants and the closure evaluation process for the regulators.

AUTHOR INDEX

This index contains names, affiliations, and book/page citations for all authors who contributed to the six books published in connection with the First International Conference on Remediation of Chlorinated and Recalcitrant Compounds, held in Monterey, California, in May 1998. Ordering information is provided on the back cover of this book.

The citations reference the six books as follows:

1(1): Wickramanayake, G.B., and R.E. Hinchee (Eds.). 1998. *Risk, Resource, and Regulatory Issues: Remediation of Chlorinated and Recalcitrant Compounds.* Battelle Press, Columbus, OH. 322 pp.

1(2): Wickramanayake, G.B., and R.E. Hinchee (Eds.). 1998. *Nonaqueous-Phase Liquids: Remediation of Chlorinated and Recalcitrant Compounds.* Battelle Press, Columbus, OH. 256 pp.

1(3): Wickramanayake, G.B., and R.E. Hinchee (Eds.). 1998. *Natural Attenuation: Chlorinated and Recalcitrant Compounds.* Battelle Press, Columbus, OH. 380 pp.

1(4): Wickramanayake, G.B., and R.E. Hinchee (Eds.). 1998. *Bioremediation and Phytoremediation: Chlorinated and Recalcitrant Compounds.* Battelle Press, Columbus, OH. 302 pp.

1(5): Wickramanayake, G.B., and R.E. Hinchee (Eds.). 1998. *Physical, Chemical, and Thermal Technologies: Remediation of Chlorinated and Recalcitrant Compounds.* Battelle Press, Columbus, OH. 512 pp.

1(6): Wickramanayake, G.B., and R.E. Hinchee (Eds.). 1998. *Designing and Applying Treatment Technologies: Remediation of Chlorinated and Recalcitrant Compounds.* Battelle Press, Columbus, OH. 348 pp.

KEYWORD INDEX

This index contains keyword terms assigned to the articles in the six books published in connection with the First International Conference on Remediation of Chlorinated and Recalcitrant Compounds, held in Monterey, California, in May 1998. Ordering information is provided on the back cover of this book.

In assigning the terms that appear in this index, no attempt was made to reference all subjects addressed. Instead, terms were assigned to each article to reflect the primary topics covered by that article. Authors' suggestions were taken into consideration and expanded or revised as necessary to produce a cohesive topic listing. The citations reference the six books as follows:

1(1): Wickramanayake, G.B., and R.E. Hinchee (Eds.). 1998. *Risk, Resource, and Regulatory Issues: Remediation of Chlorinated and Recalcitrant Compounds*. Battelle Press, Columbus, OH. 322 pp.

1(2): Wickramanayake, G.B., and R.E. Hinchee (Eds.). 1998. *Nonaqueous-Phase Liquids: Remediation of Chlorinated and Recalcitrant Compounds*. Battelle Press, Columbus, OH. 256 pp.

1(3): Wickramanayake, G.B., and R.E. Hinchee (Eds.). 1998. *Natural Attenuation: Chlorinated and Recalcitrant Compounds*. Battelle Press, Columbus, OH. 380 pp.

1(4): Wickramanayake, G.B., and R.E. Hinchee (Eds.). 1998. *Bioremediation and Phytoremediation: Chlorinated and Recalcitrant Compounds*. Battelle Press, Columbus, OH. 302 pp.

1(5): Wickramanayake, G.B., and R.E. Hinchee (Eds.). 1998. *Physical, Chemical, and Thermal Technologies: Remediation of Chlorinated and Recalcitrant Compounds*. Battelle Press, Columbus, OH. 512 pp.

1(6): Wickramanayake, G.B., and R.E. Hinchee (Eds.). 1998. *Designing and Applying Treatment Technologies: Remediation of Chlorinated and Recalcitrant Compounds*. Battelle Press, Columbus, OH. 348 pp.

A

B

F

G

H

I

K

L

M

N

O

P

U

V

W

Z